I0036750

Engineering Fluid Dynamics

Special Issue Editor

Bjørn H. Hjertager

MDPI • Basel • Beijing • Wuhan • Barcelona • Belgrade

MDPI

Special Issue Editor
Bjørn H. Hjertager
University of Stavanger
Norway

Editorial Office
MDPI AG
St. Alban-Anlage 66
Basel, Switzerland

This edition is a reprint of the Special Issue published online in the open access journal *Energies* (ISSN 1996-1073) from 2016–2017 (available at: http://www.mdpi.com/journal/energies/special_issues/eng_fluid_dyn).

For citation purposes, cite each article independently as indicated on the article page online and as indicated below:

Author 1; Author 2. Article title. *Journal Name*. **Year**. Article number/page range.

First Edition 2018

ISBN 978-3-03842-668-4 (Pbk)
ISBN 978-3-03842-669-1 (PDF)

Table of Contents

About the Special Issue Editor

Bjørn H. Hjertager, Professor, Dr. got his PhD from the University of Trondheim, Norway (NTH), in 1979, within the topic: Combustion, Heat Transfer and Fluid Flow. After one year at the Norwegian Underwater Institute, he stayed almost 10 years at Chr Michelsen Institute, Bergen Norway. Then from the late 80's and up to the late 90's, he worked at Telemark Institute of Technology (HIT-TF), and at the research institute Tel-Tek, Porsgrunn, Norway. He then stayed 11 years abroad, at Aalborg University Esbjerg in Denmark. Since 2008 he has been back to Norway, as a Professor in Fluid dynamics at the University of Stavanger. He has published more than 170 papers in the field of fluid flow, heat transfer, combustion, gas explosions and chemical reactors, and he has supervised 21 PhD candidates in Norway and Denmark.

energies

Editorial

Engineering Fluid Dynamics

Bjørn H. Hjertager

Department of Mechanical and Structural Engineering and Materials Science, University of Stavanger, N-4016 Stavanger, Norway; bjorn.hjertager@uis.no

Academic Editor: Enrico Sciubba
Received: 4 September 2017; Accepted: 12 September 2017; Published: 22 September 2017

Over the last few decades, the use of computational fluid dynamics (CFD) and experimental fluid dynamics (EFD) methods has penetrated into all fields of engineering. CFD is now becoming a routine analysis tool for design in some fields (e.g., aerodynamics of vehicles), and its implementation in other fields (e.g., chemical and marine applications) is being quickly adopted. Additionally, in the last decade, open source software has had a tremendous impact in the use of CFD. Laser-based methods have also made significant improvements in methods to obtain data for the validation of the CFD codes.

This book contains the successful submissions [1–12] to a Special Issue of *Energies* on the subject area of "Engineering Fluid Dynamics". The topic of engineering fluid dynamics includes both experimental as well as computational studies. Of special interest were submissions from the fields of mechanical, chemical, marine, safety, and energy engineering. We welcomed both original research articles as well as review articles. After one year, 22 papers were submitted and 12 were accepted for publication. The average processing time was 65.2 days. The authors had the following geographical distribution: China (four); Italy (two); Korea (one); Germany (one); UK (one); Ireland (one); Australia (one); Sweden (one); Japan (one); Spain (one); Norway (one).

Papers covered topics such as heat transfer in shell and helically coiled tube heat exchangers [1], the multiphase modeling of sprays [2], flashing flows [4], as well as mixing in a bubbling fluidized bed [8]. Two papers related to heating ventilation and air condition (HVAC) are included, namely evaporation and condensation in the underfloor space of detached houses [9] and air distribution in a railway vehicle [10]. Three papers dealt with various aspects of pumps and turbines: a performance prediction method for pumps as turbines [3]; noise radiation in a centrifugal pump [5]; periodic fluctuations in energy efficiency in centrifugal pumps [7]; and study of a high-pressure external gear pump [11]. One paper used both laser doppler velocimetry (LDV) and CFD in the study of flow behind a semi-circular step cylinder [6]. Finally, a paper investigated the influence of the equivalence ratio (ER) and feedstock particle size on birch wood gasification [12].

I found the task of editing and selecting papers for this collection to be both stimulating and rewarding. I would also like to thank the staff and reviewers for their efforts and input.

Conflicts of Interest: The author declares no conflict of interest.

References

1. Kim, S.; Jo, J.; Lee, Y.; Yoo, Y. Comparative Study of Shell and Helically-Coiled Tube Heat Exchangers with Various Dimple Arrangements in Condensers for Odor Control in a Pyrolysis System. *Energies* **2016**, *9*, 1027. [CrossRef]
2. Qian, L.; Lin, J.; Bao, F. Numerical Models for Viscoelastic Liquid Atomization Spray. *Energies* **2016**, *9*, 1079. [CrossRef]
3. Frosina, E.; Buono, D.; Senatore, A. A Performance Prediction Method for Pumps as Turbines (PAT) Using a Computational Fluid Dynamics (CFD) Modeling Approach. *Energies* **2017**, *10*, 103. [CrossRef]
4. Liao, Y.; Lucas, D. Possibilities and Limitations of CFD Simulation for Flashing Flow Scenarios in Nuclear Applications. *Energies* **2017**, *10*, 139. [CrossRef]

5. Gao, M.; Dong, P.; Lei, S.; Turan, A. Computational Study of the Noise Radiation in a Centrifugal Pump When Flow Rate Changes. *Energies* **2017**, *10*, 221. [CrossRef]
6. Sayeed-Bin-Asad, S.; Lundström, T.; Andersson, A. Study the Flow behind a Semi-Circular Step Cylinder (Laser Doppler Velocimetry (LDV) and Computational Fluid Dynamics (CFD)). *Energies* **2017**, *10*, 332. [CrossRef]
7. Zhang, H.; Deng, S.; Qu, Y. Numerical Investigation of Periodic Fluctuations in Energy Efficiency in Centrifugal Pumps at Different Working Points. *Energies* **2017**, *10*, 342. [CrossRef]
8. Zhao, X.; Eri, Q.; Wang, Q. An Investigation of the Restitution Coefficient Impact on Simulating Sand-Char Mixing in a Bubbling Fluidized Bed. *Energies* **2017**, *10*, 617. [CrossRef]
9. Oh, W.; Kato, S. Study on the Effects of Evaporation and Condensation on the Underfloor Space of Japanese Detached Houses Using CFD Analysis. *Energies* **2017**, *10*, 798. [CrossRef]
10. Suárez, C.; Iranzo, A.; Salva, J.; Tapia, E.; Barea, G.; Guerra, J. Parametric Investigation Using Computational Fluid Dynamics of the HVAC Air Distribution in a Railway Vehicle for Representative Weather and Operating Conditions. *Energies* **2017**, *10*, 1074. [CrossRef]
11. Frosina, E.; Senatore, A.; Rigosi, M. Study of a High-Pressure External Gear Pump with a Computational Fluid Dynamic Modeling Approach. *Energies* **2017**, *10*, 1113. [CrossRef]
12. Jayathilake, R.; Rudra, S. Numerical and Experimental Investigation of Equivalence Ratio (ER) and Feedstock Particle Size on Birchwood Gasification. *Energies* **2017**, *10*, 1232. [CrossRef]

energies

Article

Comparative Study of Shell and Helically-Coiled Tube Heat Exchangers with Various Dimple Arrangements in Condensers for Odor Control in a Pyrolysis System

Sun-Min Kim, Jun-Ho Jo, Ye-Eun Lee and Yeong-Seok Yoo *

Division of Environmental and Plant Engineering, Korea Institute of Civil Engineering and Building Technology, 283, Goyang-daero, Ilsanseo-gu, Goyang-si, Gyeonggi-do 10223, Korea; sunminkim@kict.re.kr (S.-M.K.); junkr@kict.re.kr (J.-H.J.); yeeunlee@kict.re.kr (Y.-E.L.)

* Correspondence: ysyoo@kict.re.kr; Tel.: +82-31-910-0298

Academic Editor: Bjørn H. Hjertager
Received: 4 October 2016; Accepted: 30 November 2016; Published: 5 December 2016

Abstract: This study performed evaluations of the shell and helically-coiled tube heat exchangers with various dimple arrangements, that is, flat, inline, staggered, and bulged, at different Dean numbers (*De*) and inlet temperatures of a hot channel. Conjugated heat transfer was analyzed to evaluate the heat transfer performance of the exchangers through temperature difference between the inlet and outlet, Nusselt number inside the coiled tube, and pressure drop of the coiled tube by using 3-D Reynolds-averaged Navier–Stokes (RANS) equations with shear stress transport turbulence closure. A grid dependency test was performed to determine the optimal number of the grid system. The numerical results were validated using the experimental data, and showed good agreement. The inline and staggered arrangements show the highest temperature differences through all *De*. The staggered arrangement shows the best heat transfer performance, whereas the inline arrangement shows the second highest performance with all ranges of *De* and the hot channel's inlet temperature. The inline and staggered arrangements show the highest pressure drop among all inlet temperatures of the hot channel.

Keywords: shell and tube; heat exchanger; odor control; heat transfer; dimple; condenser; pyrolysis; RANS

1. Introduction

Pyrolysis is defined as one of the most effective and efficient processes of thermal decomposition to obtain energy in the form of char from resources in an anoxic condition. This process has the ability to convert sewage sludge, biomass, and other carbonaceous materials to reusable fuels [1]. Biomass pyrolysis results in three products: biochar, bio-oil, and gas [2]. Biochar is a solid carbon-rich byproduct obtained through the thermal stabilization of biomass or any other organic matter [3]. Bio-oil has a good feedstock for power generation as it contains high amounts of energy, which is, in some cases, comparable to that of the fossil fuels after upgrading [4]. Moreover, gas produced through pyrolysis has high energy, which is, in some cases, comparable to that of the coal used in industries as feedstocks for fuel [5,6]. Food wastes are known as good resources of pyrolysis among various raw materials.

Food wastes have a high energy content. Consequently, they offer a good potential as feedstocks for pyrolysis in power plants [7]. Therefore, many researchers and industry officials have focused on food-waste pyrolysis as a future source of energy supply. However, in the pyrolysis of food wastes, numerous types of odors are generated and the dust formed in the process, including the dry process, can be a factor of ignition [8,9]. One of the best and commonly used methods to solve this problem is condensation [10]. By condensing, the byproducts (such as the odors or dust) of the pyrolysis process are reduced and removed. The shell and tube heat exchanger (STHX) systems have played an important role to realize condensation [11], and it is important to increase the heat transfer performance for the best condensation conditions.

Therefore, many researchers focus on the STHX systems, especially on the heat transfer performance. Selbaş et al. [12] estimated the minimum heat transfer area required for a given heat duty, as it governs the overall cost of the heat exchanger, by using the genetic algorithms (GA) with varying design variables, such as the outer tube diameter, tube layout, number of tube passes, outer shell diameter, baffle spacing, and baffle cut. In addition, they concluded that the combinatorial algorithms, such as GAs, provide significant improvement in the optimal designs than the conventional designs. Babu and Munawar [13] applied differential evolution (DE) and its various strategies for the optimal design of STHXs. They determined that the DE, an exceptionally simple evolution strategy, is significantly faster than the GA and yields the global optimum for several key parameters. Xie et al. [14] experimentally investigated three heat exchangers by applying an artificial neural network (ANN) to the heat transfer analysis of STHXs with segmental or continuous helical baffles, and predicted outlet temperature differences on each side and overall heat transfer rates. As a result, the maximum deviation between the predicted results and experimental data was less than 2%, and the ANN showed superiority for prediction than correlation. Fesanghary et al. [15] explored the use of global sensitivity analysis (GSA) and a harmony search algorithm (HSA) for design optimization of STHXs from the economic viewpoint; total cost of STHXs are identified using GSA. They compared the HSA results with those obtained using GA, and revealed that the HSA can converge to an optimum solution with higher accuracy. Guo et al. [16] applied the field synergy principle to the optimization design of an STHX with segmental baffles by using the GA to solve the heat exchanger optimization problems by using multiple design variables. The comparison with the conventional heat exchanger optimization design approach, with the total cost as the objective function, shows that the field synergy number maximization approach is more advantageous. Zhang et al. [17] experimentally compared several STHXs: one with segmental baffles and four with helical baffles at helix angles of 20°, 30°, 40°, and 50°. The results show that, based on the same shell-side flow rate, the heat transfer coefficient of the heat exchanger with helical baffles is lower than that of the heat exchanger with segmental baffles, while the shell-side pressure drop of the exchanger with helical baffles is even lower.

As mentioned above, many researchers have focused on the STHX systems to enhance the heat transfer performance during condensation of odorized materials. However, there has been no attempt to develop a new combined system with the dimpled helical tube that enhances the heat transfer performance of the STHX systems. The tube was devised to intentionally enhance the heat transfer performance. Thus, the superiority of the dimpled helical tube was evaluated. However, the various arrangements of the dimples, such as staggered, in-lined, and alternating arrangements, must be assessed. Therefore, the present study suggests several arrangements of the dimples on the helical tube; the heat transfer performance was evaluated in terms of the Nusselt number. In addition, the effect of the Dean number (De) on the heat transfer performance was estimated. Furthermore, the performance of the proposed STHX systems was comparatively assessed using 3-D Reynolds-averaged Navier–Stokes (RANS) equations.

2. Configurations of the Shell and Helically-Coiled Tube Heat Exchanger Systems

Figure 1 presents dimensions of the shell and helically-coiled dimpled-tube heat exchanger system. The overall system has a length of 230 mm and diameter of 120 mm. The coolant enters through the inlet of 12 mm diameter, and exits through the same diameter outlet. The outer and inner diameters of the helical tube are 12 and 9 mm, respectively. The helical tube has 10 turns with the coil pitch of 17 mm and a curvature radius of $2R_c$, as shown in Figure 1a. The height (H_d) and diameter (D_d) of the dimple are 1.336 and 6.903 mm, respectively; the pitch of the dimple (p) is 5.628 mm, as shown in Figure 1b.

Figure 2 shows the schematic of the dimple arrangement. The dimples on the helical tube are arranged in the following three arrangements: staggered, inline, and bulged. Further, this tube's heat transfer performance was compared with that of the flat helical tube. The heat transfer enhancement structure consists of two dimples: concave and convex dimples. However, the bulged arrangement consists of convex dimples.

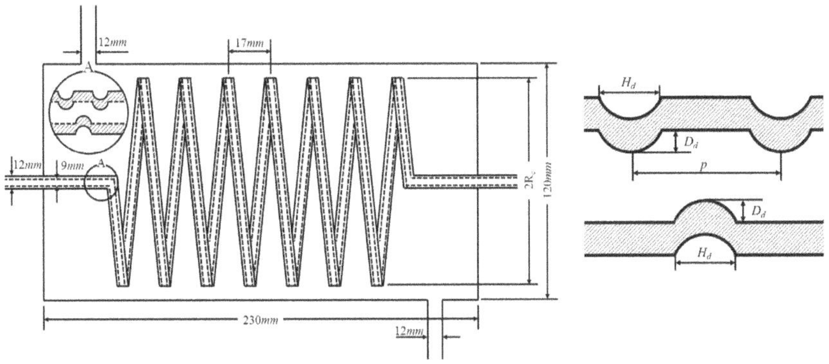

Figure 1. Dimensions of the shell and helically-coiled dimpled tube heat exchanger system: (**a**) overall system; and (**b**) dimple.

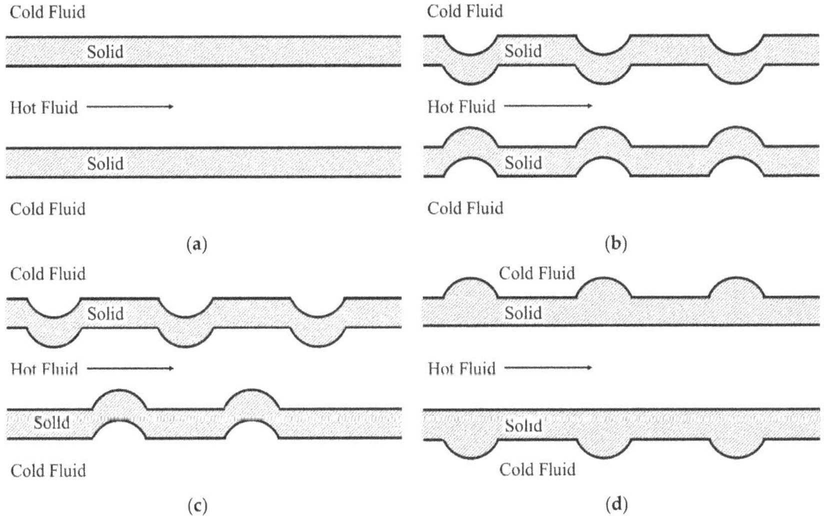

Figure 2. Schematic of the dimple arrangement: (**a**) flat; (**b**) inline; (**c**) staggered; and (**d**) bulged.

3. Numerical Analysis

The present numerical analysis was performed using commercial computational fluid dynamics code ANSYS CFX 17.0 (© 2016 ANSYS, Inc., Canonsburg, PA, USA) [18] to solve the governing equations for 3-D steady turbulent flow and conjugate heat transfer in the heat exchanger system.

For a steady, incompressible, turbulent flow, the continuity and Reynolds-averaged Navier-Stokes equations are given as:

$$\frac{\partial}{\partial x_i}(\rho u_i) = 0 \tag{1}$$

$$\frac{\partial}{\partial x_j}(\rho u_i u_j) = -\frac{\partial p}{\partial x_i} + \frac{\partial}{\partial x_j}\left[\mu\left(\frac{\partial u_i}{\partial x_j} + \frac{\partial u_j}{\partial x_i} - \frac{2}{3}\frac{\partial u_k}{\partial x_k}\delta_{ij}\right) - \rho\overline{u_i' u_j'}\right] + \overline{S}_i^u \tag{2}$$

where u_i and u_i' are mean and fluctuating velocities, respectively, and $\overline{S}_i^u$ is a production term. The shear stress transport (SST) model [19] with automatic wall treatment is used as a turbulence closure model. In addition, a second-order-accurate numerical scheme was used to discretize the governing equations. Moreover, a high-resolution scheme that uses a special nonlinear procedure at each node was used to discretize the advection terms.

Figure 3 shows the computational domain and boundary conditions. Water and Behran Hararat oil were respectively used as working fluids for the shell-side fluid and tube-side coolant. The thermophysical properties of the oil [20] were varied to state the dependency of the working fluid on temperature with an accuracy of 1.0%. The importance of temperature-dependent properties was already emphasized by many researchers [21,22]. The properties were defined as follows:

$$\begin{aligned} \mu &= -1.31 \times 10^{-7}T^3 + 3.2813 \times 10^{-5}T^2 - 2.916875 \times 10^{-3}T + 9.9418750 \times 10^{-2} \\ C_p &= 3.7093T + 1814.3575 \\ k &= -8.7 \times 10^{-5}T + 0.164570 \\ \rho &= -0.725T + 877.350 \end{aligned} \tag{3}$$

where T (°C) is the caloric mean temperature of oil through the coiled heat. At the shell-side inlet, a mass flow rate of 0.06 kg/s was applied with a temperature of 287.15 K; at the tube-side inlet, a mass flow rate of 0.008–0.054 kg/s was assigned to maintain the De in the range of 17–195 with a temperature of 45 °C, 55 °C, and 75 °C. The De is defined as follows:

$$De = Re_i\left(\frac{d}{2R_c}\right)^{1/2} \tag{4}$$

where Re_i is the Reynolds number at the inside of the coiled tube, d is the diameter of the coiled tube, and R_c is the curvature radius. A relative pressure of 0 Pa was applied at both the shell- and tube-side outlets. At the contact surface between the solid and fluid domains, an interface condition was applied to estimate the conjugate heat transfer by using a general grid interface connection [18], which has been applied to many numerical analyses of different physical domains, such as rotating and stationary domains [23], or fluid and solid domains [24]. All of the wall surfaces on the outside of the heat exchanger and calculation domain were assumed to be nonslip and adiabatic.

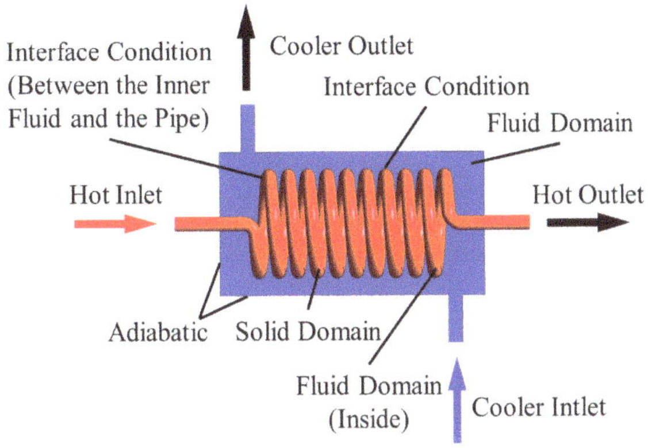

Figure 3. Computational domain and boundary conditions.

Figure 4 shows an example of the computational grid system, which comprises a tetrahedral mesh in most parts, and a hexahedral mesh near the inlet and outlet, to obtain more accurate inlet and outlet flow velocity profiles and reduce the overall number of grids. The mesh was finer in the helically-coiled tube and its surroundings for a complex flow phenomenon. A growth factor of 1.2 was adopted from the concentrated region to have a proper expansion factor for calculation stability.

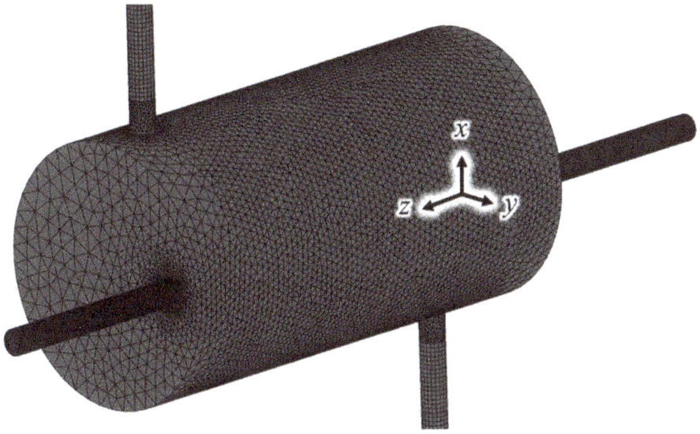

(a)

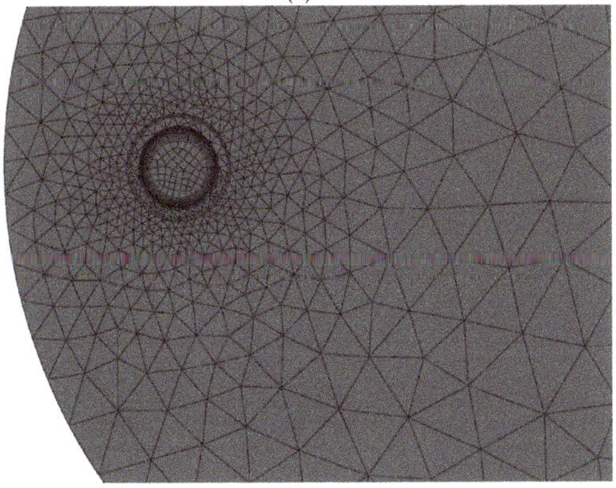

(b)

Figure 4. Example of the computational grid system: (**a**) overall and (**b**) wall treatment.

As the convergence criteria, the root-mean-square relative residual values of all the fluid and solid parameters were set to 1.0×10^{-5}. The computational time per analysis was about 12 h, and the time was dependent on the complexity of the geometry and the number of the grid system.

4. Result and Discussion

To determine the optimal number of the grid elements, three nodes of the grid system in the range of 900,000 to 3,300,000 were tested; Figure 5 illustrates the test result for a flat helical tube at $De = 111$. The Nusselt number on the inner surface of the coiled tube (Nu_i) was used as a criterion for the comparison and is defined as $Nu_i = h_i d_i / k$, where h_i and k are the area-averaged convective heat transfer coefficient at the surface of the inner fluid domain and the thermal conductivity of the inner fluid, as shown in Equation (3), respectively. The deviation of the coarsest grid system (0.9 million) shows a relative error of 12.22% compared with the finest grid system (3.3 million), while the deviation of the second finest grid system (1.9 million) shows a relative error of 2.19%. Resultantly, approximately 1.9 million nodes was decided to be the optimum number of nodes for calculation; however, the number of nodes varied with the shape of the coiled tube from 1.9 to 2.1 million.

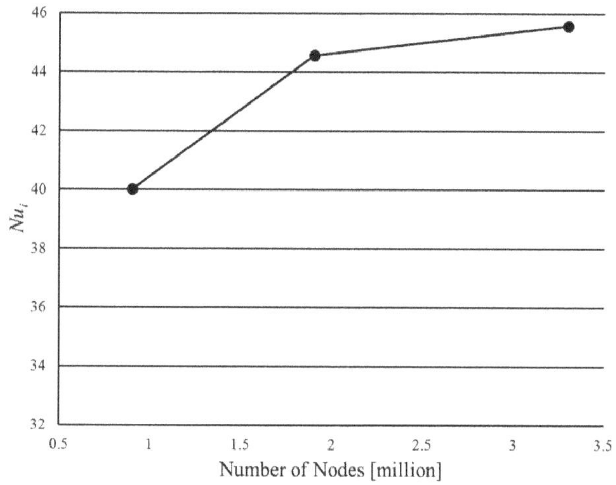

Figure 5. Grid dependency test.

To assess the validity of the numerical solutions, the results were compared with the experimental data of the flat coiled tube, as shown in Figure 6. The validation was performed for Nu_i with various De and two temperatures at the inlet of the tube-side. The results of $T_i = 55\ ^\circ\mathrm{C}$ showed a deviation of 14.0% relative error at $De = 31.8$, and 1.7% averaged relative error at $De > 66.7$. The results of $T_i = 70\ ^\circ\mathrm{C}$ showed 5.7% averaged relative error at $De > 140$, and 11.5% averaged relative error at $De < 98.3$. Overall, the present numerical results show good agreement with the experimental data.

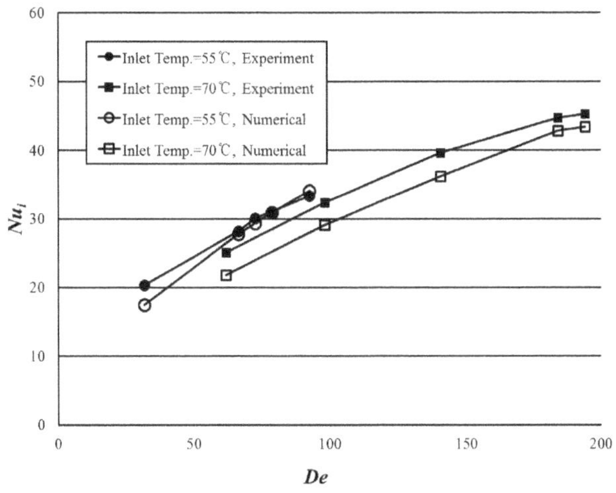

Figure 6. Validation of the numerical results with the experimental data [20] for the Nusselt number inside the helically-coiled tube.

Figure 7 presents the temperature difference charts of the hot channel with various *De*. The difference was calculated as $\Delta T = T_{hot,in} - T_{hot,out}$. The deviation in each case shows an almost similar trend. The inline and staggered arrangements show the highest temperature differences through all *De* values, whereas the staggered arrangement shows a slightly higher temperature difference. In addition, the bulged arrangement shows a higher temperature difference at lower *De* (<40) with $T_{hot,in}$ = 45 and 55 °C; however, as *De* increases, results similar with those of the flat case are observed. At $T_{hot,in}$ = 70 °C with a higher *De* (>100), the flat case shows a slightly higher temperature difference. It was obvious that increase in the residence time of the hot channel fluid increased the temperature difference of the hot channel.

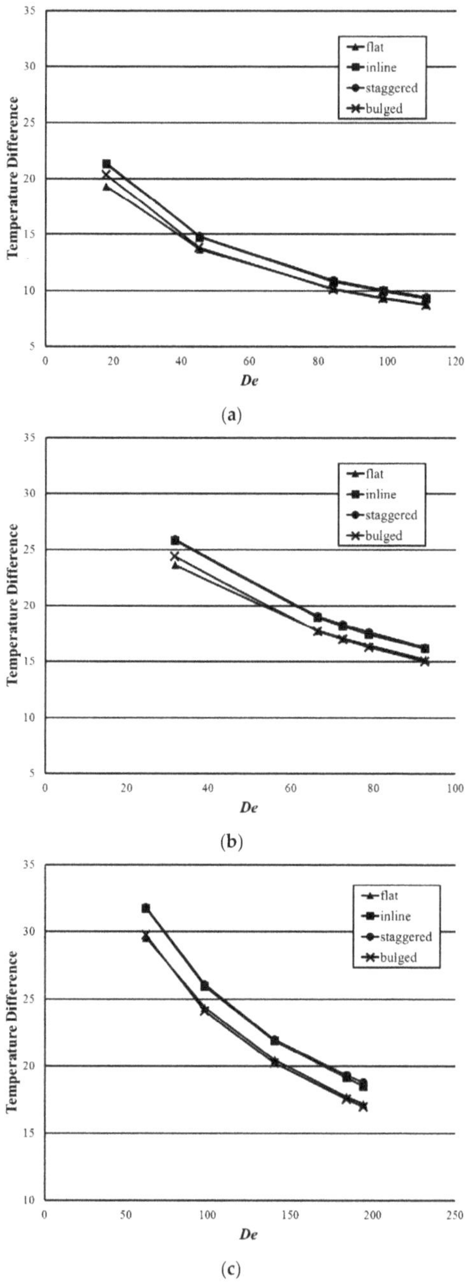

Figure 7. Temperature difference charts of the hot channel with various *De*: (**a**) $T_{hot,\ in}$ = 45 °C; (**b**) $T_{hot,in}$ = 55 °C; and (**c**) $T_{hot,in}$ = 70 °C.

Figure 8 shows temperature contours inside the helically-coiled tube with the velocity vector of the cold channel on the y–z plane at $x = 0$; $De = 17.87$ and inlet temperature of the hot channel is 45 °C. The results were focused on the temperature variation inside the coiled tube with a circumferential fluid flow on the shell-side. The hot fluid was affected by the centric fluid flow rather than outer fluid flow on the shell-side; therefore, the thermal layer inside the tube started at the inner surface for all cases. The high temperature fluid flows along the shape of the helically-coiled tube and the fluid that is closer to the center of the shell flows more slowly than the fluid that is farther from the center of the shell due to the centrifugal effect. Moreover, the fluid inside the shell-side flows faster at the inner area than at the outer (near wall) area. As the inline and staggered arrangements show rapid temperature variation in the coiled tube, the thermal layer was observed simultaneously from the inner and outer surfaces, and a lower temperature of the hot channel was observed near the exit, as shown in Figure 7.

Figure 9 presents the Nusselt number inside the helically-coiled tube with various *De* values and inlet temperatures of the hot channel. The staggered arrangement shows the best heat transfer performance, while the inline arrangement shows the second best performance for all ranges of *De* and inlet temperature. The bulged arrangement shows a higher than the flat arrangement at a lower *De* (<30); however, the flat arrangement shows a higher Nu_i value at a higher *De* (>40) at $T_{hot,in}$ = 45 °C. A similar result can be observed for $T_{hot,in}$ = 55 °C. The deviation between each case was increased with a higher *De* value at all inlet temperatures.

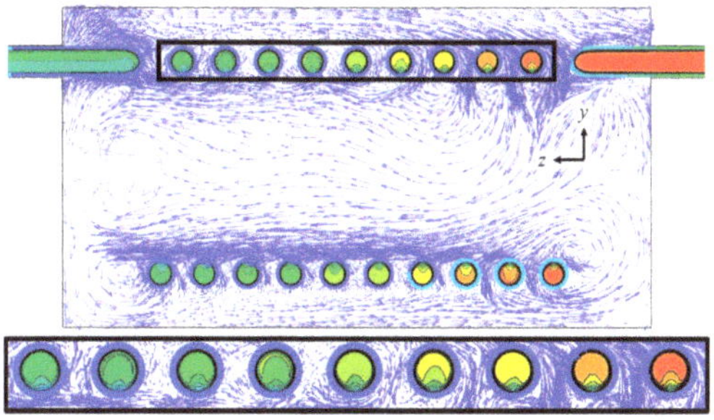

(a)

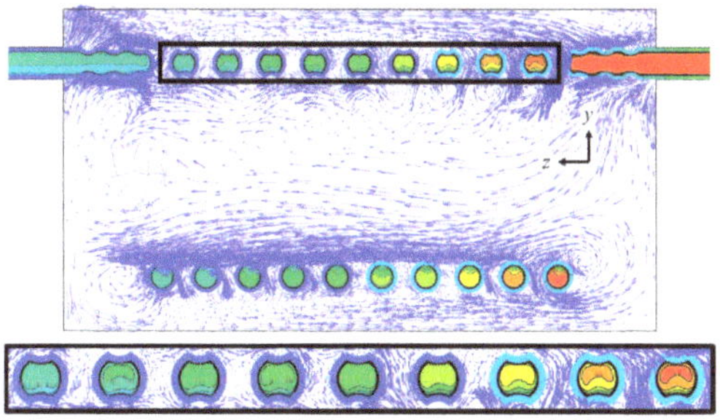

(b)

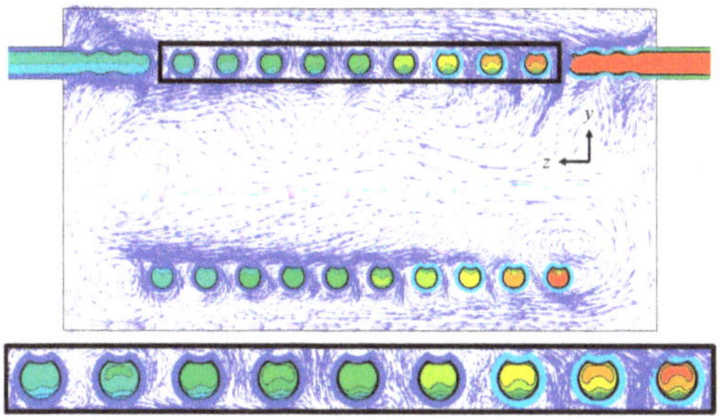

(c)

Figure 8. *Cont.*

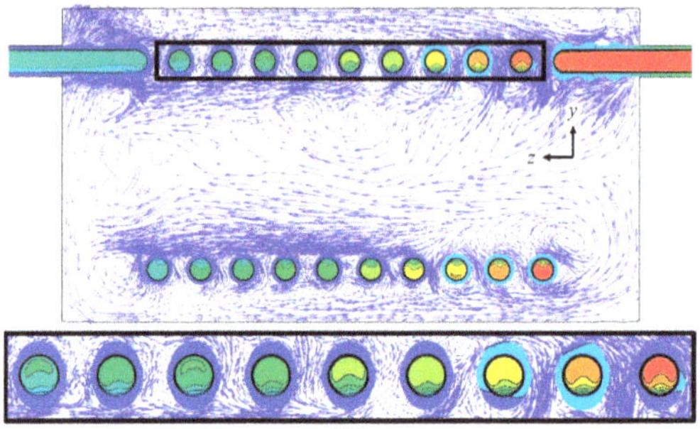

(**d**)

Figure 8. Temperature contours inside the helically-coiled tube with a velocity vector of the cold channel on the y–z plane at $x = 0$: (**a**) flat; (**b**) inline; (**c**) staggered; and (**d**) bulged.

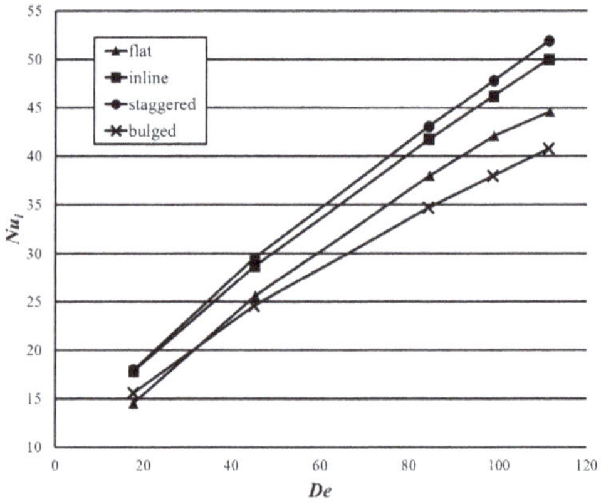

(**a**)

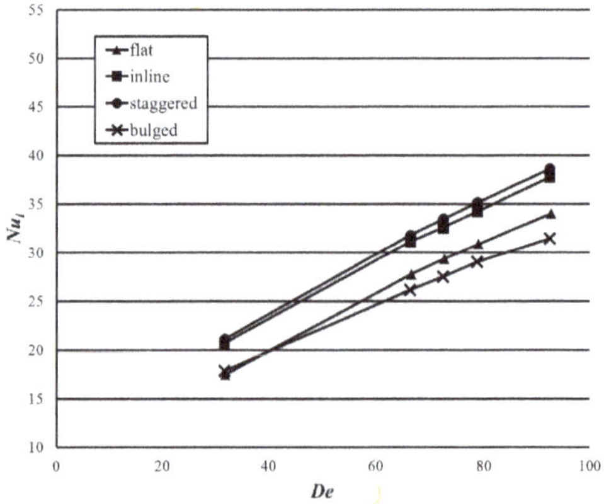

(**b**)

Figure 9. *Cont.*

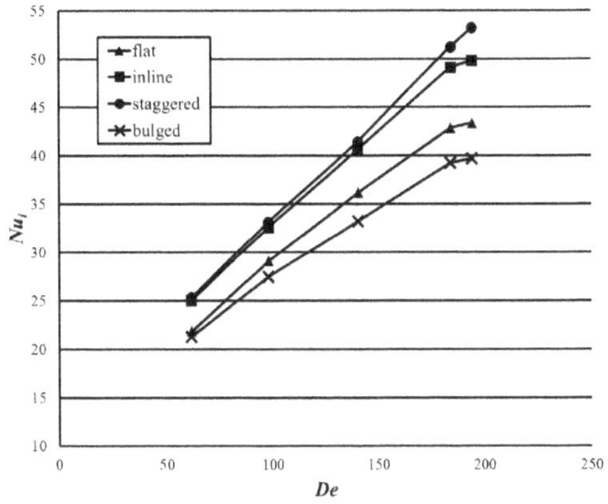

(c)

Figure 9. Nusselt number inside the helically-coiled tube with various *De*: (**a**) $T_{hot,\,in}$ = 45 °C; (**b**) $T_{hot,in}$ = 55 °C; and (**c**) $T_{hot,in}$ = 70 °C.

Figure 10 shows pressure drops at the helically-coiled tube with various *De*. The inline and staggered arrangements with the best heat transfer performance show the highest pressure drop for all inlet temperatures of the hot channel, and the flat arrangement shows the lowest pressure drop for all inlet temperatures. The deviations between each case increased with the *De* for all inlet temperatures.

Overall, introducing the staggered arrangement structure to the shell and helically-coiled tube system efficiently enhanced the heat transfer performance; however, the pressure drop increased through the system.

It is generally known that the performance of the dimple, as a device of heat transfer augment, can be influenced by the shape of the dimple, such as the dimple height, dimple width, and dimple periodic arrangement. In a further study, the effect of geometric variables of the dimple will be investigated to optimize the system and to generalize the results to other dimple systems.

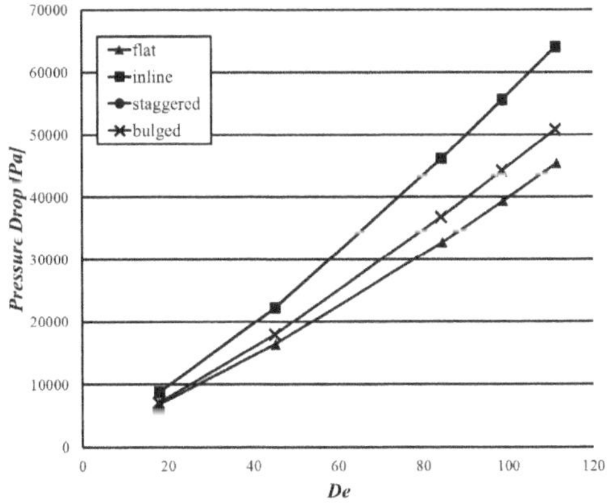

(a)

Figure 10. *Cont.*

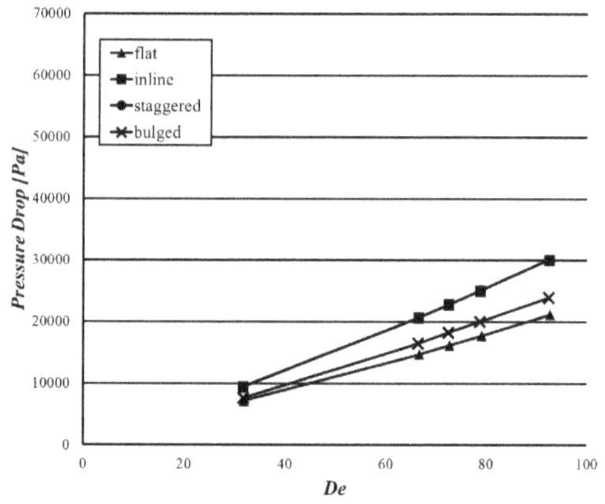

(**b**)

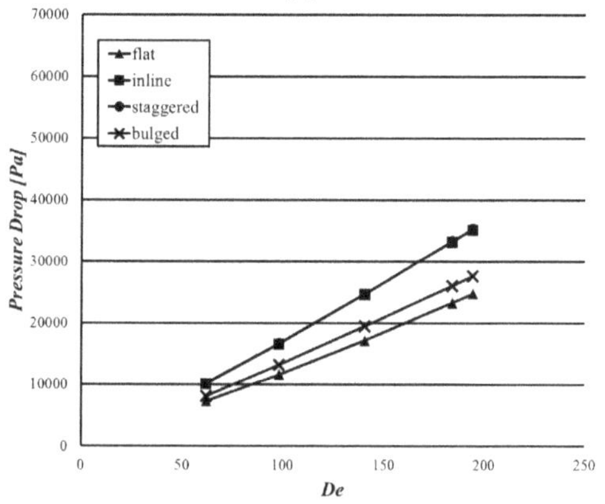

(**c**)

Figure 10. Pressure drop at the helically-coiled tube with various *De*: (**a**) $T_{hot,in}$ = 45 °C; (**b**) $T_{hot,in}$ = 55 °C; and (**c**) $T_{hot,in}$ = 70 °C.

5. Conclusions

This paper described the evaluations of the shell and helically-coiled tube heat exchangers with various dimple arrangements, flat, inline, staggered, and bulged, at different *De* and inlet temperatures of the hot channel. These evaluations were performed for determining the temperature difference between the inlet and outlet, Nusselt number inside the coiled tube, and pressure drop of the coiled tube by using 3-D RANS equations with SST turbulence closure. A grid dependency test was performed to determine the optimal number of the grid system; approximately 190 million nodes were found to be optimal for calculation. The numerical results were validated using the experimental data, and showed good agreement within the average relative error of 1.7%–5.7%. The inline and staggered arrangements show the highest temperature differences for all values of *De*, and the bulged and flat arrangements show lower temperature differences. The staggered arrangement shows the best heat transfer performance, followed by the inline arrangement for all ranges of *De* and inlet temperature of the hot channel. The inline and staggered arrangements show the highest pressure drop and the flat arrangement shows the lowest pressure drop through all inlet temperatures of the hot channel. The performance is expected to be enhanced further by optimizing the geometric configuration through a range of optimization techniques [25–28].

Acknowledgments: This research was supported by a grant (16AUDP-B083704-03) from Architecture & Urban Development Research Program funded by Ministry of Land, Infrastructure and Transport of Korean government.

Author Contributions: Jun-Ho Jo and Ye-Eun Lee supported the numerical investigation, and the revision of the manuscript. Yeong-Seok Yoo supervised the project, and provided professional technical guidelines and insights. Sun-Min Kim designed the project, analyzed the data, wrote the draft of the paper, and carried out the final revision of the paper.

Conflicts of Interest: The authors declare no conflict of interest.

Nomenclature

C_p	specific heat (J/kg·K)
d	diameter (m)
De	Dean number
h	convective heat transfer coefficient (W/m^2·K)
H	height (m)
Nu	Nusselt number
p	pitch (m)
R_c	curvature radius (m)
Re	Reynolds number
T	temperature (K)
x, y, z	Cartesian coordinates (m)

Greek symbols

μ	dynamic viscosity (Pa·s)
k	thermal conductivity
ρ	density (kg/m^3)
ν	kinematic viscosity (m^2/s)
τ	shear stress (N/m^2)

Subscripts

d	dimple
hot	hot channel
cold	cold channel
in	inner
out	outer

References

1. Doshi, V.A.; Vuthaluru, H.B.; Bastow, T. Investigations into the control of odour and viscosity of biomass oil derived from pyrolysis of sewage sludge. *Fuel Process. Technol.* **2005**, *86*, 885–897. [CrossRef]
2. Tripathi, M.; Sahu, J.N.; Ganesan, P. Effect of process parameters on production of biochar from biomass waste through pyrolysis: A review. *Renew. Sustain. Energy Rev.* **2016**, *55*, 467–481. [CrossRef]
3. Mohanty, P.; Pant, K.K.; Naik, S.N.; Parikh, J.; Hornung, A.; Sahu, J.N. Synthesis of green fuels from biogenic waste through thermochemical route–The role of heterogeneous catalyst: A review. *Renew. Sustain. Energy Rev.* **2014**, *38*, 131–153. [CrossRef]
4. Mašek, O.; Budarin, V.; Gronnow, M.; Crombie, K.; Brownsort, P.; Fitzpatrick, E.; Hurst, P. Microwave and slow pyrolysis biochar: Comparison of physical and functional properties. *J. Anal. Appl. Pyrolysis* **2013**, *100*, 41–48. [CrossRef]
5. Park, H.J.; Dong, J.I.; Jeon, J.K.; Park, Y.K.; Yoo, K.S.; Kim, S.S.; Kim, J.; Kim, S. Effects of the operating parameters on the production of bio-oil in the fast pyrolysis of Japanese larch. *Chem. Eng. J.* **2008**, *143*, 124–132. [CrossRef]
6. Fu, P.; Hu, S.; Xiang, J.; Sun, L.; Su, S.; Wang, J. Evaluation of the porous structure development of chars from pyrolysis of rice straw: Effects of pyrolysis temperature and heating rate. *J. Anal. Appl. Pyrolysis* **2012**, *98*, 177–183. [CrossRef]

7. Ahmed, I.I.; Gupta, A.K. Pyrolysis and gasification of food waste: Syngas characteristics and char gasification kinetics. *Appl. Energy* **2010**, *87*, 101–108. [CrossRef]
8. Digman, B.; Kim, D.S. Review alternative energy from food processing wastes. *Environ. Prog.* **2008**, *27*, 524–537. [CrossRef]
9. Rappert, S.; Müller, R. Odor compounds in waste gas emissions from agricultural operations and food industries. *Waste Manag.* **2005**, *25*, 887–907. [CrossRef] [PubMed]
10. Burgess, J.E.; Parsons, S.A.; Stuetz, R.M. Developments in odour control and waste gas treatment biotechnology—A review. *Biotechnol. Adv.* **2001**, *19*, 35–63. [CrossRef]
11. Wang, Q.; Chen, G.; Chen, Q.; Zeng, M. Review of improvements on shell-and-tube heat exchangers with helical baffles. *Heat Transf. Eng.* **2010**, *31*, 836–853. [CrossRef]
12. Selbas, R.; Kızılkan, Ö.; Reppich, M. A new design approach for shell-and-tube heat exchangers using genetic algorithms from economic point of view. *Chem. Eng. Process.* **2006**, *45*, 268–275. [CrossRef]
13. Babu, B.V.; Munawar, S.A. Differential evolution strategies for optimal design of shell-and-tube heat exchangers. *Chem. Eng. Sci.* **2007**, *62*, 3720–3739. [CrossRef]
14. Xie, G.N.; Wang, Q.W.; Zeng, M.; Luo, L.Q. Heat transfer analysis for shell-and-tube heat exchangers with experimental data by artificial neural networks approach. *Appl. Therm. Eng.* **2007**, *27*, 1096–1104. [CrossRef]
15. Fesanghary, M.; Damangir, E.; Soleimani, I. Design optimization of shell and tube heat exchangers using global sensitivity analysis and harmony search algorithm. *Appl. Therm. Eng.* **2009**, *29*, 1026–1031. [CrossRef]
16. Guo, J.; Xu, M.; Cheng, L. The application of field synergy number in shell-and-tube heat exchanger optimization design. *Appl. Energy* **2009**, *86*, 2079–2087. [CrossRef]
17. Zhang, J.F. Experimental performance comparison of shell-side heat transfer for shell-and-tube heat exchangers with middle-overlapped helical baffles and segmental baffles. *Chem. Eng. Sci.* **2009**, *64*, 1643–1653. [CrossRef]
18. ANSYS Inc. *ANSYS CFX-Solver Theory Guide, ANSYS CFX Release 17.0*; ANSYS Inc.: Canonsburg, PA, USA, 2016.
19. Menter, F.R. Two-equation eddy-viscosity turbulence models for engineering applications. *AIAA J.* **1994**, *32*, 1598–1605. [CrossRef]
20. Salimpour, M.R. Heat transfer characteristics of a temperature-dependent-property fluid in shell and coiled tube heat exchangers. *Int. Commun. Heat Mass Transf.* **2008**, *35*, 1190–1195. [CrossRef]
21. Kim, S.M.; Kim, K.Y. Optimization of a Hybrid Double-Side Jet Impingement Cooling System for High-Power Light Emitting Diodes. *J. Electron. Packag.* **2014**, *136*. [CrossRef]
22. Husain, A.; Kim, S.M.; Kim, K.Y. Performance analysis and design optimization of micro-jet impingement heat sink. *Heat Mass Transf.* **2013**, *49*, 1613–1624. [CrossRef]
23. Heo, M.W.; Seo, T.W.; Shim, H.S.; Kim, K.Y. Optimization of a regenerative blower to enhance aerodynamic and aeroacoustic performance. *J. Mech. Sci. Technol.* **2016**, *30*, 1197–1208. [CrossRef]
24. Kim, S.M.; Kim, K.Y. Microcooling system with impinging jets and a stalactite structure. *Numer. Heat Transf. Part A Appl.* **2016**, *69*, 1376–1389. [CrossRef]
25. Kim, K.Y.; Lee, Y.M. Design optimization of internal cooling passage with V-shaped ribs. *Numer. Heat Transf. Part A Appl.* **2007**, *51*, 1103–1118. [CrossRef]
26. Lee, K.D.; Kim, K.Y. Shape optimization of a fan-shaped hole to enhance film-cooling effectiveness. *Int. J. Heat Mass Transf.* **2010**, *53*, 2996–3005. [CrossRef]
27. Li, P.; Kim, K.Y. Multiobjective optimization of staggered elliptical pin-fin arrays. *Numer. Heat Transf. Part A Appl.* **2008**, *53*, 418–431. [CrossRef]
28. Husain, A.; Kim, K.Y. Enhanced multi-objective optimization of a microchannel heat sink through evolutionary algorithm coupled with multiple surrogate models. *Appl. Therm. Eng.* **2010**, *30*, 1683–1691. [CrossRef]

energies

Article

Numerical Models for Viscoelastic Liquid Atomization Spray

Lijuan Qian [1,*], Jianzhong Lin [1,2,*] and Fubing Bao [1]

1 College of Mechanical and Electrical Engineering, China Jiliang University, Hangzhou 310018, China; dingobao@cjlu.edu.cn
2 State Key Laboratory of Fluid Power and Mechatronic Systems, Zhejiang University, Hangzhou 310027, China
* Correspondence: qianlj@cjlu.edu.cn (L.Q.); mecjzlin@zju.edu.cn (J.L.); Tel.: +86-571-8683-6009 (J.L.)

Academic Editor: Bjørn H. Hjertager
Received: 28 October 2016; Accepted: 5 December 2016; Published: 17 December 2016

Abstract: Atomization spray of non-Newtonian liquid plays a pivotal role in various engineering applications, especially for the energy utilization. To operate spray systems efficiently and well understand the effects of liquid rheological properties on the whole spray process, a comprehensive model using Euler-Lagrangian approaches was established to simulate the evolution of the atomization spray for viscoelastic liquid. Based on the Oldroyd model, the viscoelastic linear dispersion relation was introduced into the primary atomization; an extended viscoelastic version of Taylor analogy breakup (TAB) model was proposed; and the coalescence criteria was modified by rheological parameters, such as the relaxation time, the retardation time and the zero shear viscosity. The predicted results are validated with experimental data varying air-liquid mass flow ratio (*ALR*). Then, numerical calculations are conducted to investigate the characteristics of viscoelastic liquid atomization process. Results showed that the evolutionary trend of droplet mean diameter, Weber number and Ohnesorge number of viscoelastic liquids along with axial direction were qualitatively similar to that of Newtonian liquid. However, the mean size of polymer solution increased more gently than that of water at the downstream of the spray, which was beneficial to stable control of the desirable size in the applications. As concerned the effects of liquid physical properties, the surface tension played an important role in the primary atomization, which indicated the benefit of selecting the solvents with lower surface tension for finer atomization effects, while, for the evolution of atomization spray, larger relaxation time and zero shear viscosity increased droplet Sauter mean diameter (*SMD*) significantly. The zero shear viscosity was effective throughout the jet region, while the effect of relaxation time became weaken at the downstream of the spray field.

Keywords: viscoelastic fluid; atomization spray; numerical modeling

1. Introduction

Atomization of non-Newtonian liquid has become increasingly prevalent in various engineering applications, such as injection of gelled fuels for propulsion system [1], spray coating for painting [2,3] manufacture of pharmaceutical tablets [4], and materials processing for suspension plasma spray [5].The non-Newtonian droplets can provide attractive and variegated properties to meet the specific requirements. The constitution relation for non-Newtonian liquids are complicated since the relationship between shear stress and rate of strain exhibits non-linear behavior. For the widespread shear-thinning liquids, there are several rheological models to describe the variation of viscosity along with strain rate [6,7], such as power-law model, Carreau model and Oldroyd model [8]. Among them, Oldroyd model with three constants can effectively describe the rheological characteristics for viscoelastic liquids.

For the atomization of non-Newtonian liquid, the external two-phase flow outside of the nozzle orifice can be generally divided into two parts, the near-field region and the far-field domain, where specific physical phenomena govern the flow.

In the near-field region, the primary atomization in which the liquid stream disintegrates into ligaments and drops by interacting with the gas takes place. The studies on primary atomization can be classified into three approaches: experimental measurements, theoretical analysis and modeling calculations.

Based on the techniques of high speed camera and phase Doppler anemometry, experiments provided visualized approaches to reveal the morphology and detailed information of the primary atomization such as breakup length, geometry, spray angle and so on [9]. Dumouchel [10] has reviewed the experimental investigate dedicated to the primary atomization step for Newtonian liquids. Aliseda et al. [11] extracted the images of liquid jet break-up which is closed to the nozzle orifice from high-speed visualizations for several Weber and Reynolds numbers for viscous and non-Newtonian liquids. They observed that at lower Weber numbers the Kelvin–Helmholtz instability grew slowly and several intact wavelengths were observed prior to break-up, In contrast at larger weber number, the Rayleigh-Taylor (R-T) instability was arresting. Acceleration of the interface resulted in the R-T instability creating ligaments of fluid which eventually break-up into droplets. This experimental finding could inspire us to use the R-T instability to describe the breakup of cylindrical ligaments. Harrison et al. [12] experimentally examined the spray cone angle with three types of polymer solutions and studied the cone angle as a function of type of polymer and reduced concentration. Chao et al. [13] evaluated the effect of polymeric additives on spray structure from photographs of the sprays. Their studies show the correlations between additive concentrations and the formation of droplets. These experimental investigations are very helpful for understanding the features of viscoelastic liquid breakup and give some guidance to build up models.

The theoretical analysis of primary atomization mainly focused on the derivation of dispersion relations between the surface wave growth rate and the wave number, which is based on two instability analysis: Rayleigh-Taylor (R-T) and the Kelvin-Helmholtz (K-H) instability. Various breakup modes resulted in different dispersion relations. Sirignano and Mehring [14] have reviewed the basic mechanism of distortion and disintegration of liquid streams. For the non-Newtonian liquid, especially viscoelastic, the linear instability dispersion relationship is more complex than the Newtonian liquid. Middleman [15] firstly performed a linearized stability analysis on viscoelastic jet to predict the stream disintegration. Goren and Gorttlieb [16] expanded the Weber's Newtonian liquid linear stability analysis by involving an equivalent liquid viscosity and including an unrelaxed axial tension into the dispersion equation which can be used to explain the stabilization of the viscoelastic jet. Joseph et al. [17] took the high relative velocity between liquid and the acceleration of droplets into consideration and obtained the dispersion relation based on R-T instability. Yang et al. [18] used linear stability analysis to investigate the instability of a three-dimensional viscoelastic liquid jet surrounded by a swirling air stream.

For the modeling of primary atomization, the direct numerical simulation (DNS) of multiphase flows which is based on the one-fluid formalism coupled with interface tracking algorithms [19–22] is a promising approach to thoroughly investigate the whole process. As reviewed by Villermaux [20] and Gorokhovski et al. [21], the DNS of turbulent primary atomization process is still in its infancy owing to the high computational cost and big numerical challenges. In addition, most of the numerical studies are focused on the Newtonian liquids, few investigations have devoted to viscoelastic liquids since the complicated rheological relations further enhanced the difficulties [22]. Instead, based on some simplifications and assumptions of the gas-liquid interface, and involving the instability analysis, we can find a compromising way to establish a viable physical model and predict the size of ligaments/droplets after primary atomization directly and effectively.

At the far-field domain, the droplets produced by primary atomization are unstable in the turbulent spray, and will undergo a series of events such as secondary breakup, collision and coalescence. For non-Newtonian spray, the effects of viscoelasticity on the droplets evolution and distribution in the downstream field have attracted numerous industrial applications [23,24]. For different applications, the requirements of droplet size and distribution are different, for instance, finer droplets are desirable for combustions and painting spray [25]. On the contrary, in pesticide sprays [26] and anti-misting jet, too small drops are undesirable. The evolution of droplets at the downstream of the spray is critical to the final outcomes. However in the downstream of the spray, there is a scarcity in literature to investigate the evolution and distribution of atomized drops, especially for the atomization of viscoelastic liquids.

In the experimental aspect, most studies devoted to determine the features of non-Newtonian droplet secondary breakup. Rivera [27] used the high-speed photography to find that Non-Newtonian drop secondary breakup mode is qualitatively similar to those observed for Newtonian drops. The only major difference is that the bag stretches more before it ruptures and the resulting breakup fragments persist longer than their Newtonian counterparts. Dechelette et al. [28] experimentally investigated the combined effect of polymers and soluble surfactants on dynamics of jet breakup and formation of satellite drop. Ma et al. [29] found the particle size distribution was fit to the Rosin-Rammler function through 3-D phase Doppler methods. Hartranft and Settles [25] indicated that extensional viscosity had the dominant influence on sheet breakup features compared to Weber number, which means the traditional Weber number may not qualify to describe the breakup of non-Newtonian fluids. These experimental investigations can support us to extend the breakup mode of Newtonian drop to the viscoelastic version.

In the modeling aspects, for the secondary atomization, some investigations have devoted to build up physical models to illustrate droplet breakup. Xue et al. [30] used a two-dimensional computational approach to predict the effect of coupled surfactant and non-Newtonian mechanisms on the formation of satellite drops. Rivera [27] extended the Taylor analogy breakup (TAB) model to describe the inelastic non-Newtonian liquid with power-law model and validated the model with experimental data. It is noted that most of previous models are unable to account for the whole spray pattern and its evolution without considering the poly-dispersion of droplets as well as collision and coalescence [31]. There is very few investigations devoted to establish the numerical models accounting for the evolution of viscoelastic liquid spray [32], due to the complexity of rheological behavior and statistical difficulties.

As such, the motivation of this paper is to logically expand the above mentioned research efforts and aim specially to shed light on the simulations of the downstream of the spray for viscoelastic liquid. The numerical model will consider the primary breakup, secondary breakup, droplet tracking and collision. For primary breakup, Goren and Joseph's dispersion relationship will be used to describe the linear instability of viscoelastic liquid. For secondary atomization, the TAB is extended to the viscoelastic version based on the rheological analysis. For droplet collision, the criterion of coalescence is modified by the rheological properties. The numerical calculations will try to give some detailed information about the evolutions of droplet distribution, size distribution and critical dimensionless numbers (such as Weber number and Ohnesorge number). Results will involve three parts. Firstly, the primary atomization model is validated and compared by experimental data. Then, the evolution of spray field is investigated in detailed by comparison of polymer solution and water. Finally, the effects of rheological properties on spray performance are discussed.

2. Mathematical Models

The mathematical model considers two parts. As shown in Figure 1, the first one is the primary breakup sub-model, which predicts gas velocity at the nozzle exit and the droplet mean size. The second sub-model calculates the evolutions of the gas flow and the droplets spray accounting for secondary breakup, collision and coalescence, momentum transfer. The initial droplets characteristics in the second sub-model are provided by the first sub-model calculation. In droplet-gas two-phase flow, the two-way coupling of gas and droplet phase is considered, while the effect of primary breakup jet on the environment gas is neglected.

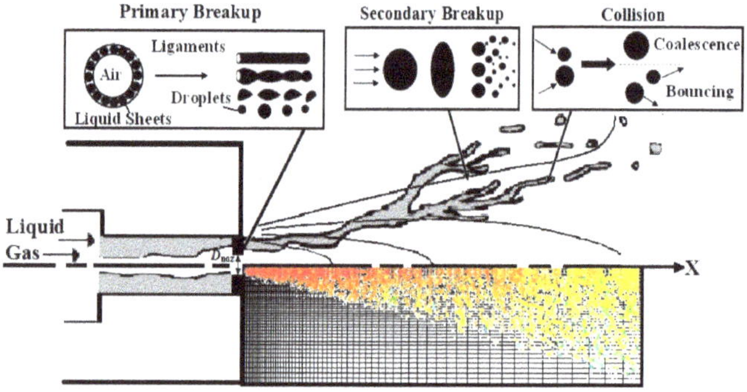

Figure 1. Schematic of the effervescent atomization spray.

2.1. *Sub-Model of Primary Atomization*

The primary atomization is a process that the liquid stream disintegrates into ligaments and drops by interacting with the gas. It is a complex gas-liquid flow. In order to simplify the process, we extracted the main features of the primary atomization from experimental observations.

At present study, effervescent atomizer is chosen since it has so far been found effective for high viscosity liquids [33], in which the gas is introduced into the liquid at low pressure to form a bubbly two-phase mixture upstream of the orifice. As indicated by experimental observation, the flow pattern closed to the nozzle exit is similar to annular. Then, as illustrated in Figure 1, the primary breakup sub-model for effervescent atomizer can be established through three steps.

The first step is estimation of the annular sheets as well as the cylindrical ligaments. The momentum equation of the annular flow combined with the state equation can be written as:

$$RT\ln\left(\rho_g RT\right)+\frac{1}{2}\left(\frac{\dot{m}_l ALR}{\rho_g \pi r_g^2}\right)^2=\text{const} \tag{1}$$

where ρ_g is the gas density, r_g is the radius of gas flow, $\dot{m}_l$ is the mass flow rate of liquid, and ALR is the air-liquid ratio by mass. The radius of gas flow can be written in terms of orifice radius r_{orific} and void fraction ϑ, $r_g=\sqrt{\vartheta}r_{\text{orifice}}$. The interface velocity slip ratio v_{slip} under different flow rate is expressed as [34]:

$$v_{\text{slip}}=\sqrt{\frac{\rho_l}{\rho_g}\frac{\sqrt{\vartheta}}{1+C(1-\vartheta)}} \tag{2}$$

where C is the experimental coefficient of the mass flow rate scaling. The interface velocity slip ratio and void fraction are related by:

$$1+\frac{\rho_g sr}{\rho_l ALR}=\frac{1}{\vartheta} \tag{3}$$

By solving Equations (1)–(3), ρ_g, ϑ and v_{slip} can be calculated for different operating conditions. The thickness of annular liquid sheet is then calculated from $\delta_{lig}=4(r_o-r_g)/\sqrt{\pi}$, which is also the diameter of the typical cylindrical ligament. Besides, the gas velocity inside the nozzle orifice can be expressed as $V_{g1}=\dot{m}_g/(A_{noz}\times\vartheta\times\rho_g)$, where A_{noz} is the area of the nozzle orifice and $\dot{m}_g$ is the gas flow rate. According to the energy conservation equation, the gas velocity at the nozzle exit can be derived from $V_{g2}=\sqrt{{V_{g1}}^2+2RT\log(P_{in}/P_{out})}$, where R is the gas constant, T is the temperature, P_{in} is the pressure inside the nozzle, and P_{out} is the pressure outside the nozzle.

Secondly, the formed ligaments are approximated by cylindrical jets. The ligaments will breakup into fragments at the wavelength of the most rapidly growing wave. The dispersion relations based on instability analysis are used to determine the critical wave number. Here, Joseph's dispersion relationship [17] is used to describe the linear instability of viscoelastic liquid.

Joseph's dispersion relationship is derived by considering the acceleration of a drop exposed to a high speed air stream. The equations of ligament motion coupled with boundary conditions are solved to deduce the expression of the disturbed interface:

$$-\left[1+\frac{1}{\alpha^2}(-\dot{a}k+\frac{\sigma k^3}{\rho_l})\right]-4\frac{k^2}{\alpha}\frac{\mu_{ve}}{\rho_l}+4\frac{k^3}{\alpha^2}(\frac{\mu_{ve}}{\rho_l})^2(\sqrt{k^2+\alpha\rho_l/\mu_{ve}}-k)=0\ \left(\rho_l>>\rho_g\right) \quad (4)$$

Equation (4) is an approximate analysis of Rayleigh-Taylor instability based on viscoelastic potential flow which gives the critical wavelength and growth rate to within less than 10% of the exact theory. In Equation (4), k is the magnitude of the wave number, α is the amplification rate, and σ is the liquid surface tension. μ_{ve} is the effective shear viscosity of the liquid, $\mu_{ve}=\mu_0(1+\alpha\lambda_2)/(1+\alpha\lambda_1)$ (Section 2.2.2), for viscoelastic liquid, μ_{ve} is large and $\alpha\rho_l/(k^2\mu_{ve})<<1$, then $(\sqrt{k^2+\alpha\rho_l/\mu_{ve}}-k)$ can be assumed to be $\alpha\rho_l/2k\mu_{ve}$. $\dot{a}$ is the acceleration of the liquid ligament:

$$\dot{a}=\frac{F_{c_d}}{m_{lig}}=\frac{c_d\rho_g(u_g-u_l)^2}{\rho_l\delta_{lig}} \quad (5)$$

where c_d is the drag coefficient. As the ligament Reynolds number $Re_{lig}=\rho_g\delta_{lig}|u_g-u_l|/\mu_g$ is between 3 and 1000, here $c_d=24(1+{Re_{lig}}^{2/3}/6)/Re_{lig}$. Then, Equation (4) reduces to be a three-degree polynomial of α as follows and can be solved analytically:

$$\lambda_1\rho_l\alpha^3+(\rho_l+2\mu_0\lambda_2k^2)\alpha^2+(2\mu_0k^2-\dot{a}k\rho_l\lambda_1+\sigma k^3\lambda_1)\alpha-\dot{a}k\rho_l+\sigma k^3=0 \quad (6)$$

Based on the above dispersion relation, the critical breakup wave number that corresponds to the maximum growth rate can be used to calculate the wavelength of a typical ligament fragment. Finally, assuming that each fragment is stabilized to one spherical droplet under the influence of surface tension, the Sauter mean drop diameter SMD (defined as the ratio of volume–surface mean diameter $\sum_i n_pd_p^3/\sum_i n_pd_p^2$) can then be calculated from the conservation of mass:

$$SMD=\left[3\pi\frac{{\delta_{lig}}^2}{k}\right]^{1/3} \quad (7)$$

2.2. Sub-Model of Droplet-Gas Two-Phase Flow

The second sub-model is based on hybrid Euler-Lagrangian coordinate system to describe the evolution of spray. The flow field is axisymmetric, and the parcels are tracked in three-dimensional coordinate.

2.2.1. Gas Phase

For the turbulent gas spray, the conservation equations of mass, momentum and energy are written as follows:

$$\begin{cases}\frac{\partial\rho_g}{\partial t}+\nabla\cdot\left(\rho_g\mathbf{u}_g\right)=0\\ \frac{\partial\left(\rho_g\mathbf{u}_g\right)}{\partial t}+\nabla\cdot\left(\rho_g\mathbf{u}_g\mathbf{u}_g\right)=-\nabla\left(p+\frac{2}{3}\rho_g\kappa\right)+\nabla\cdot[\sigma_s]+\mathbf{F}\\ \frac{\partial\left(\rho_g e\right)}{\partial t}+\nabla\cdot\left(\rho_g e\mathbf{u}_g\right)=-p\nabla\cdot\mathbf{u}_g-\nabla\cdot\mathbf{q}+\rho_g\varepsilon+Q\end{cases} \quad (8)$$

where $\mathbf{u}_g$ is the gas velocity vector, $\mathbf{F}$ is the momentum transfer between the gas and the discrete liquid phase, $\mathbf{K}$ is the turbulent kinetic energy per unit mass, σ_s is the viscous stress tensor, $\mathbf{q}$ is the heat flux vector, ε is the viscous dissipation rate, and Q is the heat transfer between the gas and the liquid phase. Turbulence effects were modeled by the standard $k-\varepsilon$ model. F and Q are defined as: $\mathbf{F} = -(\alpha_g V_{cell})^{-1}\sum_P N_P \mathbf{F}_\mathrm{p}$, $Q = (\alpha_g V_{cell})^{-1}\sum_p N_p[Q_p + \mathbf{F}_\mathrm{p}(\mathbf{u}_\mathrm{g} - \mathbf{u}_\mathrm{l})]$, where α_g is the fraction of gas in the computational cell, V_{cell} is the cell volume, N_p is the number of droplets in this parcel, and F_p is the force acting on each droplet, Q_p is the heat flux through surface of each droplet, and $\mathbf{u}_\mathrm{l}$ is the droplet velocity.

For the discrete droplets, the momentum equation can be given as: $\mathbf{F}_\mathrm{p} = \pi r_l^2 \rho_l C_D |\mathbf{u}_\mathrm{g} - \mathbf{u}_\mathrm{l}|(\mathbf{u}_\mathrm{g} - \mathbf{u}_\mathrm{l})/2$, where $\mathbf{F}_\mathbf{p}$ is the force exerted on droplet, m_l is the mass of droplet, r_l is the radius of droplet, $\mathbf{g}$ is the gravity force, and C_D is the coefficient of the drag force and can be expressed as: $C_D = 24/Re_l + 6(1 + Re_l^{-1/2}) + 0.4$. Here the droplets are assumed to be spherical, and the effect of droplet deformation on drag coefficient is not taken into consideration.

2.2.2. Liquid Phase

Secondary Atomization

In this section, we first analyze the constitutive relations of the Newtonian and viscoelastic liquids, introducing more physical properties to describe the viscoelastic contributions. Then, the TAB model which is based on Taylor's analogy between an oscillating droplet and a spring-mass system was extend to a non-Newtonian version to describe the deformation and distortion of viscoelastic liquids.

For the Newtonian fluid, the shear stress τ_{ij} can be expressed as:

$$\tau_{ij} = -\mu d_{ij} \tag{9}$$

where μ is the Newtonian viscosity, and d_{ij} is the rate of deformation tensor. In cylindrical coordinates, d_{ij} with (i,j) component can be taken as: $d_{ij} = \partial u_r/\partial z + \partial u_z/\partial r$.

For the viscoelastic fluid, according to Jeffreys' study [35], the rheological model can be expressed with three constants $(\mu, \lambda_1, \lambda_2)$ as follows:

$$\tau_{ij} + \lambda_1 \frac{\partial \tau_{ij}}{\partial t} = -\mu(d_{ij} + \lambda_2 \frac{\partial d_{ij}}{\partial t}) \tag{10}$$

Using the kinetic theory of transient mechanical response in small amplitude motions, the variables (u_r, u_z) in Equations (9) and (10) can be written as $f(r,z,t) = F(r)\mathrm{e}^{ikz+\alpha t}$, where k is the wave number, and α is the growth rate of a perturbation. As deduced by Middleman [15], Equations (9) and (10), respectively become the following:

$$T_{rz}(r) = -\mu(ikU_r(r) + \frac{\mathrm{d}U_z(r)}{\mathrm{d}r}) \tag{11}$$

$$T_{rz}(r) = -\mu\frac{1+\alpha\lambda_2}{1+\alpha\lambda_1}(ikU_r(r) + \frac{\mathrm{d}U_z(r)}{\mathrm{d}r}) \tag{12}$$

By comparing the above two dynamical equations, it concludes that they are identical, the Newtonian viscosity μ can be replaced by $\mu_{ve} = \mu(1+\alpha\lambda_2)/(1+\alpha\lambda_1)$ for the viscoelastic model. Based on the above analysis, the original TAB model proposed by Amsden et al. [36] can be used as a foundation for the viscoelastic version. The derivation of viscoelastic version of TAB model include three aspects as below.

First is the modification of deformation equation. The TAB model is based on the Raleigh–Taylor instability analysis, describing the droplet distortion by a forced, damped, harmonic oscillator in which the forcing term is given by the aerodynamic interaction between the droplet and gas; the damping is due to the liquid viscosity, and the restoring force is supplied by the surface tension. Droplet distortion is denoted by the deformation parameter, $\zeta = 2\tau/r$, where τ denotes the change of radial cross-section from its equilibrium position and r is the initial drop radius. The deformation equation of ζ in terms of the normalized distortion parameter is:

$$\ddot{\zeta} + \frac{5\mu_l}{\rho_l r^2}\dot{\zeta} + \frac{8\sigma}{\rho_l r^3}\zeta = \frac{2\rho_g |U_{rel}|^2}{3\rho_l r^2} \tag{13}$$

where ρ_l is the density, μ_l is the viscosity, σ is the surface tension, U_{rel} is the relative drop–gas velocity, and the subscripts g and l denote the gas and liquid, respectively. It is assumed that the external aerodynamic force is normal to the drop and the turbulence fluctuations of the surrounding gas are ignored. In reality, there are various modes when droplets breakup as they are subjected to the stresses of the surrounding turbulence gas. However, it is impossible to describe all secondary breakup modes in a single model, therefore only the fundamental mode of oscillation is considered in TAB model.

For a viscoelastic version of TAB model, the constant Newtonian shear viscosity μ_l was replaced by a viscoelastic viscosity μ_{ve}. In the expression of μ_{ve}, the growth rate of a perturbation α can be taken by the rate of strain $\dot{\gamma}$:

$$\mu_{ve} = \mu_0 \frac{1 + \dot{\gamma}\lambda_2}{1 + \dot{\gamma}\lambda_1} \tag{14}$$

where λ_1 is the fluid relaxation time, λ_2 is the retardation time, and μ_0 is the zero shear viscosity. The rate of strain $\dot{\gamma}$ is approximated as $\dot{\gamma} = U_{rel}/2r_0$. As indicated in Equation (14), on the one hand, μ_{ve} will approach zero shear viscosity μ_0 when the rate of strain $\dot{\gamma}$ is close to 0. On the other hand, μ_{ve} will approach $\mu_0\lambda_2/\lambda_1$ when $\dot{\gamma}$ becomes larger. Therefore, when $\lambda_2 < \lambda_1$ and the rate of strain is increasing, $\mu_{ve} < \mu_0$, which means that the liquid is shear thinning. For dilute polymer solutions, the relationship of retardation time and relaxation time is $\lambda_2 = \lambda_1\eta_{solvent}/\eta_{solution}$ [37], where $\eta_{solvent}$ and $\eta_{solution}$ are the viscosity of solvent and solution, respectively.

Second, the breakup criterion is changed. In TAB model, drop breakup occurs when the normalized drop distortion $\zeta(t)$ exceeds the critical value of 1. For Newtonian droplet, the stability limit has been determined experimentally to be $We_g = \rho_g U_{rel}^2 r/\sigma = 6$. For high viscous liquid, the effect of viscosity on the stability should be related to the breakup criterion. As proposed by Brodkey [38], the critical Weber number can be modified by introducing the Ohnesorge number:

$$We_{g.crit} = 6(1 + 1.077Oh^{1.6}),\ Oh = \mu_l/\sqrt{2\rho_l \sigma r} \tag{15}$$

when $Oh \leq 0.1$, the value of this expression approximates to 6, which means the case of low viscosity for Newtonian liquid is recovered. When Oh is higher, the stability limit is increased and scales as $\mu_l^{1.6}$ in the asymptotic limit. Based on the Brodkey's empirical relation, an effective Weber number is introduced to instead of the regular one [38]:

$$We_g^{eff} = \frac{We_g}{1 + 1.077Oh^{1.6}} \tag{16}$$

for the viscoelastic liquid, μ_{ve} is used instead of μ_l in Oh number.

Third, the creation of the product droplets is also revised. The population dynamics is used to define the rate of droplet creation. For each breakup event, with the mass conservation principle, it is assumed that the number of product droplets is proportional to the number of critical parent drops, where the proportionality constant depends on the drop breakup regime:

$$\frac{\mathrm{d}}{\mathrm{d}t}\overline{m}(t) = -3K_{bu}\overline{m}(t) \tag{17}$$

where $\overline{m}(t)$ denotes the mean mass of the product drop distribution, and the breakup frequency K_{bu} depends on the drop breakup regimes. Breakup frequency $K_{bu} = 0.05\omega$ as suggested by Amsden et al. [36]

is used. The drop oscillation frequency ω is given by $\omega^2 = 8\sigma/(\rho_l r^3) - 25\mu_l^2/(4\rho_l^2 r^4)$, where μ_l is also replaced by μ_{ve} for viscoelastic version.

Droplets Collision

Droplet collision is calculated by a statistical method, rather than a deterministic method since the real collisions process is highly stochastic. All the droplets in one computational parcel behave in the same manner, they either do or do not collide, which is depended on the probability of the collision whether larger than a random number. For collision result, an impact parameter, $\psi = \chi/(r_1 + r_2)$ is involved to judge the outcomes, where χ is the distance between the center of one drop and the relative velocity vector U_{rel}, r_1 and r_2 are the radii of small and large droplets, respectively. The criteria, χ_{cr} determines the transition boundary: drops coalescence when $\chi \leq \chi_{cr}$, and bounce when $\chi > \chi_{cr}$. χ_{cr} can be expressed as:

$$\chi_{cr}{}^2 = (r_1 + r_2)^2 \min[1.0, 2.4f(\gamma)/We_l] \tag{18}$$

where the collision Weber number for Newtonian liquid defined as: $We_l = \rho_l |\boldsymbol{U}_{rel}| r_1/\sigma_l$. For viscoelastic liquid, $We_l^{eff} = We_l/(1 + 1.077Oh^{1.6})$ is used instead of We_l. $f(\gamma)$ is a function of the radius ratio γ: $f(\gamma) = \gamma^3 - 2.4\gamma^2 + 2.7\gamma$ $(\gamma = r_2/r_1 \geq 1)$.

According to the conservation of momentum, the drop velocity after a bouncing collision is [36]: $u_{1new} = \{u_1 m_1 + u_2 m_2 + m_2(u_1 - u_2)[(\chi - \chi_{cr})/(r_1 + r_2 - \chi_{cr})]\}/(m_1 + m_2)$. The droplet velocity after coalescence is calculated as $u_{new} = (u_1 m_1 + u_2 m_2)/(m_1 + m_2)$. m_1 and m_2 are the mass of the small and large droplets, respectively; and u_1 and u_2 are the velocity of the small and large droplets, respectively. The new radius after droplet coalescence is $r_{new} = (r_1^3 + r_2^3)^{1/3}$.

3. Numerical Setup

Figure 2 shows the computational mesh and boundary conditions of numerical model. The radius and axial lengths of the computational regime are 6 cm and 15 cm respectively, and 2π in the circular direction. The computational domain is in a cylindrical coordinate system with size of 56 (radial) × 65 (axial) × 32 (azimuthal). Finer meshes are used for the core region of the spray. The grid dependency has been validated in our previous studies [39,40] and the computational results are convergence under present mesh size.

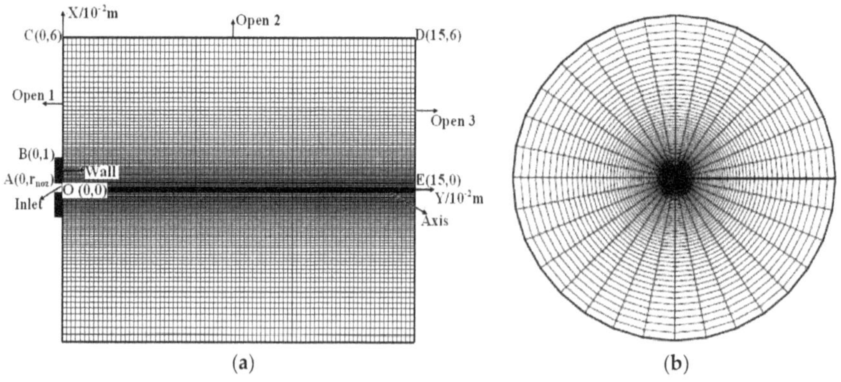

Figure 2. Geometry and computational mesh: a middle section plane. (**a**) Middle section plane; and (**b**) Cross section plane.

As presented in Table 1, there are five types of boundary conditions involved for the gas field. The parcels are introduced in the system at the nozzle exit, which has the original size calculated from primary breakup sub-model with Rosin-Rammler distribution: $F_{R-R}(D) = 1 - \exp\{-(D/SMD_0)^{q_{co}}\}$, where q_{co} indicates the value of the width of the distribution. The Rosin-Rammler distribution is a prevalent size distribution used in the atomization field [41], and the experimental data also supported the Rosin-Rammler function for non-Newtonian fluid atomization spray [31]. The number of parcels should be large enough to reduce statistical fluctuation. Usually, a total number of 10,000 computational parcels are injected into the flow for each case.

Table 1. Five types of boundary conditions involved in the model.

Type		Location	Conditions
Open	1	Line BC	$v(x,0,\theta) > 0$: P, T, u are assigned the same as that in ambient $v(x,0,\theta) \leq 0$: $\partial u/\partial y = 0, \partial\phi/\partial y = 0$ ($\phi = P, T$)
	2	Line CD	$u(6,y,\theta) < 0$: P, T, v are assigned the same as that in ambient $u(6,y,\theta) \geq 0$: $\partial\phi/\partial x = 0, \partial v/\partial y = 0$
	3	Line DE	$v(x,15,\theta) < 0$: P, T, u are assigned the same as that in ambient $v(x,15,\theta) \geq 0$: $\partial u/\partial y = 0, \partial\phi/\partial y = 0$
Wall		Line AB	$u, v, w = 0$; $T = 300$ K; turbulent wall function
Inlet		Line OA	Data from primary breakup sub-model
Axis		$x = 0$	$\xi\|_{x=0} = (\sum_{i=1}^{N} \xi_i)/N$ ($\xi = u, v, T$, ξ_i: ξ at Δx from the centerline); $w = 0$
Periodic		$\theta = 2\pi$	Circular direction: periodic boundary conditions

4. Results and Discussion

4.1. Validation of Primary Atomization

Figure 3 shows a comparison of experimental data [42] and model predictions for polymer solution atomization at $P_{inj} = 0.35$ MPa, $D_{noz} = 1$ mm and *ALR* ranging from 0.02 to 0.11. The physical and rheological properties are listed in Table 2. The polymer solution were formulated by adding poly(ethylene oxide) to solvent having a common composition of 60% glycerine–40% water solution by mass. Polyethylene oxide (PEO) molecular weights of 100,000 were used at mass concentrations of 0.001%, 0.01% and 0.15% [42]. The dispersion relations proposed by Joseph are solved analytically. Results show that, with an increase in air–liquid mass flow ratio (*ALR*), droplet *SMD* will decline significantly, especially at lower *ALR*. The agreement between the predictions and measurements is qualitative achieved within 5%–20%. Joseph's model is effective to capture the influence of rheological properties on the breakup process of polymer solution and to quantitatively predict the resulting droplet sizes. However the shortcoming is that the divergence may occur when the liquid ligament is not similar to the flattened drop. Therefore, at lower *ALR*, the divergence is obviously since the liquid ligaments is slender.

Table 2. Physical properties of the fluids in Figure 3 [42].

Code	Polymer Concentration	Zero Shear Viscosity η_0	Relaxation Time λ_1	Surface Tension σ
		mPa·s	s	10^{-3} J/m^2
1	0.001%	11	2.8×10^{-7}	65.1
2	0.01%	11.1	1.4×10^{-5}	62.6
3	0.15%	15.4	2.06×10^{-4}	62.6

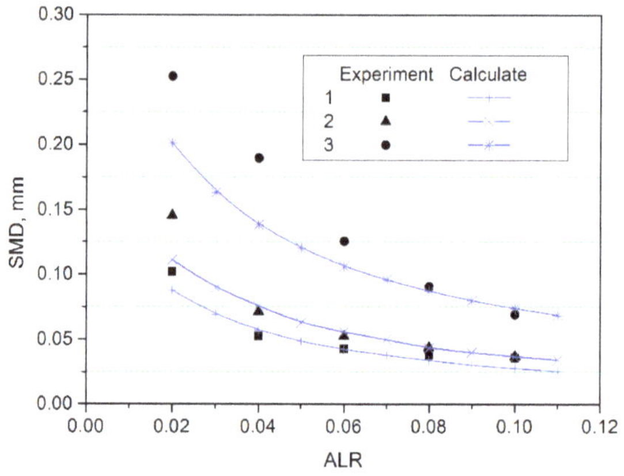

Figure 3. Comparison of experimental data and model predictions for the Sauter mean diameter (*SMD*) versus air-liquid mass flow ratio (*ALR*) at $P_{inj} = 0.35$ MPa, $D_{noz} = 1$ mm.

4.2. Evolution of Spray Field

Among the investigations of the far-field region, the particle size evolution along with the axial distance is the primary interest. Figure 4 firstly validates the numerical results of water's *SMD* with experimental data [43] and then compares the downstream variation of *SMD* for water and polymer solution. Both experiment and simulations were conducted based on the same operating conditions. The operating parameters and liquid physical properties are listed in Table 3.

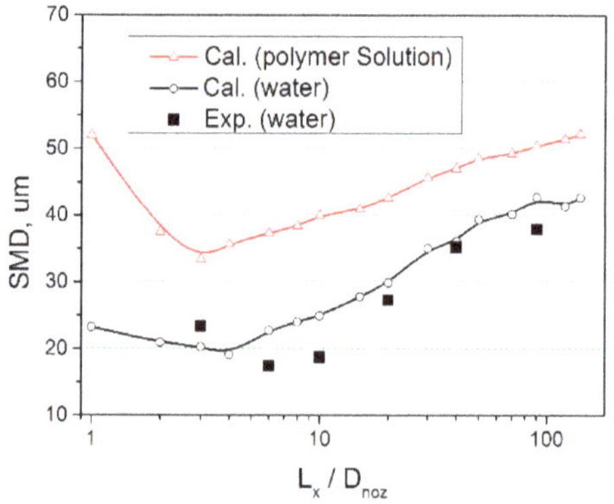

Figure 4. Droplet Sauter mean diameter along with the dimensionless axial distance for water and polymer solution.

Table 3. Operating parameters and liquid physical properties of baseline case.

Parameters	Value	Parameters	Value
D_{noz} (mm)	1	η_0 (10^{-3}Pa·s)	15
σ (10^{-3}J/m^2)	65	$\eta_{solvent}$ (10^{-3}Pa·s)	10
ALR	0.06	λ_1 (s)	2.0×10^{-6}
$\dot{m}_l$ (g/s)	10	P_{inj} (10^6 Pa)	0.35

As shown in Figure 4, an agreement between the simulation results and experimental data has been obtained for the case of water, and the divergence of the experiment and simulation is between 2.5% and 25%. In Figure 4, the viscoelastic liquid added the difficulties in the breakup of ligaments and droplets, impairing the atomization effect by increasing the drop size in the whole process, which is consistent with the results of other researchers [12,44]. Further, in both the cases of water and polymer solution, the droplet diameter first decreases and then increases with the axial distance. This phenomenon can be explained by the competition between breakup and coalescence effects. Figure 5 shows the occurrence number of droplets breakup and collision. Close to the nozzle exit, with high kinetic energy and large velocity difference between the droplet and atomization gas, turbulence breakup dominates, causing the drop size to rapidly decrease, while, further downstream, as the kinetic energy decay and small velocity difference, turbulence breakup ends and droplet coalescence plays an important role, causing the drop size to gradually increase.

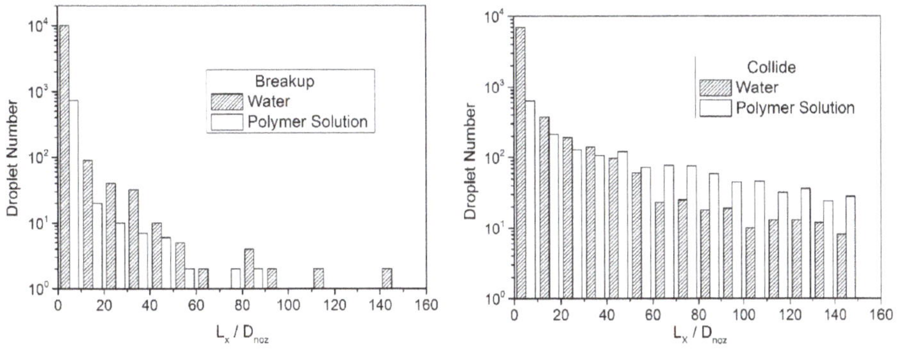

Figure 5. Occurrence number of droplets breakup and collision.

Besides the common and well-known points, there are two noteworthy differences between these two types of liquids. First, in Figure 4, the gap between the mean size of water and polymer solution is shortened along with the development of the jet. Before the nadir of the curve, the case of polymer solution decays more steeply than that of water. In contrast, after the nadir of the curve, the case of water rises more obviously than that of polymer solution. On the rising part of the curves, the increasing degree of water is about 123%, while the case of polymer solution is less than 57% (this value is calculated from $(SMD_{L_x/D_{noz}=140} - SMD_{L_x/D_{noz}=4})/SMD_{L_x/D_{noz}=4}$). This phenomenon indicates that the mean size of polymer solution increases more gently than that of water at the axial distance $(5\text{–}100)L_x/D_{noz}$, which is benefit to control the desirable size in industrial applications. Secondly, in Figure 5, the occurrence numbers for water are higher than polymer solutions for both breakup and collision process in the upstream of the spray. This is because droplets breakup time and frequency depend on the competition between the aerodynamic force and restoring force. An increase in liquid viscosity will delay and weaken the breakup process, resulting in higher droplet size. The decrease of the frequency of droplet secondary breakup would also lead to the number of product droplets decreasing, therefore the collision occurrence number of polymer solutions is obviously less than that of water in the upstream region. However, when the axial distance beyond $40L_x/D_{noz}$, the collision occurrence number for polymer solution exceeds that of water. Since the droplets collision process were influenced by the droplet size and concentration rate. Concentrated drop distribution causes the collision happening frequently. The outcome of collision was also affected by the liquid viscosity. For the polymer solution droplets, impact parameter increased due to higher viscosity, leading to an increase in possibility of droplet coalescence after collision. However, the change of collision occurrence number did not influence the trend of *SMD* in Figure 4 significantly, due to the occurrence number is too small compared to the total particle number.

The most important factors influencing the droplet breakup can be grouped into two dimensionless numbers, the Weber number (*We*) and the Ohnesorge number (*Oh*). The average Weber number is defined as $We = \rho_g \left|\overline{u}_{rel}\right|^2 SMD/\sigma_l$. Here the modified weber number (Equation (16)) is used for the polymer solution. The average Ohnesorge number is expressed as $Oh = \mu_l/\sqrt{\rho_l \sigma_l SMD}$. For viscoelastic liquid, the constant viscosity μ_l was replaced by a viscoelastic viscosity μ_{ve}.

The evolutions of Weber and Ohnesorge number along with the axial direction are illustrated in Figure 6. Weber number indicates the ratio of the inertial force to the surface tension force, in which the trend is basically determined by the gas-liquid relative velocity. At the nozzle exit, *We* is higher due to the large difference between the velocity of droplet and gas, causing the secondary breakup of droplets. In this period, *We* of polymer solution is larger than that of water due to the larger *SMD* at the beginning. Then at downstream of the spray, *We* dramatically declines as the kinetic energy dissipated and *We* of polymer solution is less than that of water since *We* for viscoelastic liquid is modified by the effective viscosity. The effective Weber number decreased as liquid viscosity increasing. The Ohnesorge number is the ratio of an internal viscosity force to an interfacial surface tension force, which significantly increases with the increase of liquid viscosity. For both water and polymer solution, the *Oh* curves have a peak at about 4 mm from the nozzle exit, which is corresponding to the nadir of the curve of droplet size (as shown in Figure 4). In short, the case of polymer solution has a greater magnitude of change for both *We* and *Oh* due to introducing the changes in viscosity.

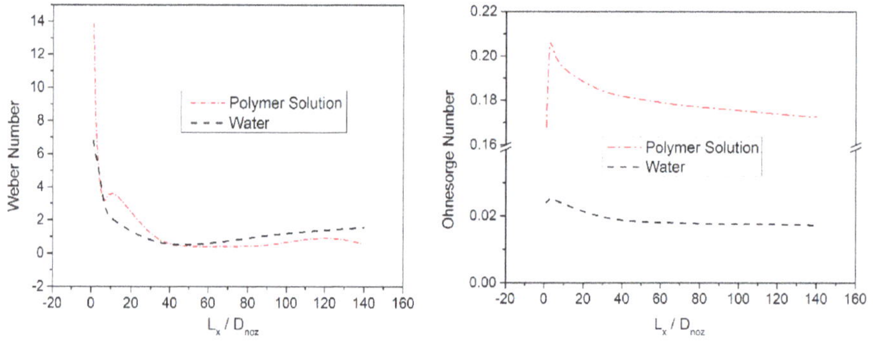

Figure 6. Evolutions of Weber and Ohnesorge number along with axial direction for water and polymer solution.

In addition to drop size and critical numbers, liquid viscosity can also affect the spray pattern. Figure 7 shows the spatial distributions of droplet size and number for water and polymer solution, respectively. The cross-sectional plane is located at the axial distance of 50 mm. It can be seen from Figure 7 that the spatial distribution of droplets becomes wider for water. The distribution range of water case is about 40% larger than the polymer solution. It can be explained as the ligaments and fragments are more likely to adhere due to viscoelastic properties. Furthermore, finer droplets are easier for the gas to accelerate and carry them outward of the centerline, resulting in the spray angle will expand. In turn, the distribution will also affect the droplet size. Concentrated drop distribution causes the collision happening frequently and the droplet size increasing, which is in accordance with that indicated in Figure 5.

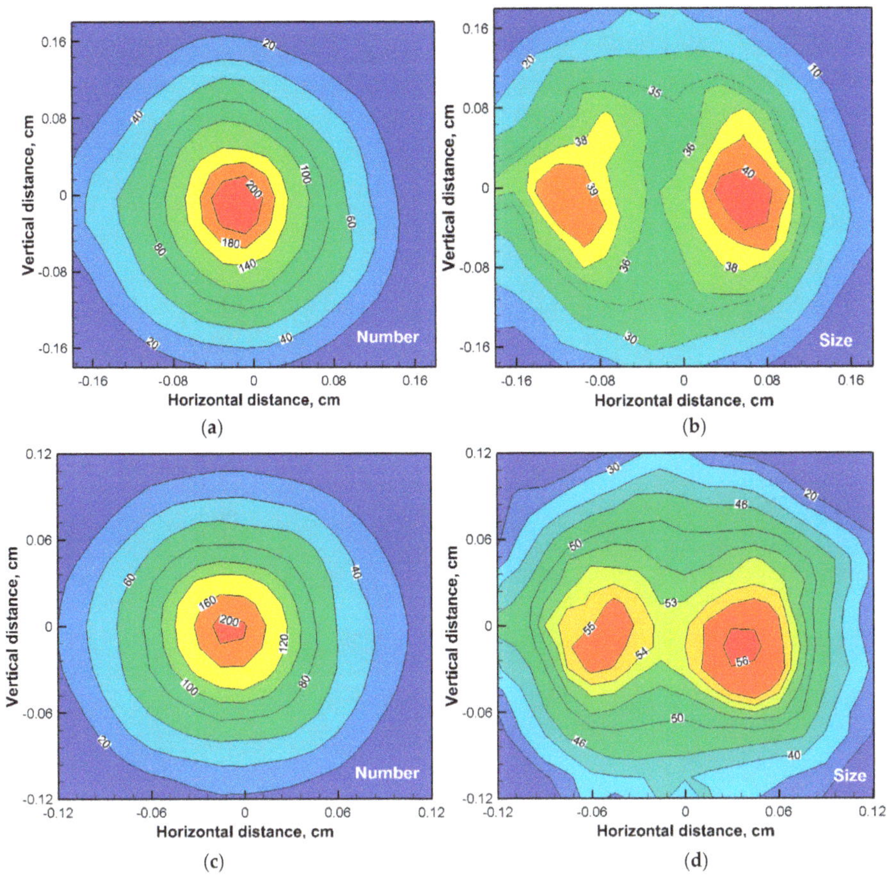

Figure 7. Spatial distribution of the drop number and size at the axial distance of 50 mm. (**a**) Number distribution of water; (**b**) Size distribution of water ; (**c**) Number distribution of polymer solution; (**d**) Size distribution of polymer solution.

4.3. Effects of Viscoelastic Parameters

Figure 8 shows the analysis of the effects of rheological properties on primary atomization. Figure 9 explores the evolutions of droplet size along with axial direction for various liquid properties. For polymer solutions, the rheology properties can be adjusted through the addition of polymer molecules. The liquid physical properties of base case are as the same as that of code 2 in Table 2. All cases in Figures 8 and 9 share the same operating conditions except the specified parameters in the legend of the figure. In order to compare the behavior of Newtonian and viscoelastic liquid, the calculated results of water spray have been added into Figures 8 and 9.

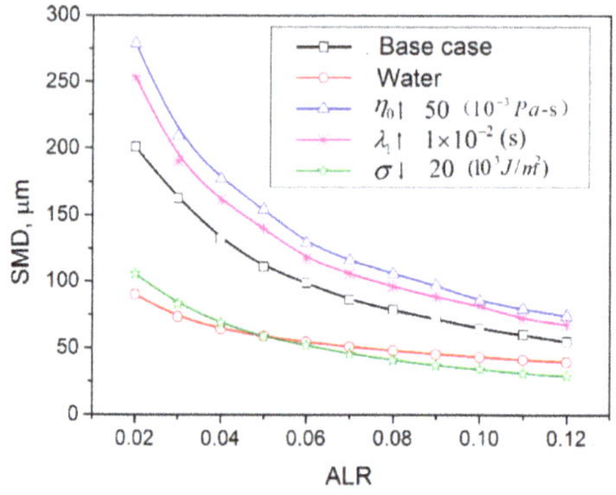

Figure 8. Droplet *SMD* versus *ALR* for various liquid properties.

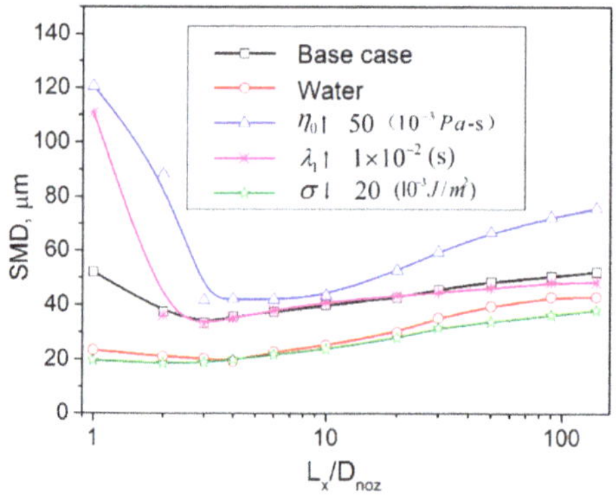

Figure 9. Droplet *SMD* along the axial distance for various liquid properties.

Obviously, from Figures 8 and 9, the increases in relaxation time and zero-shear viscosity cause the atomization quality deteriorates since the growth rate of the disturbance wave reduces and the liquid becomes stringier to resist the disruptive force, while small surface tension brings the droplet size to decrease significantly. The case with surface tension of 20×10^{-3} J/m^2 is almost identical to the case of water. In both instability analysis and TAB model, the surface tension can resist the occurrence and development of instability and smooth the disturbance. That is, a decrease in surface tension brings the ligaments and droplet less stability. This finding shows the importance of selecting solvents with lower surface tension for the polymer solution when finer atomization effects are desirable. Moreover, the effects of relaxation time become weaken at the downstream of the spray field. The case with shorter relaxation time approaches to the base case since the influence of relaxation time depends on the gas-liquid relative velocity. When the relative velocity declined rapidly, the effective viscosity of viscoelastic liquid is mainly up to zero shear viscosity.

5. Conclusions

A comprehensive mathematical model has been proposed to simulate the outflow of the effervescent atomization for viscoelastic liquid, accounting for primary atomization droplet tracking, secondary breakup and droplet collision. The computational models have included two parts. The first one is the primary breakup sub-model. With the simplification and assumptions of the atomization of liquid stream, the primary breakup sub-model has been established based on the viscoelastic linear dispersion relation. The second sub-model can calculate the evolutions of the droplets spray, in which the gas-liquid two-phase flow has been simulated by the Euler-Lagrangian approach. Both the droplet secondary atomization model and collision model have been extended to the viscoelastic version by introducing rheological parameters. The main findings of the present study can be summarized in four points:

(1) Based on the comparison with experimental data and the predicted results of other dispersion relation, the primary model using Joseph's relation effectively captured the influence of rheological properties on the breakup process of polymer solution and to quantitatively predict the resulting droplet sizes.
(2) The evolutionary trend of droplet mean diameter, Weber number and Ohnesorge number of viscoelastic liquids along with axial direction were qualitatively similar to that of Newtonian liquids. The *SMD* for the case of polymer solutions were obviously larger than that of water throughout the spray field. While, the gap of *SMD* between water and polymer solution became shorten along with the development of the jet. This phenomenon indicated that the mean size of polymer solution increased more gently than that of water at the axial distance $(5\text{–}100) L_x / D_{noz}$, which benefited the stable control of the desirable size in the applications.
(3) The changes of Weber and Ohnesorge number for polymer solution were more dramatic than that of water, since the viscosity of the viscoelastic liquid was influenced by the relative velocity of gas and droplets. Besides, the distribution range of water was about 40% larger than that of polymer solution in present study, which indicated the spray angle became narrow for the viscoelastic liquids.
(4) In primary atomization, the surface tension was the dominating physical property for viscoelastic liquid rather than relaxation time and zero shear viscosity. When the finer atomization effects were desirable, it was important to select solvents with lower surface tension for the polymer solutions, while, concerning the evolution of atomization spray, larger relaxation time and zero shear viscosity increased droplet *SMD* significantly. The zero shear viscosity was effective throughout the jet region, while the effect of relaxation time became weaken at the downstream of the spray field.

Acknowledgments: This work was supported by the Major Program of the National Natural Science Foundation of China with Grant No. 11632016, and the National Natural Science Foundation of China with Grant No. 11372006 and 11402259. Lijuan Qian would also like to thank Zhejiang Provincial Natural Science Foundation (LY17A020008) and Young Talent Cultivation Project of Zhejiang Association for Science and Technology (2016YCGC013).

Author Contributions: Lijuan Qian performed the codes and wrote the paper; Fubing Bao analyzed the data; Jianzhong Lin contributed ideas and suggestions.

Conflicts of Interest: The authors declare no conflict of interest.

References

1. Mallory, J.A. Jet Impingement and Primary Atomization of Non-Newtonian Liquids. Ph.D. Thesis, Purdue University, West Lafayette, IN, USA, 2012.
2. Yoo, H.; Kim, C. Generation of inkjet droplet of non-Newtonian fluid. *Rheol. Acta* **2013**, *52*, 313–325. [CrossRef]
3. Hoath, S.D.; Vadillo, D.C.; Harlen, O.G.; McIlroy, C.; Morrison, N.F.; Hsiao, W.; Tuladhar, T.R.; Jung, S.; Martin, G.D.; Hutchings, I.M. Inkjet printing of weakly elastic polymer solution. *J. Non-Newton. Fluid Mech.* **2014**, *205*, 1–10. [CrossRef]
4. Petersen, F.J.; Wørts, O.; Schæfer, T.; Sojka, P.E. Effervescent atomization of aqueous polymer solutions and dispersions. *Pharm. Dev. Technol.* **2001**, *6*, 201–210. [CrossRef] [PubMed]
5. Pawlowski, L. Suspension and solution thermal spray coatings. *Surf. Coat. Technol.* **2009**, *203*, 2807–2829. [CrossRef]
6. Bartolo, D.; Boudaoud, A.; Narcy, G.; Bonn, D. Dynamics of non-newtonian droplets. *Phys. Rev. Lett.* **2007**, *99*, 174502. [CrossRef] [PubMed]

7. Schowalter, W.R. *Mechanics of Non-Newtonian Fluids*; Pergamon Press: Oxford, UK, 1978.
8. Oldroyd, O.G. Non-Newtonian effects in steady motion of some idealized elastic viscous fluids. *Proc. R. Soc.* **1958**, *279*, A245.
9. Li, L.K.B.; Dressler, D.M.; Green, S.I.; Davy, M.H.; Eadie, D.T. Experiments on air-blast atomization of viscoelastic liquids, Part 1: Quiescent conditions. *At. Sprays* **2009**, *19*, 157–190. [CrossRef]
10. Dumouchel, C. On the experimental investigation on primary atomization of liquid streams. *Exp. Fluids* **2008**, *45*, 371–422. [CrossRef]
11. Aliseda, A.; Hopfinger, E.J.; Lasheras, J.C.; Kremer, D.M.; Berchielli, A.; Connolly, E.K. Atomization of viscous and non-newtonian liquids by a coaxial, high-speed gas jet: Experiments and droplet size modeling. *Int. J. Multiph. Flow* **2008**, *34*, 161–175. [CrossRef]
12. Harrison, G.M.; Mun, R.P.; Cooper, G.; Boger, D.V. A note on the effect of polymer rigidity and concentration on spray atomization. *J. Non-Newton. Fluid Mech.* **1999**, *85*, 93–104. [CrossRef]
13. Chao, K.K.; Child, C.A.; Grens, E.A., II; Williams, M.C. Anti-misting action of polymeric additives in jet fuels. *AIChE J.* **1984**, *30*, 111–120. [CrossRef]
14. Sirignano, W.A.; Mehring, C. Review of theory of distortion and disintegration of liquid streams. *Prog. Energy Combust. Sci.* **2000**, *26*, 609–655. [CrossRef]
15. Middleman, S. Stability of a viscoelastic jet. *Chem. Eng. Sci.* **1965**, *20*, 1037–1040. [CrossRef]
16. Goren, S.L.; Gorttlieb, M. Surface-tension driven breakup of viscoelastic liquid threads. *J. Fluid Mech.* **1982**, *120*, 245–266. [CrossRef]
17. Joseph, G.E.; Funda, B.T. Rayleigh-Taylor instability of viscoelastic drops at high Weber number. *J. Fluid Mech.* **2002**, *453*, 109–132. [CrossRef]
18. Yang, L.J.; Tong, M.X.; Fu, Q.F. Linear stability analysis of a three-dimensional viscoelastic liquid jet surrounded by a swirling air stream. *J. Non-Newton. Fluid Mech.* **2013**, *191*, 1–13. [CrossRef]
19. Fuster, D.; Bagué, A.; Boeck, T.; Le Moyner, L.; Leboissetier, A.; Popinet, S.; Scardovelli, R.; Zaleski, S. Simulation of primary atomization with an octree adaptive mesh refinement and VOF method. *Int. J. Multiph. Flow* **2009**, *35*, 550–565. [CrossRef]
20. Villermaux, E. Fragmentation. *Annu. Rev. Fluid Mech.* **2007**, *39*, 419–446. [CrossRef]
21. Gorokhovski, M.; Herrmann, M. Modeling primary atomization. *Annu. Rev. Fluid Mech.* **2008**, *40*, 343–366. [CrossRef]
22. Zhu, C.; Ertl, M.; Weigand, B. Numerical investigation on the primary breakup of an inelastic non-Newtonian liquid jet with inflow turbulence. *Phys. Fluids* **2013**, *25*, 083102. [CrossRef]
23. Stelter, M.; Brenn, G.; Durst, F. The influence of viscoelastic fluid properties on spray formation from flat-fan and pressure-swirl atomizers. *At. Sprays* **2002**, *12*, 299–327. [CrossRef]
24. Dexter, R.W. Measurement of extensional viscosity of polymer solutions and its effects on atomization from a spray nozzle. *At. Sprays* **1996**, *6*, 167–191. [CrossRef]
25. Hartranft, T.J.; Settles, G.S. Sheet atomization of non-Newtonian liquids. *At. Sprays* **2003**, *13*, 191–221. [CrossRef]
26. Mun, R.P.; Young, B.W.; Boger, D.V. Atomization of dilute polymer solutions in agricultural spray nozzles. *J. Non-Newton. Fluid Mech.* **1999**, *83*, 163–178. [CrossRef]
27. Rivera, C.L. Secondary Breakup of Inelastic Non-Newtonian Liquid Drops. Ph.D. Thesis, Purdue University, West Lafayette, IN, USA, 2010.
28. Dechelette, A.; Campanellab, O.; Corvalanc, C.; Sojka, P.E. An experimental investigation on the breakup of surfactant-laden non-Newtonian jets. *Chem. Eng. Sci.* **2011**, *66*, 6367–6374. [CrossRef]
29. Ma, Y.; Bai, F.; Chang, Q.; Yi, J.; Jiao, K.; Du, Q. An experimental study on the atomization characteristics of impinging jets of power law fluid. *J. Non-Newton. Fluid Mech.* **2015**, *217*, 49–57. [CrossRef]
30. Xue, Z.; Carlos, M.C.; Dravid, V.; Sojka, P.E. Breakup of shear-thinning liquid jets with surfactants. *Chem. Eng. Sci.* **2008**, *63*, 1842–1849. [CrossRef]
31. Mallory, J.; Sojka, P.E. On the primary atomization of non-Newtonian imping jets: Volume II linear stability theory. *At. Sprays* **2014**, *24*, 525–554. [CrossRef]
32. Ashgriz, N. *Handbook of Atomization and Sprays: Theory and Applications*; Springer: New York, NY, USA, 2011.
33. Broniarz-Press, L.; Ochowiak, M.; Woziwodzki, S. Atomization of PEO aqueous solutions in effervescent atomizers. *Int. J. Heat Fluid Flow* **2010**, *31*, 651–658. [CrossRef]

34. Lund, M.T.; Jian, C.Q. The influence of atomizing gas molecular weight on low mass flow-rate effervescent atomization performance. *J. Fluids Eng.* **1998**, *120*, 750–754. [CrossRef]
35. Bird, R.B. Useful non-Newtonian models. *Annu. Rev. Fluid Mech.* **1976**, *8*, 13–34. [CrossRef]
36. Amsden, A.A.; O'Rourke, P.J.; Butler, T.D. *KIVA-II: A Computer Program for Chemically Reactive Flows with Sprays*; Technical Report LA-11560-MS; Los Alamos National Laboratory: Los Alamos, NM, USA, 1980.
37. Denn, M.M. Issues in viscoelastic fluid mechanics. *Annu. Rev. Fluid Mech.* **1999**, *22*, 13–32. [CrossRef]
38. Brodkey, R.S. *The Phenomena of Fluid Motions, Addison-Wesley Series in Chemical Engineering*; Addison-Wesley: New York, NY, USA, 1967.
39. Xiong, H.B.; Lin, J.Z.; Zhu, Z.F. Three-dimensional simulation of effervescent atomization spray. *At. Sprays* **2009**, *19*, 1–16. [CrossRef]
40. Lin, J.Z.; Qian, L.J.; Xiong, H.B.; Chan, T.L. Effects of operating conditions on droplet deposition onto surface of atomization impinging spray. *Surf. Coat. Technol.* **2009**, *203*, 1733–1740. [CrossRef]
41. Lefebvre, A.H. *Atomization and Sprays*; Hemisphere Publishing Corporation: Carlbad, CA, USA, 1989.
42. Geckler, S.C.; Sojka, P.E. Effervescent atomization of viscoelastic liquids: Experiments and modeling. *J. Fluids Eng.* **2008**, *130*, 061303. [CrossRef]
43. Liu, L.S. Experimental and Theoretical Investigation on the Characteristics and Two-Phase Spray Flow Field of Effervescent Atomizers. Ph.D. Thesis, Tianjing University, Tianjin, China, 2001.
44. Mansour, A.; Chigier, N. Air-Blast atomization of Non-Newtonian liquids. *J. Non-Newton. Fluid Mech.* **1995**, *58*, 161–194. [CrossRef]

energies

MDPI

Article

A Performance Prediction Method for Pumps as Turbines (PAT) Using a Computational Fluid Dynamics (CFD) Modeling Approach

Emma Frosina *, Dario Buono and Adolfo Senatore

Department of Industrial Engineering, University of Naples Federico II, Via Claudio, 21-80125 Naples, Italy; darbuono@unina.it (D.B.); senatore@unina.it (A.S.)
* Correspondence: emma.frosina@unina.it; Tel.: +39-081-768-32-85

Academic Editor: Bjørn H. Hjertager
Received: 18 October 2016; Accepted: 6 January 2017; Published: 16 January 2017

Abstract: Small and micro hydropower systems represent an attractive solution for generating electricity at low cost and with low environmental impact. The pump-as-turbine (PAT) approach has promise in this application due to its low purchase and maintenance costs. In this paper, a new method to predict the inverse characteristic of industrial centrifugal pumps is presented. This method is based on results of simulations performed with commercial three-dimensional Computational Fluid Dynamics (CFD) software. Model results have been first validated in pumping mode using data supplied by pump manufacturers. Then, the results have been compared to experimental data for a pump running in reverse. Experimentation has been performed on a dedicated test bench installed in the Department of Civil Construction and Environmental Engineering of the University of Naples Federico II. Three different pumps, with different specific speeds, have been analyzed. Using the model results, the inverse characteristic and the best efficiency point have been evaluated. Finally, results have been compared to prediction methods available in the literature.

Keywords: energy saving; PAT; urban hydraulic network; numerical modeling

1. Introduction

Electricity generation presents many issues and is studied with different techniques in order to reduce its production cost and environmental impact. Conventional production with fossil fuels presents problems associated with the high cost, rapid depletion and detrimental environmental effects of these fuels. Renewable energy is probably the best solution for environmental issues, and many solutions have been developed since the last century, such as hydropower, hydrogen, fuel cells, biofuels, and solar power generation.

Among the renewable sources, small hydropower represents a very attractive source of energy generation. In many countries, small and micro hydropower systems are an important means of electricity generation. An efficient solution, from the point of view of energy efficiency, is the adoption of a turbine, but the purchase and maintenance costs of turbines make their implementation economically unattractive, especially for small hydropower [1–6].

Reverse-running centrifugal pumps (also called pumps as turbines or PAT) are a solution for generating and recovering power in small and micro hydropower situations. Pumps are relatively simple machines, inexpensive (compared to a hydraulic turbines), and readily available worldwide. It has been estimated that the capital payback period of a reverse-running pump in the range of 5–50 kW is less than two years [7,8]. Moreover, the use of PAT could be suitable because manufacturers of turbines worldwide are less numerous than pump producers, the market for turbines is smaller compared to pumps, and pumps are mechanically simple and require less maintenance. Moreover, an integral pump and electric motor can be purchased for use as a turbine and generator set; pumps are available in a wide range of heads and flows and in a large number of standard sizes. Generally, pumps have short delivery times, spare parts (such as seals and bearings) are easily available and the installation can be done using standard pipes and fittings.

The use of a pump running in reverse mode to generate electricity is not new; the first applications started almost 80 or 90 years ago and many theoretical and experimental studies have been done [2–6,9,10]. Much research is still being conducted, especially to predict the operating conditions and the efficiency of centrifugal pumps running in reverse [11].

The selection of a proper PAT for an existing site represents a critical issue because pump manufacturers do not supply the characteristic of the pump running in reverse. Many methods have been used to predict the inverse characteristic of a pump, based on numerical models, experiments, or theoretical procedures [4–10].

This research has demonstrated that these methods can be used only for a limited set of pumps. None of them, in fact, allows prediction for the reverse running conditions for all geometries and over a wide range of pump specific speeds. Several studies, based on a modeling approach with CFD code, are available and generally show good correspondence with the available experimental data [4,10].

A study [4] carried out with a computational model of a PAT is based on the concept called "flow zone". The flow regime within a PAT is divided into four major flow regions (volute casing, impeller, casing outlet and draft tube). A comparison has been made between the experimental and numerical results of a single stage end suction centrifugal pump that was operated in turbine mode at a speed of 800 rpm. CFD predictions of the hydraulic parameters were in good correspondence with the experimental results, but deviations (within 5% to 10%) have been found at certain load regions.

Nautiyal et al. [5,6] carried out a study on the application of CFD and its limitations for PAT using cases reported by previous researchers [4,9,10]. The study reported that CFD analysis was an effective design tool for predicting the performance of centrifugal pumps in turbine mode and for identifying the losses in turbo-machinery components such as the draft tube, impeller and casing, but there was some deviation between the experimental results and the CFD modeling results. Barrio et al. [11] carried out a numerical investigation on the unsteady flow in commercial centrifugal pumps operating in direct and reverse mode with the help of CFD code. The results of their simulation were in good correspondence with the experimental results. The study revealed that in the reverse mode, the flow only matched the geometry of the impeller at nominal conditions; re-circulating fluid regions developed at low flow rates (near the discharge side of the blades) and high flow rates (near the suction side).

Many correlations based on theoretical approaches are available to predict the performance of a PAT. Several researchers (Stepanoff, Childs, Sharma, Wong, Williams, Alatorre-Frenk, and others) have presented correlations for predicting the performance of a pump-as-turbine [5]. These correlations were based upon either pump efficiency or specific speed. However, deviations of more than 20% have been found between the experimental and predicted reverse operation of standard pumps [12]. The objective of these correlations is to calculate the best efficiency point (BEP) of pumps for operation in turbine mode by using the pump operation data provided by the manufacturer.

In 1962, Childs [13] presented a PAT prediction method based on the efficiency of the pump. A similar approach was then presented by McClaskey and Lundquist [14] and Lueneburg and Nelson [15] in 1976 and 1985, respectively.

Hancock [16] stated that for most pumps the turbine BEP lies within 2% of the pump mode BEP. Grover and Hergt [17,18] proposed a PAT prediction method based on specific speed for the turbine mode (obtained similar to the specific speed for a pump). Grover's method is applicable for the turbine mode specific speed range between 10 and 50 [17]. A comparison between experimental results and the methods proposed by the above researchers show relatively large deviations; therefore, the use of these formulae must be confined to an approximate selection of PATs.

Finally, a large number of experimental studies can be used to evaluate the inverse characteristics. These are often limited to the specific pumps tested, so that they cannot serve as a valid tool for pump selection, but are very useful for tuning and validating theoretical and modeling analyses.

In this paper, authors present a methodology for obtaining the reverse characteristics of a pump, starting from the results of three-dimensional CFD models. After a description of all prediction methods available in the literature, in the third section the adopted modeling approach is described. Three pumps have been studied and modeled using the three-dimensional CFD commercial code PumpLinx®, developed by Simerics Inc.® (1750 112th Ave NE, Ste C250, Bellevue, WA, USA). In the fourth section, the test bench layout is shown with all the transducers' characteristics.

Numerical models have been validated with experimental data obtained on a dedicated test bench installed in the Department of Civil Construction and Environmental Engineering of the University of Naples Federico II. Simulations have been run in both direct (as pump) and reverse (as turbine) modes with good accuracy.

In the fifth section, the models' results have been used to predict the efficiency curves of the three analyzed pumps in both modes. At the end of the paper, results of the proposed methodology have been compared to the prediction methods described in the second section. Analysis has demonstrated that the existing prediction methods underestimate or overestimate the real operation.

In this paper, the authors have shown only the first step of a research done on PAT. Other pumps are under study using the same modeling approach in order to realize a macro database for the prediction of pump performance as turbines. The final aim is the identification of a new prediction method more accurate than the others already available in literature. Using the new prediction method, a reduction of the experimentation will then be realized allowing an easy and fast choice of the PAT for each application.

2. Literature Overview on Prediction Methods

A methodology to calculate the inverse characteristic of a commercial pump is presented in this paper. The proposed approach is based on results of CFD modeling using a commercial code developed to simulate centrifugal machines. Therefore, it is important to describe prediction methods already available in literature. In fact, these methods, will be used in the last section of the paper to analyze the proposed methodology and to discuss our results. The following equations summarize different methods to predict the pump inverse characteristics. They are based on theoretical or experimental analyses [13–24]. Stepanoff [19] calculates the head, flow rate at the BEP in reverse mode using the efficiency, head and flow rate value at the BEP in direct mode. All relations between head and flow rate are reported in Equation (1).

$$\frac{H_t}{H_p} = \frac{1}{\eta_p}; \quad \frac{Q_t}{Q_p} = \frac{1}{\sqrt{\eta_p}}; \quad \eta_T = \eta_P; \quad N_{st} = N_s\eta_p \qquad (1)$$

Alatorre-Frenk [20] calculates the head, flow rate and efficiency at the BEP in reverse mode using the efficiency, head and flow rate value at the BEP in direct mode. Correlations are presented by following:

$$\frac{H_t}{H_p} = \frac{1}{0.85{\eta_p}^5+0.385}; \quad \frac{Q_t}{Q_p} = \frac{0.85{\eta_p}^5+0.385}{2{\eta_p}^{9.5}+0.205}; \quad \eta_t = \eta_p - 0.03 \qquad (2)$$

The prediction method developed by Sharma [21] calculates the head and flow rate at the BEP in reverse mode using the efficiency, head and flow rate value at the BEP in direct mode:

$$\frac{H_t}{H_p} = \frac{1}{{\eta_p}^{1.2}}; \quad \frac{Q_t}{Q_p} = \frac{1}{{\eta_p}^{0.8}}; \quad P_t = P_p; \quad \eta_t = \eta_p \qquad (3)$$

Schmiedl [22] as Sharma [21] calculates the head and flow rate at the BEP in reverse mode using the efficiency, head and flow rate value at the BEP in direct mode. Correlations are reported by following:

$$\frac{H_t}{H_p} = -1.4 + \frac{2.5}{\eta_p}; \quad \frac{Q_t}{Q_p} = -1.5 + \frac{2.4}{\eta_p{}^2} \frac{\eta_t}{p} = 1.158 - 0.265 N_{st} \tag{4}$$

Head and flow rate at the BEP in reverse mode can be evaluated with correlations of Grover [17] using the specific speed value of the PAT at the BEP ($N_{st} = N_s \eta_p$). All relations between head and flow rate are listed by following.

$$\begin{cases} \frac{H_t}{H_p} = 2.693 - 0.0229 N_{st} \\ \frac{Q_t}{Q_p} = 2.379 - 0.0264 N_{st} \\ \frac{\eta_T}{\eta_P} = 0.893 - 0.0466 N_{st} \end{cases} \tag{5}$$

As Grover [17], knowing that $N_{st} = N_s \eta_p$, Hergt [18] calculates the head and flow rate at the BEP in reverse mode using the specific speed value of the PAT at the BEP:

$$\frac{H_t}{H_p} = 1.3 - \frac{6}{N_{st} - 3}; \quad \frac{Q_t}{Q_p} = 1.3 - \frac{1.6}{N_{st} - 5} \tag{6}$$

Relations of Childs [13] evaluate the head and the flow rate at the BEP in reverse mode using the efficiency, head and flow rate value at the BEP in direct mode. Equations are reported by following:

$$\frac{H_t}{H_p} = \frac{1}{\eta_p}; \quad \frac{Q_t}{Q_p} = \frac{1}{\eta_p}; \quad \eta_t = \eta_p \tag{7}$$

Moreover, Derakhshan and Nourbakhsh [24] introduced a method based on theoretical analysis to evaluate the BEP of an industrial centrifugal pump. This method is based on the geometrical and hydraulic characteristics of the pump in direct mode. The final formula to evaluate the turbine's maximum efficiency is:

$$\eta_t = \frac{P_{nt}}{\gamma \times Q_t \times H_t} = \frac{\gamma \times Q_t \times H_t - P_{vt} - P_{lt} - P_{et} - P_{it} - P_{mt}}{\gamma \times Q_t \times H_t} \tag{8}$$

All of presented methods are based on different hypotheses. In Section 6 of this paper, all methods will be compared with the results of the proposed modeling methodology. The comparison has confirmed that each prediction method can be used only for a limited set of pumps. None of them, in fact, allows the prediction in the reverse running conditions for all geometries and over a wide range of pump specific speeds. In some cases, performance is underestimated or overestimated by as much as 30%.

3. Simulation Model

Three different centrifugal pumps have been modeled in order to obtain the necessary data to predict the performance of the pumps by the described procedures. The analyzed pumps are commercial ones and have three different specific speeds. The main characteristics are summarized in Table 1. It was decided to use pumps with different heads (from 3.9 to 60 m) and flow rates (from 45.4 to 148 m^3/s) to have different geometries and operating conditions to better test the prediction method.

Table 1. Pumps characteristics.

	Impeller Diameter (mm)	Delivery Outlet Diameter (mm)	H_{bep} (m)	Q_{bep} (m^3/h)
(*Ns* 37.6)	190	80	39	148
(*Ns* 20.5)	200	70	60	45.4
(*Ns* 64.0)	120	80	3.9	54

On the top side of Figure 1, the disassembled pump with a N_S = 37.6 is shown. This pump is a shrouded one with one- channel impeller and six blades and is called Pump 1. Starting from the real geometry in a .step format, the fluid volume has been extracted. In Figure 1, the fluid-volume is colored in green while the solid impeller is in blue and the solid rotor in red. In the same way, fluid volumes of others two pumps have been extracted and then modelled using a 3D-CFD approach.

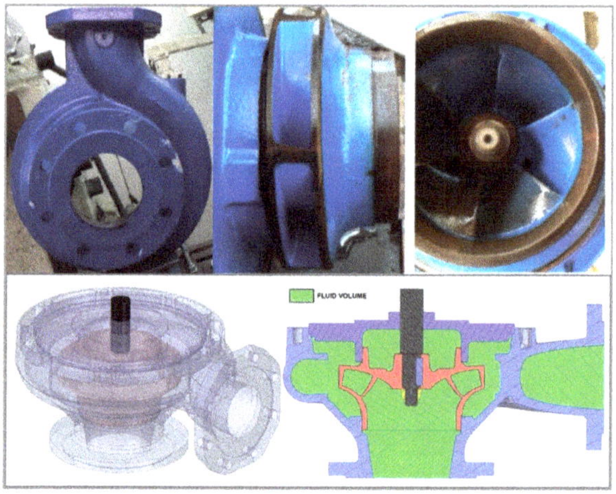

Figure 1. Geometry and fluid volume.

The study has been approached using the commercial code PumpLinx®. PumpLinx® is a three-dimensional CFD software developed by Simerics Inc. (1750 112th Ave NE, Ste C250, Bellevue, WA 98004, USA) [25–29]. It numerically solves the fundamental conservation equations of mass, momentum and energy and includes robust models of turbulence and cavitation.

The fluid volume of each pump has then been meshed with the PumpLinx® grid generator using a body-fitted binary tree approach. These grids have been demonstrated to be extremely accurate and efficient [25–29]. In fact, the parent-child tree architecture allows for an expandable data structure with reduced memory storage, the binary refinement is optimal for transitioning between different length scales and resolutions within the model, the majority of cells are cubes, and, since the grid is created from a volume, it can tolerate inaccurate CAD surfaces with small gaps and overlaps. It is important to underline that, cells are hexagonal not deformed therefore the skewness is zero [30].

Figure 2 shows the binary tree mesh of three pumps under study: Pump 1 (*Ns* = 37.6), Pumps 2 (*Ns* = 20.5) and Pump 3 (*Ns* = 64).

It is important to underline that using the binary tree approach in the boundary layer can easily increase the grid density on the surface without excessively increasing the total cell count. In this way, the grid has been subdivided and cut to conform it to the surface in regions of high curvature and small details [27]. Impellers and rotors fluid volumes of each model have been meshed separately. A maximum cell size of 0.025 has been chosen, where no cell in the volume can have a cell side larger than the maximum cell size. The minimum cell size has been fixed at 0.0001. The minimum cell size is a parameter used to limit how small cells can be in attempting to resolve the geometry using the general mesher. No cell in the volume can have a cell side smaller than the minimum cell size. The cell size on surfaces has also been fixed at 0.00625. This parameter is used to control the size of the cells for all surfaces of a mesh volume. Using a pure Eulerian approach the mesh of rotors is deformed by squeezing and expanding the cells. The rotor mesh is colored in green in Figure 2.

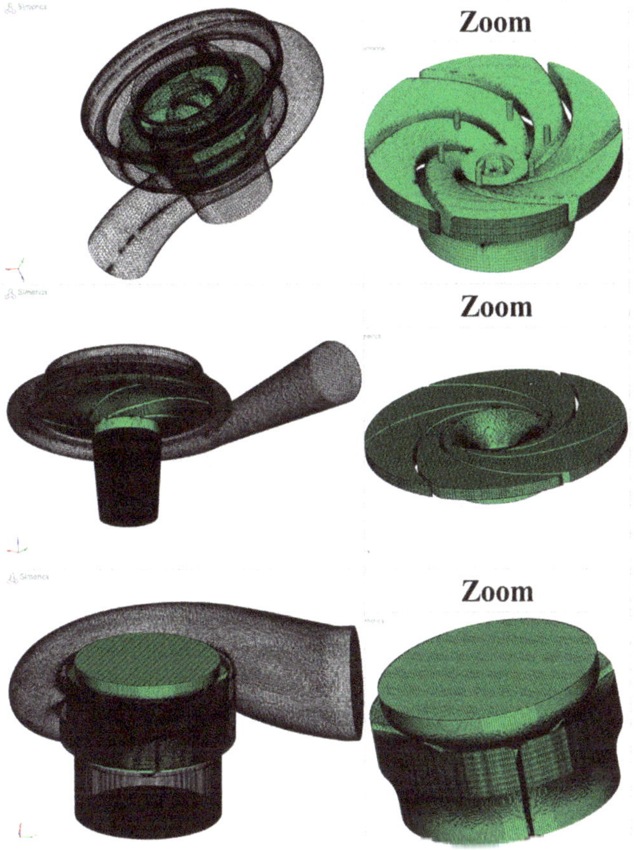

Figure 2. Binary tree mesh for three pumps.

A mesh sensitivity, as well known, is fundamental in studying any new problem with a CFD solver. The target is to allow excellent accuracy (in comparison with experimental tests) and computational efficiency. Pumps fluid volumes have been meshed increasing and decreasing the already described parameters: maximum cell size, minimum cell size and the cell size on surfaces. For each case the best compromise between results accuracy and computational time has been found and by following all mesh characteristic are listed:

PUMP 1 ($N_S = 37.6$): Total number of cells: 851.673

Total number of faces: 3.383.745

Total simulation time: 8.9 h as Pump, 9 h as PAT

PUMP 2 ($N_S = 20.5$): Total number of cells: 1.039.450

Total number of faces: 3.926.412

Total simulation time: 4.2 h as Pump, 4.8 h as PAT

PUMP 3 (N_S = 64): Total number of cells: 324.596

Total number of faces: 3.348.318

Total simulation time: 4.5 h as Pump, 4.8 h as PAT

Simulations have been run with an Intel(R) Xeon(R) CPU 2.66 GHz (two processors). It is important to underline that models are transient, and during simulations there is a simultaneous treatment of moving and stationary fluid volumes. In particular, each volume connects to the others via an implicit interface. Mismatched grid interfaces can identify overlap areas and match them without interpolation. These faces, during the simulation process, are treated no differently than an internal face between two neighboring cells in the same grid domain.

Using the models described it has been possible to study the internal fluid dynamics of each pump working in the direct (as pump) and reverse (as turbine) modes. A mature turbulence model has been implemented. It has been demonstrated that for these applications one of the more accurate model to study turbulence is the *k*–ε model and RNG *k*–ε model. Other turbulence models, such as LES and RNG *k*–ε model, might also provide good results but the adoption of higher order turbulence models would have increased the computational time with no relevant improvement of the results. Authors have also applied the presented strategy confirming the solution accuracy [25–29,31–33] in other analyses.

Therefore, in this research, the *k*-epsilon model has been used providing a good accuracy and computationally efficiency. This model is used by the adopted CFD code since it has been available for more than a decade and has been widely demonstrated to provide good engineering results for a wide range of applications [29,31–33]. The standard *k*–ε model, used for the simulations presented in this paper is based on the following two equations [25–29,31,32]:

$$\frac{\partial}{\partial t}\int_{\Omega(t)}^{0} \rho k d\Omega + \int_{\sigma}^{0} \rho((v-v_{\sigma})\times n)k d\sigma = \int_{\sigma}^{0}\left(\mu+\frac{\mu_t}{\sigma_k}\right)(\nabla k\times n)d\sigma + \int_{\Omega}^{0}(G_t-\rho\epsilon)d\Omega \quad (9)$$

$$\frac{\partial}{\partial t}\int_{\Omega(t)}^{0} \rho\epsilon d\Omega + \int_{\sigma}^{0} \rho((v-v_{\sigma})\times n)\epsilon d\sigma = \int_{\sigma}^{0}\left(\mu+\frac{\mu_t}{\sigma_\varepsilon}\right)(\nabla\epsilon\times n)d\sigma + \int_{\Omega}^{0}\left(c_1 G_t\frac{\varepsilon}{k}-c_2\rho\frac{\varepsilon^2}{k}\right)d\Omega \quad (10)$$

where c_1 = 1.44, c_2 = 1.92, σ_k = 1, σ_ε = 1.3, where σ_k and σ_ε are the turbulent kinetic energy and the turbulent kinetic energy dissipation rate Prandtl numbers.

The turbulent kinetic energy, *k*, is defined as [25–29,31–33]:

$$k = \frac{1}{2}(v'\cdot v') \quad (11)$$

where v' is the turbulent fluctuation velocity, and the dissipation rate, ε, of the turbulent kinetic energy is defined as [25–29]:

$$\varepsilon = 2\frac{\mu}{\rho}\overline{\left(S'_{ij}S'_{ij}\right)} \quad (12)$$

In which the strain tensor is [29,31]:

$$S'_{ij} = \frac{1}{2}\left(\frac{\partial u'_i}{\partial x_j}+\frac{\partial u'_j}{\partial x_i}\right) \quad (13)$$

with u_i' (i = 1, 2, 3) being components of v'.

The turbulent viscosity μ_t is calculated by [16,23,24,32,33]:

$$\mu_t = \rho C_\mu \frac{k^2}{\epsilon} \quad (14)$$

with $C_\mu = 0.09$.

The turbulent generation term G_t can be expressed as a function of velocity and the shear stress tensor as [16,23,24]:

$$G_t = -\rho\overline{u'_i u'_j}\frac{\partial u'_i}{\partial x_j} \tag{15}$$

where $\tau'_{ij} = \rho\overline{u'_i u'_j}$ is the turbulent Reynolds stress, which can be modelled by the Boussinesq hypothesis [15,16,23,24,32,33]:

$$\tau'_{ij} = \mu_t\left(\frac{\partial u_i}{\partial x_j} + \frac{\partial u_j}{\partial x_i}\right) - \frac{2}{3}\left(\rho k + \frac{\partial u_k}{\partial x_k}\right)\delta_{ij} \tag{16}$$

With the built models, simulations have first been run comparing the results in the direct and reverse working modes. Model results for Pump 1 at 2900 rpm are shown in Figure 3; in Figure 3a the pressure distribution at the BEP in the 0–9 bar pressure range is presented. The fluid properties of water have been used in all the simulations.

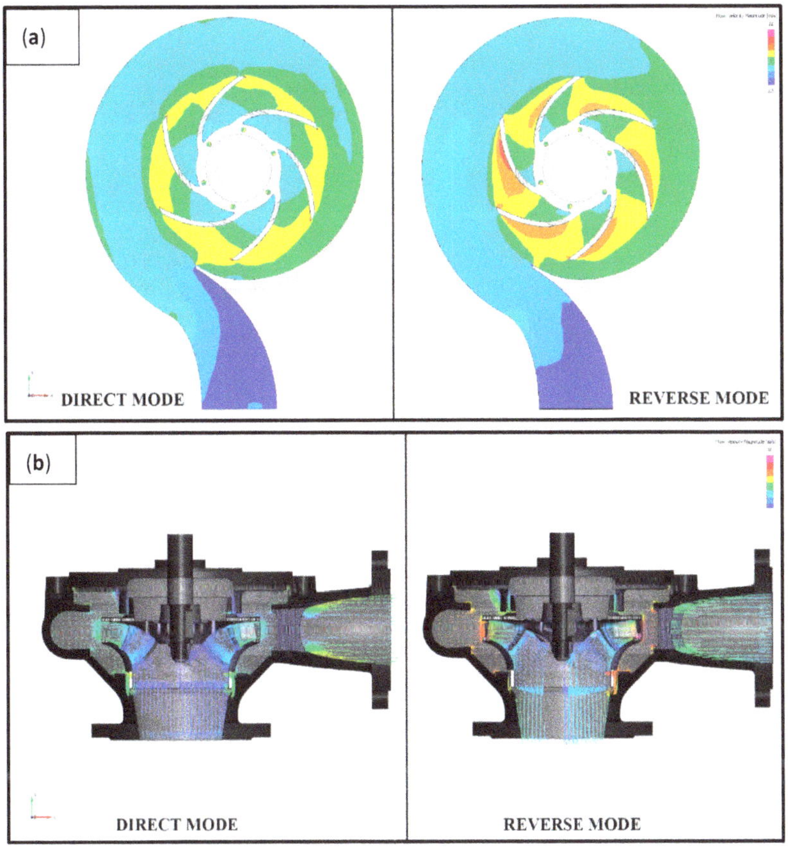

Figure 3. Model results for Pump 1 (Ns = 37.6) at 2900 rpm. (**a**) Pressure distribution; (**b**) velocity vectors.

Figure 3b shows the velocity vectors in the fluid volume; the velocity range (0–32 m/s) is the same for the direct and reverse modes. In this picture, it is possible to visualize the flow evolution inside the machine and the acceleration/deceleration of the fluid. Both figures confirm that the velocity is higher in the reverse mode.

Similarly, Figure 4 shows the pressure distributions for pump 2 (N_S = 20.5) and pump 3 (N_S = 64) at 2900 rpm.

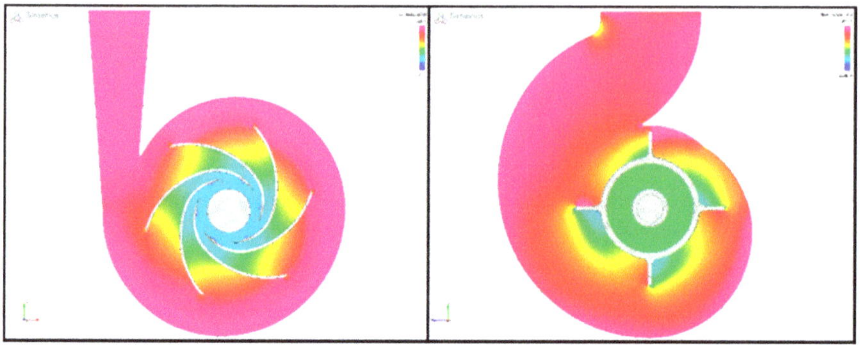

(a) (b)

Figure 4. Pressure distribution in the fluid volume of pumps 2 and 3 at 2900 rpm. (**a**) *Ns* = 20.5; (**b**) *Ns* = 64.

For the direct mode, CFD models have been validated using the data supplied by the pump manufacturers. In Figure 5, the head vs. flow rate plots (as the blue curves) are shown. Across the range of flow rates (30–207) m^3/h, the head varies from 47 m to 3 m. In the plots in Figure 5, the model results are shown in red. The comparison in Figure 5 demonstrates the accuracy of the adopted methodology; in fact, the percentage error is always less than 4% while for many points the error is near zero.

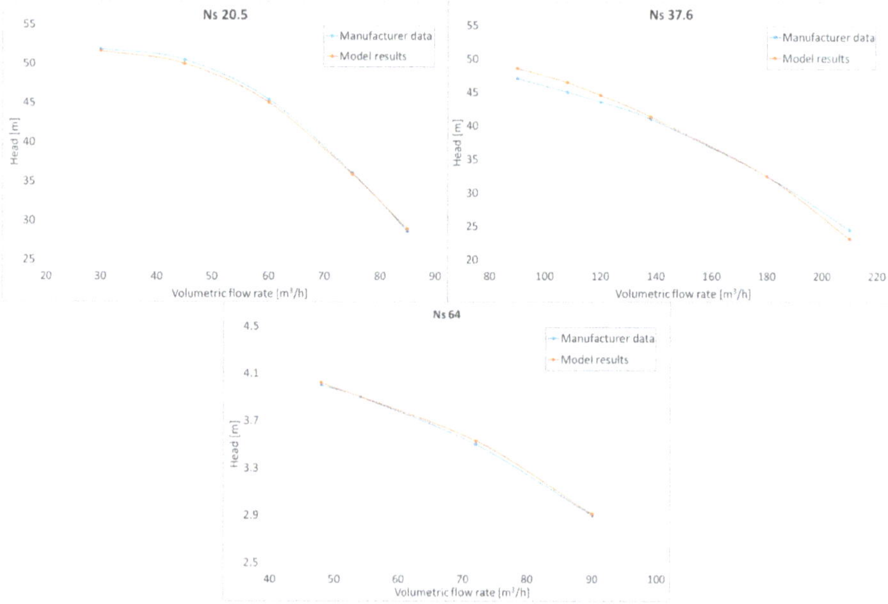

Figure 5. Model validation: Comparison between manufacturer's data and model results.

4. *Ns* 37.6 Pump Model Validation with Experimental Data

Once the model had been validated in the pumping mode, it was decided to also validate it in the reverse mode, to assess whether the model reproduces the turbine mode well. Because the proposed methodology is based only on the results of the CFD model under reverse conditions, the validation under reverse conditions was necessary to confirm the entire methodology.

The model of centrifugal Pump 1 has been validated with data from an experiment performed on a dedicated test bench of the Department of Civil Construction and Environmental Engineering of the University of Naples Federico II. The bench enables testing a centrifugal pump running in reverse mode. The aim of this activity was to further validate the simulation model under reverse conditions. The test bench reproduces a full-scale hydraulic network, made up of four nodes (Figure 6). An external pump increases the water pressure to simulate the behavior of a real urban network while an air chamber stabilizes the flow rate. The tested pump has been installed in one node where two pressure-reducing valves (PRV) regulate the water flow rate and the pressure at the inlet and outlet of the pump. The pressure is regulated with a valve installed at the network inlet, while the flow rate at the outlet is varied with a butterfly valve. Valves are remotely controlled with electronic actuators controlled by a dedicated homemade software program, with an external PLC [34–40].

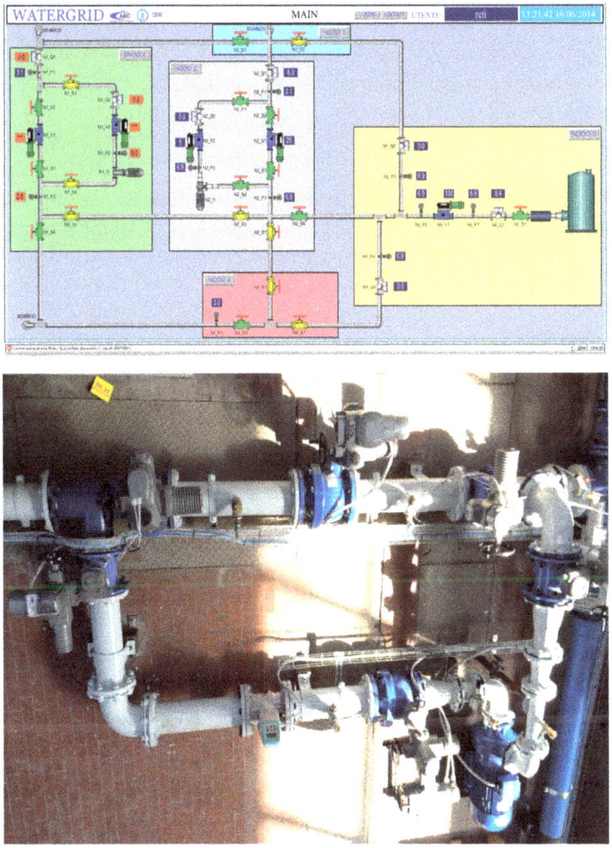

Figure 6. Test bench—water grid of the Department of Civil Construction and Environmental Engineering of the University of Naples Federico II.

The electric motor of the pump is linked to an inverter and the produced electrical power is connected to the urban power grid. In the node, two pressure transducers, P_1 and P_2 (Burkert® model 8314), and a flow meter Q (Siemens® mag 500) have been installed. All test bench data have been acquired by a homemade acquisition system. Furthermore, a 360-tooth encoder has been installed on the electrical motor to acquire the shaft speed.

The hardware system is based on a data acquisition board NI DAQ Card (12-bit ADC converter resolution, 16 input channel, two 24 bit counters), a 68-pin shielded desktop connector block (NI TBX-68). NI LabView performs a homemade software.

The pressure is acquired with Buckert transducers. Sensors are installed upstream and downstream of the PAT, as it is shown in Figure 6. The characteristics of the pressure transducers are as follows:

- Ceramic technology
- 0–10 bar pressure range
- ±0.25% accuracy
- 2 ms response time

The electric motor shaft speed is acquired with a BAUMER BHK 16.05A.360-I2-5 incremental encoder, with 360 teeth, while, as already said, the flow rate is measured with an electromagnetic Siemens mag 500 transducer, (accuracy 0.25%, response time 1 s). The sample frequency was 1 Hz.

The experiments have been performed only in steady-state conditions, varying the water flow rate and the pressure at the inlet of the pump, for different shaft speeds. In particular, the flow rate has been varied between 8 and 21 L/s, and the shaft speed between 300 and 2200 rpm. During the test, pressure at the inlet and outlet of the pump have been acquired at a sample frequency of 1 Hz.

As stated above, the tests have been done in steady-state conditions, running the pump in reverse mode. The flow rate, the pressures at the inlet and outlet of the pump and the shaft rpm were measured. In Figure 7, all results of the experimental campaign are shown.

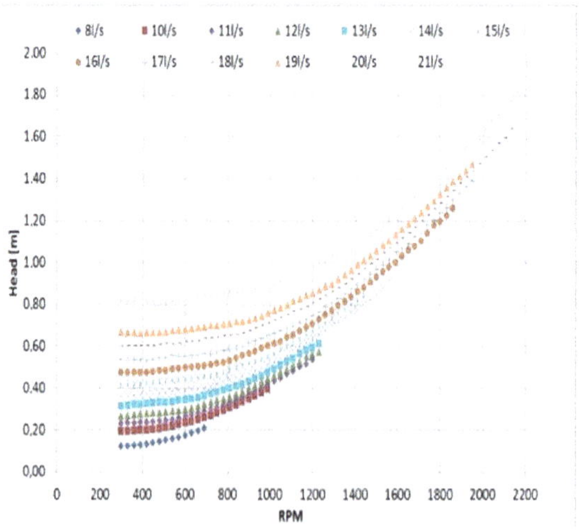

(a)

Figure 7. *Cont.*

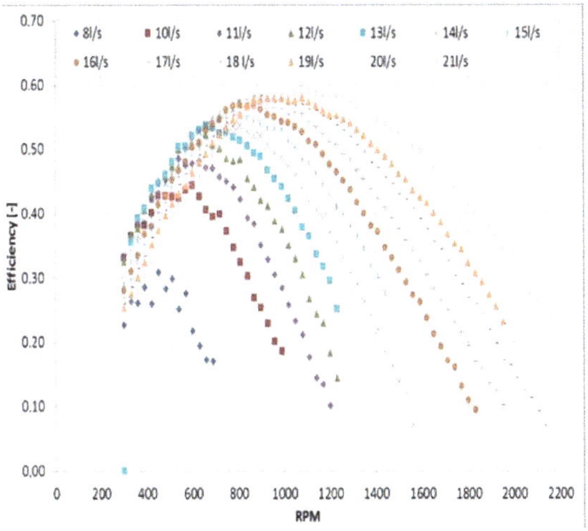

(b)

Figure 7. Experimental results. (**a**) Head; (**b**) Efficiency.

To examine the PAT performance, the total head (m) versus shaft rpm is reported, varying the water flow rate for all the examined conditions. Results confirm what is known from the literature: the PAT head increases with the rpm and with the flow—rate, and it can be easily noted that, for the tested conditions, the head varies between 0.1 and 1.8 m.

In Figure 8, the whole validation of the simulation model is presented: it shows that the model reproduces the experimental data well with very small differences between the experimental and the model results for all the running conditions that were analyzed.

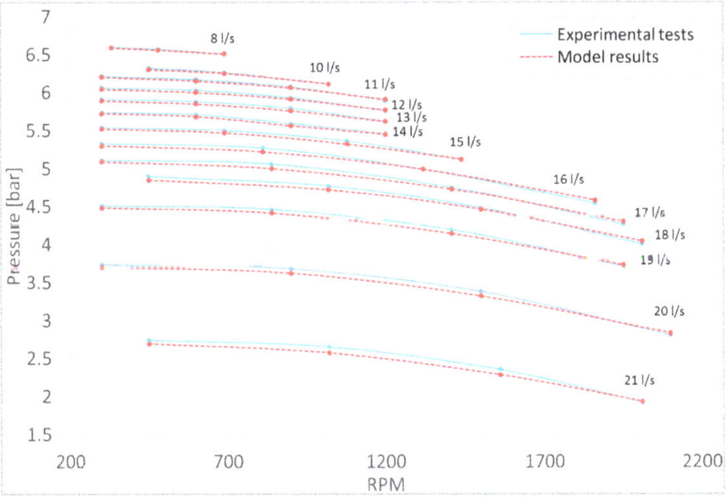

Figure 8. Model validation.

5. Model Results

After the validation phase in pump and turbine mode, simulation models have been used to predict the efficiency curves of the three analyzed pumps. Then, all the simulations have been performed to obtain the data necessary to evaluate the inverse characteristics.

The specific head ψ can be evaluated as:

$$\psi = \frac{gH}{n^2 D^2} \tag{17}$$

The specific capacity φ depends by the flow Q and the impeller diameter D:

$$\varphi = \frac{Q}{nD^3} \tag{18}$$

The specific power is defined as:

$$\pi = \frac{P}{\rho n^3 D^5} \tag{19}$$

While the efficiency can be evaluated as:

$$\eta = \frac{P}{\rho Q H} \tag{20}$$

where H (m), Q (m^3/s), and P (W) are the head, flow rate and power, respectively. The rotational speed is n (RPS) and D (m) is the impeller diameter. In the reverse mode simulation, the boundary conditions in pump and PAT mode were the same (declared data in pump mode). The boundary conditions in reverse mode are summarized in Table 2.

Table 2. Boundary conditions.

Boundary Conditions	Pump 1	Pump 2	Pump 3
Outlet pressure	1.9 bar	1.9 bar	1.9 bar
Inlet Volumetric Flow	90/210 m^3/h	30/85 m^3/h	48/90 m^3/h
T_{in}	293.15 K	293.15 K	293.15 K
P_{sat}	2886 Pa	2886 Pa	2886 Pa

The specific head, the specific power and the efficiency have been evaluated for both pump and turbine mode and for all the studied pumps. These are plotted versus specific capacity in Figure 9.

It is clear that at high capacity the specific head in reverse mode is always higher than in direct mode. In reverse mode, pumps have a larger power range than in direct mode. The trends are quite different and the curves arising at low flow rate and at higher capacities have a higher power value than in direct mode. In pump mode, the efficiency has a typical "bell shape" while in PAT mode its profile resembles that of a Francis turbine: it increases at low flow rates and reaches a maximum value at high flow rates. Moreover, the pump with the low specific speed works with low flow rates but high heads in direct mode. In reverse mode at high flow rates, the head is higher than the direct-mode value. The pump with the high specific speed works at high flow rates but low heads. In reverse mode, working with the same flow rates, the maximum head is also higher than the direct-mode value. For all pumps, the maximum value is lower than the direct-mode value. For low specific speed pumps, this maximum value is approximately equal to the direct-mode value, while for high specific speed pumps, it is lower.

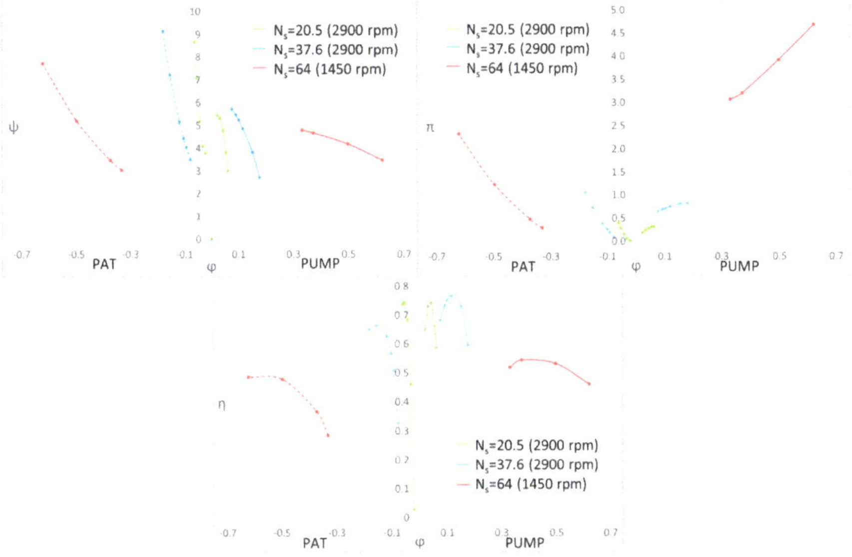

Figure 9. Specific head, specific power and efficiency.

In conclusion, in Table 3, the BEP values are summarized:

Table 3. Boundary conditions.

Boundary Conditions	Pump 1		Pump 2		Pump 3	
	Direct Mode	Reverse Mode	Direct Mode	Reverse Mode	Direct Mode	Reverse Mode
Head (m)	39	61	45.4	67	3.9	4.6
Capacity (m^3/s)	0.041	0.05	0.017	0.021	0.015	0.022
Power (kW)	20.5	19.98	10.01	10.24	1.05	0.75
Efficiency	0.787	0.663	0.743	0.741	0.543	0.487

6. Comparison of Prediction Methods

After the evaluation of the inverse characteristics, the results of the proposed methodology have been compared to the prediction methods available in literature. To this end, all the previously discussed methods have been applied to the three analyzed pumps. Some methods predict only the head and flow rate, while others also predict power and efficiency. In Table 4, all the results are shown for Pump 1.

The flow rates calculated using the methods of Sharma and of Hergt and Schmiedl are very close to those of the proposed CFD methodology. Grover's and Alatorre-Frenk's methods overestimate this value while the other methods underestimated the value by a margin of 10%–15%. All methods underestimate the head value, except for Grover's. These values diverge with errors of up to 30%.

Evaluating the percentage deviation as:

$$\text{percentage deviation} = \frac{\text{predicted value} - cfd\text{ value}}{cfd\text{ value}} \times 100 \tag{21}$$

In Figure 10 the deviations between the predictions of these methods and the simulated data are plotted vs. the pump specific speeds for all three pumps.

Table 4. Comparison for pump 1.

Methods	*H* (m)	*Q* (m^3/s)	*P* (kW)	η
Model results	61.42	0.05	19.98	0.663
Stepanoff	49.55	0.0463	22.5	0.787
Alatorre-Frenk	42.79	0.091	28.92	0.757
Sharma	51.99	0.0498	20.5	0.807
Schmiedl	43.94	0.0514	16.82	0.759
Grover	78.59	0.0657	36.92	0.729
Hergt	41.90	0.0508	-	-
Childs	49.55	0.0522	19.96	0.787
D&N	58.56	0.0411	18.17	0.769

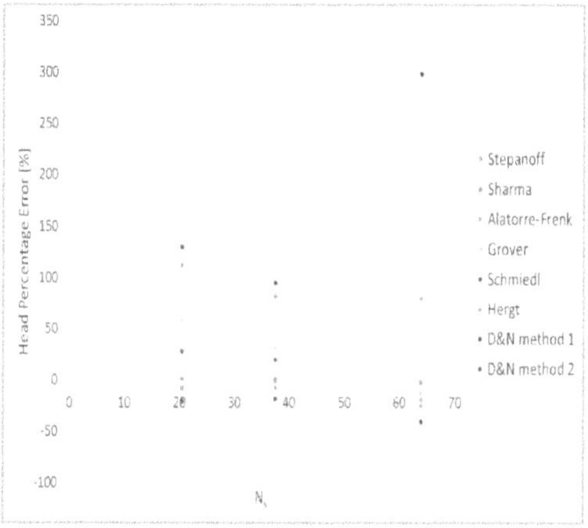

(**a**)

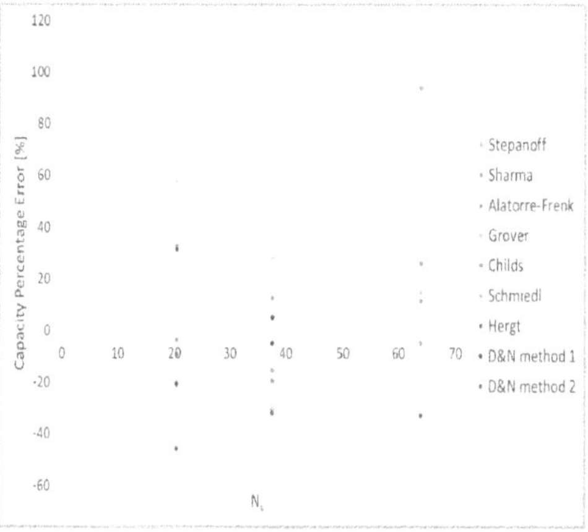

(**b**)

Figure 10. Prediction methods comparison. (**a**) Head Percentage Error; (**b**) Capacity Percentage Error.

It easy to observe that in some cases the deviation is very high. To evaluate the Derakhshan and Nourbakhsh efficiency [24], the CFD model results were used as shown in Figure 11. In this figure, the relative velocity magnitude (m/s) is shown and the angle between the relative and absolute velocity is highlighted in red, at BEP conditions and in pump mode.

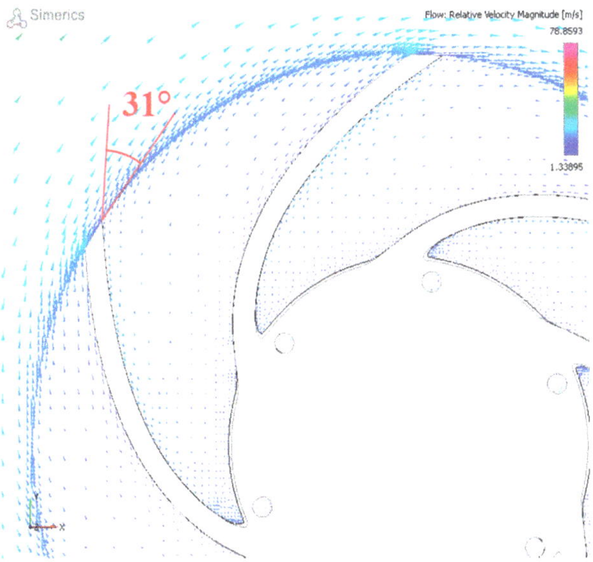

Figure 11. Relative velocity magnitude.

7. Conclusions

In this paper a methodology to predict the inverse characteristic of a centrifugal pump has been presented. This methodology is based on the results of a three-dimensional simulation model built with a commercial CFD code. Three industrial pumps have been analyzed, with different specific speeds. First, the simulation models have been validated with data supplied by the pump manufacturers. Then the results of an experimental campaign have been used to validate a model simulating the pump working in reverse conditions.

Starting from the CFD model results, the specific head, capacity, power and efficiency have been evaluated and the best efficient point of all the analyzed pumps was found. Furthermore, several prediction methods have been applied to the tested pumps and their predicted values were compared with those of the proposed methodology. Some methods (e.g., Childs' method) are not in accord while others (e.g., Stepanoff's method) show small relative differences.

Acknowledgments: This research has been supported by the Department of Industrial Engineering of the University of Naples "Federico II". We appreciate the technical contribution from Gennaro Stingo and Giuseppe Iovino of the Department of Industrial Engineering and from colleagues of the Department of Civil Construction and Environmental Engineering of the University of Naples "Federico II".

Author Contributions: Emma Frosina, Dario Buono and Adolfo Senatore conceived and designed the experiments; Dario Buono performed the experiments; Emma Frosina and Dario Buono analyzed the data; Adolfo Senatore contributed reagents/materials/analysis tools; Emma Frosina wrote the paper.

Conflicts of Interest: The authors declare no conflict of interest.

Nomenclature

PRV	Pressure Reducing Valve
PAT	Pump as Turbine
CFD	Computational Fluid Dynamic
BEP	Best Efficiency Point
H_p	Pump head
Q_p	Pump flow rate
P_p	Pump power

η_p	Pump overall efficiency
H_t	Turbine head
Q_t	Turbine flow rate
P_t	Turbine power
η_t	Turbine overall efficiency
ψ	Specific head
φ	Specific capacity
π	Specific power
η	Efficiency
N_s	Pump specific speed
N_{st}	Turbine specific speed
z	Impeller's blade number
P_{it}	Power losses due to leakage
P_{vt}	Volute power losses
P_{et}	Kinetic energy losses
P_{nt}	Turbine net power
P_{it}	Hydraulic losses of the impeller in turbine mode
n	Surface normal
k	Turbulence kinetic energy
p	Pressure (Pa)
Q	Flow rate (m^3/h)
rpm	Revolution per minute
U	Initial velocity
u	Velocity component (m/s)
u'	Component of v'
v	Velocity vector
v'	Turbulent fluctuation velocity
μ	Fluid viscosity (Pa-s)
ρ	Fluid density (kg/m^3)
$\tilde{\tau}$	Shear stress tensor
c_1, c_2	Constant
σ_k	Turbulent kinetic energy
σ_ε	Turbulent kinetic energy dissipation
S'_{ij}	Strain tensor
μ_t	Turbulent viscosity
G_t	Turbulent generation
τ_{ij}	Turbulent Reynolds stress
ε	Turbulence dissipation
Ω	Control volume

References

1. Williams, A. *Pumps as Turbines: A User's Guide*; Intermediate Technology Publications: London, UK, 1995.
2. Arriaga, M. Pump as turbine—A pico-hydro alternative in Lao People's Democratic Republic. *Renew. Energy* **2010**, *35*, 1109–1115. [CrossRef]
3. Derakhshan, S.; Nourbakhsh, A. Experimental study of characteristic curves of centrifugal pumps working as turbines in different specific speeds. *Exp. Therm. Fluid Sci.* **2008**, *32*, 800–807. [CrossRef]
4. Williams, A.A.; Rodrigues, A.; Singh, P.; Nestmann, F.; Lai, E. Hydraulic analysis of a pump as a turbine with CFD and experimental data. In Proceedings of the IMechE Seminar, Computational Fluid Dynamics for Fluid Machinery, London, UK, 18 November 2003.
5. Nautiyal, H.; Varun, A.K. Reverse running pumps analytical, experimental and computational study: A review. *Renew. Sustain. Energy Rev.* **2010**, *14*, 2059–2067. [CrossRef]
6. Nautiyal, H.; Varun, K.A.; Yadav, S. 2010 CFD analysis on pumps working as Turbines. *Hydro Nepal* **2010**, *6*, 35–37.

7. Carravetta, A.; Del Giudice, G.; Fecarotta, O.; Ramos, H. PAT design strategy for energy recovery in water distribution networks by electrical regulation. *Energies* **2013**, *6*, 411–424. [CrossRef]
8. Carravetta, A.; Del Giudice, G.; Fecarotta, O.; Ramos, H. Pump as Turbine (PAT) design in water distribution network by system effectiveness. *Water* **2013**, *5*, 1211–1225. [CrossRef]
9. Derakhshan, S.; Nourbakhsh, A.; Mohammadi, B. Efficiency improvement of centrifugal reverse pumps. *ASME J. Fluids Eng.* **2009**, *131*. [CrossRef]
10. Natanasabapathi, S.R.; Kshirsagar, J.T. *Pump as Turbine—An Experience with CFX 5.6*; Kirloskar Bros. Ltd., Corporate Research and Engineering Division: Pune, India, 2004.
11. Barrio, R.; Fernández, J.; Parrondo, J.; Blanco, E. Performance prediction of a centrifugal pump working in direct and reverse mode using Computational Fluid Dynamics. In Proceedings of the International Conference on Renewable Energies and Power Quality (ICREPQ'10), Granada, Spain, 23–25 March 2010.
12. Chapallaz, J.M.; Eichenberger, P.; Fischer, G. Manual on pumps used as turbines. In *Friedr Vieweg Sohn Verlagsgesellschaft, Braunschweig*; Informatica International, Inc.: Braunschweig, Germany, 1992.
13. Childs, S.M. Convert pumps to turbines and recover HP. *Hydro Carbon Process. Pet. Refin.* **1962**, *41*, 173–174.
14. McClaskey, B.M.; Lundquist, J.A. Can You Justify Hydraulic Turbines? *Hydrocarb. Process.* **1976**, *55*, 163–166.
15. Lobanoff, V.S.; Ross, R.R. Hydraulic power recovery turbines. In *Centrifugal Pumps: Design and Applications*; Gulf Publishing Company: Houston, TX, USA, 1985; pp. 246–282.
16. Sanjay, V.J.; Rajesh, N.P. Investigations on pump running in turbine mode: A review of the state-of-the-art. *Renew. Sustain. Energy Rev.* **2014**, *30*, 841–868.
17. Grover, K.M. *Conversion of Pumps to Turbines*; GSA Inter Corp.: Katonah, NY, USA, 1980.
18. Williams, A. The turbine performance of centrifugal pumps: A comparison of prediction methods. *J. Power Energy* **1994**, *208*, 59–66. [CrossRef]
19. Stepanoff, A.J. *Centrifugal and Axial Flow Pumps*; John Wiley: New York, NY, USA, 1957.
20. Alatorre-Frenk, C. Cost Minimization in Micro Hydro Systems Using Pumps-Asturbines. Ph.D. Thesis, University of Warwick, Coventry, UK, 1994.
21. Sharma, K. *Small Hydroelectric Project-Use of Centrifugal Pumps as Turbines*; Technical Report; Kirloskar Electric Co.: Bangalore, India, 1985.
22. Schmiedl, E. *Serien-Kreiselpumpen im Turbinenbetrieb*; Pumpentagung: Karlsruhe, Germany, 1988.
23. Lewinsky, K.; Heinz, P. Pumpen als Turbinen fur Kleinkraftwerke. *Wasserwirtschaft* **1987**, *77*, 531–537.
24. Derakhshan, S.; Nourbakhsh, A. Theoretical, numerical and experimental investigation of centrifugal pumps in reverse operation. *Exp. Therm. Fluid Sci.* **2008**, *32*, 1620–1627. [CrossRef]
25. Frosina, E.; Senatore, A.; Buono, D.; Santato, L. Analysis and Simulation of an Oil Lubrication Pump for the Internal Combustion Engine. In Proceedings of the Fluids Engineering Systems and Technologies, San Diego, CA, USA, 15–21 November 2013.
26. Frosina, E.; Senatore, A.; Buono, D.; Olivetti, M. A Tridimensional CFD Analysis of the Oil Pump of a High Performance Engine. In Proceedings of the SAE 2014 World Congress and Exhibition, Detroit, MI, USA, 8–10 April 2014.
27. Frosina, E.; Buono, D.; Senatore, A. Modeling methodology to study the internal fluid-dynamic of a gas filter. *Int. Rev. Model. Simul.* **2015**, *8*, 533. [CrossRef]
28. Frosina, E.; Senatore, A.; Buono, D.; Stelson, K.A.; Wang, F.; Mohanty, B.; Gust, M.J. Vane pump power split transmission: Three dimensional computational fluid dynamic modeling. In Proceedings of the ASME/BATH 2015 Symposium on Fluid Power and Motion Control, Chicago, IL, USA, 12–14 October 2015.
29. Frosina, E.; Senatore, A.; Buono, D.; Stelson, K.A. A Mathematical Model to Analyze the Torque Caused by Fluid-Solid Interaction on a Hydraulic Valve. *ASME J. Fluids Eng.* **2016**, *138*, 061103. [CrossRef]
30. Automated Meshing with Binary Tree. Available online: https://www.simerics.com/download/automated-meshing-with-binary-tree/ (accessed on 11 January 2017).
31. Frosina, E.; Buono, D.; Senatore, A.; Costin, I.J. A Simulation Methodology Applied on Hydraulic Valves for High Fluxes. *Int. Rev. Model. Simul.* **2016**, *9*, 217. [CrossRef]
32. Frosina, E.; Buono, D.; Senatore, A.; Stelson, K.A. A modeling approach to study the fluid dynamic forces acting on the spool of a flow control valve. *J. Fluids Eng.* **2016**, *139*, 11103–11115. [CrossRef]
33. Pellegri, M.; Vacca, A.; Frosina, E.; Buono, D.; Senatore, A. Numerical analysis and experimental validation of Gerotor pumps: A comparison between a lumped parameter and a computational fluid dynamics-based approach. *Inst. Mech. Eng. Part C J. Mech. Eng. Sci.* **2016**, *1989–1996*, 203–210. [CrossRef]

34. Pugliese, F.; De Paola, F.; Fontana, N.; Giugni, M.; Marini, G. Experimental characterization of two pumps as turbines for hydropower generation. *Renew. Energy* **2016**, *99*, 180–187. [CrossRef]
35. Su, X.; Huang, S.; Zhang, X.; Yang, S. Numerical research on unsteady flow rate characteristics of pump as turbine. *Renew. Energy* **2016**, *94*, 488–495. [CrossRef]
36. De Marchis, M.; Milici, B.; Volpe, R.; Messineo, A. Energy saving in water distribution network through pump as turbine generators: Economic and environmental analysis. *Energies* **2016**, *9*, 877. [CrossRef]
37. Wang, T.; Kong, F.; Chen, K.; Duan, X.; Gou, Q. Experiment and analysis of effects of rotational speed on performance of pump as turbine. *Trans. Chin. Soc. Agric. Eng.* **2016**, *32*, 67–74.
38. Frosina, E.; Senatore, A.; Buono, D.; Monterosso, F.; Olivetti, M.; Arnone, L.; Santato, L. A tridimensional CFD analysis of the lubrication circuit of a non-road application diesel engine. In Proceedings of the 11th International Conference on Engines and Vehicles, ICE 2013, Naples, Italy, 15–19 September 2013.
39. Venturini, M.; Alvisi, S.; Simani, S. Energy potential of pumps as turbines (PATs) in water distribution networks. In Proceedings of the 28th International Conference on Efficiency, Cost, Optimization, Simulation (ECOS) and Environmental Impact, Pau, France, 29 June–3 July 2015.
40. Barbarelli, S.; Amelio, M.; Florio, G. Predictive model estimating the performances of centrifugal pumps used as turbines. *Energy* **2016**, *107*, 103–121. [CrossRef]

energies

MDPI

Article

Possibilities and Limitations of CFD Simulation for Flashing Flow Scenarios in Nuclear Applications

Yixiang Liao [1,2,*] and Dirk Lucas [2]

1 Gesellschaft für Anlagen-und Reaktorsicherheit, Boltzmannstraße 14, 85748 Garching bei München, Germany

2 Helmholtz-Zentrum Dresden-Rossendorf, Institute of Fluid Dynamics, Bautzner Landstraße 400, 01328 Dresden, Germany; D.Lucas@hzdr.de

* Correspondence: Yixiang.Liao@grs.de or Y.Liao@hzdr.de; Tel.: +49-351-260-2389

Academic Editor: Bjørn H. Hjertager

Received: 19 August 2016; Accepted: 12 January 2017; Published: 23 January 2017

Abstract: The flashing phenomenon is relevant to nuclear safety analysis, for example by a loss of coolant accident and safety release scenarios. It has been studied intensively by means of experiments and simulations with system codes, but computational fluid dynamics (CFD) simulation is still at the embryonic stage. Rapid increasing computer speed makes it possible to apply the CFD technology in such complex flow situations. Nevertheless, a thorough evaluation on the limitations and restrictions is still missing, which is however indispensable for reliable application, as well as further development. In the present work, the commonly-used two-fluid model with different mono-disperse assumptions is used to simulate various flashing scenarios. With the help of available experimental data, the results are evaluated, and the limitations are discussed. A poly-disperse method is found necessary for a reliable prediction of mean bubble size and phase distribution. The first attempts to trace the evolution of the bubble size distribution by means of poly-disperse simulations are made.

Keywords: flashing; computational fluid dynamics (CFD) simulation; two-fluid-model; mono-disperse; poly-disperse

1. Introduction

Flash boiling (flashing) is a process of phase change from liquid to vapor. It distinguishes itself from traditional boiling by the way the liquid gets superheated. It is caused by depressurization or pressure drop instead of heating. For this reason, flash boiling is sometimes also called "adiabatic boiling", namely without external heat sources. Another more familiar phenomenon of the same type is "cavitation". Under the term of cavitation, we often understand an isothermal process, where the growth or collapse of pre-existing vapor nuclei is an inertia-controlled process driven by the pressure difference across the gas-liquid interface. The phase change is controlled by pressure drop, and its recovery while liquid temperature remains constant. In this case, the vapor generation rate can be approximated by using the Rayleigh-Plesset equation. On the other hand, the phase change process in flash boiling is similar to that in traditional boiling with external heat sources. It is characterized by nucleation and the inter-phase heat transfer rate, namely by thermal non-equilibrium. The superheated liquid is cooled down by giving up surplus energy to vapor generation. The effect of pressure non-equilibrium between the inside and outside of a vapor bubble is often neglected. Nevertheless, an actual phase change process taking place in superheated liquid due to depressurization is controlled by both thermal (temperature difference) and mechanical (i.e., pressure difference) effects. The terms of "cavitation" and "flashing" above represent only simplification assumptions, i.e., neglecting thermal or mechanical effects, in theoretical treatments. The ratio of thermal and mechanical non-equilibrium contribution to bubble growth and phase change is dependent on the temperature (or pressure) level at which the process is taking place. At a high pressure level, especially in low depressurization rate situations, the phase change process is dominated by the inter-phase thermal heat transfer and is often treated as a flash boiling process.

The flash boiling phenomenon has a fundamental and decisive presence in many industrial and technical applications, where significant pressure drop is present. In nuclear engineering, as an example, it can take place in the following scenarios:

1. Large break loss-of-coolant accidents (LOCA) of pressurized water nuclear reactors;
2. Pressure release through blow-off valves at the pressurizer or steam generator;
3. Two-phase critical flow problem through nozzles;
4. Flashing-induced instability in natural circulation systems.

Under these circumstances, the properties of the flashing flow, such as the discharge flow rate, vapor generation rate, void hold-up, as well as two-phase morphology, are of key safety and economic importance. All of these quantities are influenced by the degree of non-equilibrium substantially.

Since the middle of the last century there have arisen many theoretical and experimental investigations on two-phase flashing flow owing to great concern in nuclear safety. With respect to the theoretical study, simplifying and empirical assumptions usually have to be adopted due to the complexity of its nature. The degree of non-equilibrium is accounted for partially or neglected fully. Therefore, the general validity of available methods is largely limited. So far, one-dimensional approaches or system codes are routinely applied to deal with such kinds of issues. However, the flashing two-phase mixture has a strongly three-dimensional nature, which is accompanied with a large heterogeneous gaseous structure and high gas volume fraction. All of these features along with the micro-scale bubble dynamic processes require a more sophisticated prediction tool with high time and space resolution, such as the CFD (computational fluid dynamics) technology. Furthermore, the drawback of increased computational effort in CFD simulations is being offset by the rapidly increasing speed and decreasing hardware costs of parallel computers. Therefore, there exists a need to assess the restrictions and update the closures for flashing situations.

2. State of the Art of CFD Simulation of Flashing Flow

Recently, promising CFD research on flashing flows has been published, such as in [1–9]. All of these simulations are based on the framework of a simplified two-fluid-model for bubbly flow, where the interfacial area density for inter-phase exchange is obtained by the particle model. Furthermore, they all assume a mono-disperse interface morphology. That means that the bubble size in each computational cell has a single value instead of a spectrum at any given time.

Laurien and his co-workers studied water evaporation and re-condensation phenomena caused by steady-state pressure variation inside a three-dimensional complex pipeline [1–4]. Frank [5] simulated the well-known Edwards pipe blow-down test [10] using a one-dimensional simplified mesh in CFX (a commercial CFD code of the company ANSYS). Both of them employed a five-equation model including two continuity equations, two momentum equations for liquid and vapor, respectively, and one energy equation for liquid. The vapor was assumed to remain always at the saturation condition corresponding to local pressure, which is uniform inside and outside the bubble. The assumption is reasonable in the case of a small depressurization rate. For the computation of interfacial area density, a constant bubble diameter, e.g., $d_g = 1$ mm, is prescribed in the whole domain. In addition, the momentum interaction between the gas and liquid phases is modelled only as a drag force while the effect of non-drag forces is ignored. However, Liao [11] showed that non-drag forces have a significant effect on the spatial distribution of phases.

Later on, Laurien [4] suggested that a model presuming bubble number density instead of bubble size, which allows bubble size to grow, is more close to the physical picture of boiling flow. This assumption is acceptable when the nucleation zone is sufficiently narrow and bubble dynamics, such as coalescence and break-up, is negligible. Otherwise, an additional transport equation for the bubble number density with appropriate source terms or even a poly-disperse method is necessary.

Maksic and Mewes [6] simulated flashing flows in pipes and nozzles by using a four-equation model, where a common velocity field is assumed for both phases. The inter-phase heat transfer is assumed to be dominated by conduction. However, it has been shown that in most flashing expansion cases, the convective contribution due to the relative motion of bubbles dominates the heat transfer [12]. Neglecting of inter-phase velocity slip obviously under-predicts the vapor generation rate [11]. Wall nucleation was considered as a unique source of bubble number density in the additional transport equation. The Jones model [13–16] was used to determine the nucleation rate. Inter-phase mass, momentum and energy transfer due to nucleation were ignored.

Marsh and O'Mahony [7] simulated the nozzle flashing flow using a six-equation model in FLUENT (a commercial CFD code of the company ANSYS), with separate mass, momentum and enthalpy balance equations for liquid and vapor. Inter-phase mass and momentum, as well as energy transfer resulting from both nucleation and phase change were accounted for. However, the effects of non-drag forces on momentum exchange and the heat transfer between vapor and vapor-liquid interface were neglected. A modified version of the Blander and Katz nucleation model [17] was employed to compute the source of bubble number density. The original model was found to create large numerical instability, which is based on the classical homogeneous nucleation theory.

Mimouni et al. [8] simulated the cavitating flow using a six-equation model in NEPTUNE_CFD. The vapor temperature was ensured to be very close to the saturation temperature by using a special heat transfer coefficient. Besides the drag, added mass and lift force were included in the inter-phase momentum transfer. The contribution of nucleation to the vapor generation rate, as well as momentum and energy transfer were considered by the slightly modified Jones model [13–16]. The original model was shown to be insufficiently general by the authors. Nevertheless, the effect of nucleation and vaporization on the mean bubble size was ignored, namely, a constant bubble size was assumed.

Janet [9] studied the performance of various nucleation models in a flashing nozzle flow by using the five-equation model in CFX Version 14.5. It was found that predictions obtained with the Jones model are more reliable than the RPI (Rensselaer Polytechnic Institute) [18] and Riznic models [19].

Besides closure models for heat transfer coefficient and nucleation rate, basic differences among the above model setups can be summarized as follows.

Although in [7,8], the energy equation of the steam phase was solved, special treatments on heat the transfer coefficient were needed to maintain numerical stability. As a consequence, steam was either near saturated or no heat transfer to the interface. Based on these considerations, the five-equation model is chosen for the present work, which is proven to be sufficient. Various mono-disperse particle models are applied to flashing scenarios relevant to nuclear safety analysis. The limitations of each approach are discussed, and subsequently, a poly-disperse model free of open parameters is established for flow conditions with a broad spectrum of bubble size.

The rest of this paper is structured as follows. Section 3 begins with a brief mathematical description of applied transport equations, as well as the main closure relations. Results and discussions of the mono-simulations are found in Section 4. Section 5 presents a poly-disperse method and the simulation results. Suggestions for future work in Section 6 conclude the paper.

3. Physical Setup

3.1. Fundamental Transport Equations

The ensemble-averaged mass, momentum and energy conservation equations for individual phases are given as follows. Steam is assumed to stay always at the saturated state corresponding to local absolute pressure, and a common pressure field p is assumed for both phases. Pressure non-equilibrium at the steam-water interface is neglected.

Water:

$$\frac{\partial}{\partial t}(\alpha_l \rho_l) + \nabla \cdot \left(\alpha_l \rho_l \vec{U}_l\right) = \Gamma_{lg}, \tag{1}$$

$$\frac{\partial}{\partial t}\left(\alpha_l \rho_l \vec{U}_l\right) + \nabla \cdot \left(\alpha_l \rho_l \vec{U}_l \vec{U}_l\right) = -\alpha_l \nabla p + \nabla \cdot \left(\alpha_l \vec{\tau}_l\right) + \alpha_l \rho_l \vec{g} + \vec{F}_{lg} + \Gamma_{lg} \vec{U}_i, \tag{2}$$

$$\frac{\partial}{\partial t}\left(\alpha_l \rho_l \left(H_{tot,l} - \frac{p}{\rho_l}\right)\right) + \nabla \cdot \left(\alpha_l \rho_l \vec{U}_l H_{tot,l}\right) = \nabla \cdot (\alpha_l \lambda_l \nabla T_l) + \Gamma_{lg} H_{l,i} + E_l. \tag{3}$$

Steam:

$$\frac{\partial}{\partial t}(\alpha_g \rho_g) + \nabla \cdot \left(\alpha_g \rho_g \vec{U}_g\right) = -\Gamma_{lg}, \tag{4}$$

$$\frac{\partial}{\partial t}\left(\alpha_g \rho_g \vec{U}_g\right) + \nabla \cdot \left(\alpha_g \rho_g \vec{U}_g \vec{U}_g\right) = -\alpha_g \nabla p + \nabla \cdot \left(\alpha_g \vec{\tau}_g\right) + \alpha_g \rho_g \vec{g} - \vec{F}_{lg} - \Gamma_{lg} \vec{U}_i, \tag{5}$$

The unknown terms Γ_{lg}, F_{lg} and E_l describe the mass, momentum and heat flux to the liquid phase from the interface, which have to be modelled by constitutive relations or closure models. U_i and $H_{l,i}$ represent interfacial values of velocity and enthalpy carried into or out of the phase due to phase change. They are determined according to an upwind formulation:

$$\vec{U}_i = \begin{cases} \vec{U}_g, & \text{if } \Gamma_{lg} > 0 \text{ (condensation)} \\ \vec{U}_l, & \text{if } \Gamma_{lg} < 0 \text{ (evaporation)} \end{cases}, \tag{6}$$

$$H_{l,i} = \begin{cases} H_{l,sat}, & \text{if } \Gamma_{lg} > 0 \text{ (condensation)} \\ H_{l,tot}, & \text{if } \Gamma_{lg} < 0 \text{ (evaporation)} \end{cases}, \tag{7}$$

where $H_{l,sat}$ is liquid saturation enthalpy. The saturation parameters are interpolated from the published IAPWS-IF97 steam-water property tables corresponding to local pressure.

3.2. Main Closure Models

3.2.1. Inter-Phase Mass Transfer

As discussed in the Introduction, phase change in a flashing process is assumed to be induced by inter-phase heat transfer, which is called "thermal phase change model" in CFX. The volumetric mass transfer rate is related to heat flux density as follows:

$$\Gamma_{lg} = \frac{E_l \cdot A_i}{H_{g,sat} - H_{l,i}}, \tag{8}$$

where A_i is the interfacial area density and $H_{g,sat}$ is the vapor saturation temperature.

3.2.2. Interfacial Area Density

The particle model is adopted for interfacial transfer between two phases, which assumes that one of the phases (here water) is continuous and the other (steam) is dispersed. The surface area per unit volume is then calculated by assuming that steam is present as spherical particles of mean diameter d_g. Using this model, the interfacial area density is:

$$A_i = 6\frac{\alpha_g}{d_g} \text{ or } (6\alpha_g)^{2/3}(\pi n_g)^{1/3}. \tag{9}$$

Either mean diameter d_g or the number density n_g has to be known. They are prescribed or obtained by solving additional transport equations; see Table 1. Depending on the way of calculating d_g or n_g, mono-disperse and poly-disperse methods are derived. In the latter case, d_g represents the Sauter mean diameter of a size spectrum.

Table 1. Model setup available for CFD simulation of flashing flow.

Papers	Number of Conservation Equations	CFD Software	Particle Model for Interfacial Area Density with
[1–3]	Five	CFX 4.2	prescribed size
[4]	Five	CFX 4.2	prescribed number density
[6]	Four	CFX 4.2	additional transport equation for number density
[7]	Six	FLUENT 6.2.16	additional transport equation for number density
[8]	Six	NEPTUNE_CFD	prescribed size
[9]	Five	CFX 14.5	additional transport equation for number density

3.2.3. Inter-Phase Heat Transfer

The sensible heat flux transferring to water from the steam-water interface, E_l, is given by:

$$E_l = h_l(T_{sat} - T_l) \tag{10}$$

where h_l is the overall heat transfer coefficient. It is approximated by the Ranz–Marshall correlation [20,21], which is applicable for spherical particles. It has been widely used such as in [1–5,8,9].

3.2.4. Nucleation Model

The effect of nucleation is investigated by considering two kinds of nucleation mechanisms, namely wall nucleation and bulk nucleation. The wall nucleation rate is computed according to the Jones model [16].

$$J_{het,1,W} = 0.25 \times 10^{-7} \frac{R_d^2}{R_{cr}^4} \cdot C_{dp}(T_l - T_{sat})^3, \tag{11}$$

where R_d and R_{cr} are bubble departure radius and critical radius, respectively. The constant $C_{dp} = 10^4$ ($K^{-3} \cdot s^{-1}$).

Heterogeneous nucleation due to impurities in the bulk flow is accounted for with the model given by Rohatgi and Reshotko [22].

$$J_{nuc,1,B} = N_{im,B} \cdot \sqrt{\frac{2\sigma}{\pi m_W}} \cdot \exp\left(-\frac{16\pi\sigma^3\varphi}{3k_B T_l}\left(\frac{T_{sat}}{T_l - T_{sat}} \frac{\rho_g{}^{-1} - \rho_l{}^{-1}}{\Delta H_L}\right)^2\right), \tag{12}$$

where m_w is the mass of a single liquid molecule, k_B the Boltzmann constant and ΔH_L latent heat. $N_{im,B}$ is the number density of impurities (nucleation sites) in the bulk, and φ is the heterogeneous factor. Both $N_{im,B}$ and φ are treated as adjustable constants; $N_{im,B} = 5 \times 10^3$ (m^{-3}) and $\varphi = 10^{-6}$ are adopted in this work.

3.2.5. Turbulence Model

Turbulence in the liquid phase is described by a standard shear stress transport (SST) model augmented by the addition of more source terms. These source terms describe the effect of bubbles, i.e., the bubble-induced turbulence (BIT).

Concerning the source for the k-equation, there is a general agreement that the bubbles' contribution to turbulence generation comes from the work done by interphase drag force, i.e.,:

$$\varphi_k = 0.75\rho_l \frac{C_D}{d_g} \alpha_g \left|\vec{U}_l - \vec{U}_g\right|^3, \tag{13}$$

Similar to single-phase dissipation, the ε-equation source is obtained by scaling φ_k with a time scale:

$$\varphi_\varepsilon = C_{\varepsilon B} \frac{\varphi_k}{\tau}, \tag{14}$$

where $C_{\varepsilon B}$ is an adjustable constant and set to one. The time scale τ is approximated with the ratio between bubble size and turbulence intensity [23]:

$$\tau = \frac{d_g}{\sqrt{k_l}}. \tag{15}$$

3.2.6. Inter-Phase Momentum Transfer

Inter-phase momentum transfer occurs due to interfacial forces acting on each phase by interaction with the other phase. In the present work, both drag force and non-drag forces like, lift force, wall lubrication force, virtual mass force and turbulent dispersion force, are considered, i.e.,:

$$\vec{F}_{lg} = \vec{F}_{lg,D} + \vec{F}_{lg,L} + \vec{F}_{lg,W} + \vec{F}_{lg,VM} + \vec{F}_{lg,TD}. \tag{16}$$

These forces have to be computed according to appropriate models. So far, identical relations are commonly used for air-water and steam-water dispersed flows. The baseline-closure models defined in the previous work [24] are adopted.

In the following text, the above transport equations and closure models will be used in the CFD simulation of various flashing scenarios. Results obtained with different assumptions for interfacial area density are presented below.

3.3. Numerical Schemes and Convergence Criteria

In the simulation the coupled volume fraction algorithm was used, which allows the implicit coupling of the velocity, pressure and volume fraction equations. The high resolution scheme was selected to calculate the advection terms in the discrete finite volume equations. The discretization algorithm for the transient term was the second order backward Euler. The upwind advection and the first order backward Euler transient scheme were used in the turbulence numerics. The convergence of the simulation was monitored by using the criterion of the root mean square residual throughout the domain of less than 10^{-4}.

4. Mono-Disperse Simulation Results

4.1. Edwards and O'Brien Blowdown Test

4.1.1. Description of the Test

In nuclear safety analysis, the Edwards and O'Brien test [10] represented a standard problem for the verification and assessment of computational programs [25,26]. It consisted of fluid depressurization studies in a horizontal straight pipe 4.096 m long with an inner diameter of 0.073 m; see Figure 1. One end of the pipe is a fixed wall, while at the other end, a glass disc was mounted, which was designed to burst to initiate the depressurization transient.

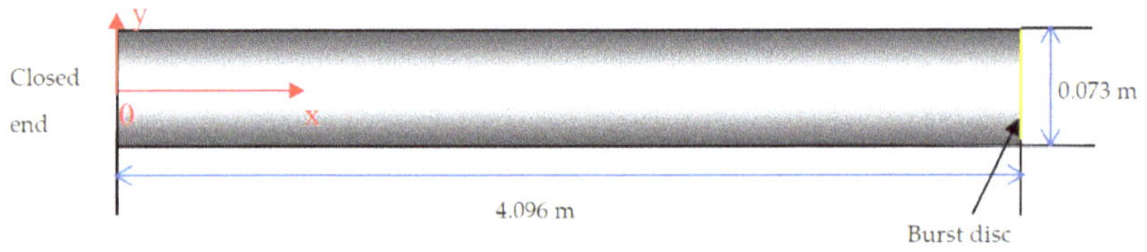

Figure 1. Edwards and O'Brien blowdown test.

The pipe was filled with sub-cooled water, whose initial conditions ranged from 3.55 MPa to 17.34 MPa and from 514.8 K to 616.5 K. At the burst of the glass disc, the depressurization wave propagates through the pipe, and water becomes superheated and vaporizes suddenly. Measurements were performed at several horizontal positions.

4.1.2. Simulation Results

The problem investigated in the present work has initial conditions of 7.00 MPa and 513.7 K. A pressure boundary was assumed for the outlet, and the background pressure was 1 atm. The simulation is done on a 3D cylinder grid, and furthermore, the five-equation model mentioned above with the prescribed bubble diameter (d_g = 1 mm) as in [3] is applied. The results of absolute pressure and void fraction at the position x = 1.469 m are presented in Figure 2a,b and compared with the experimental data.

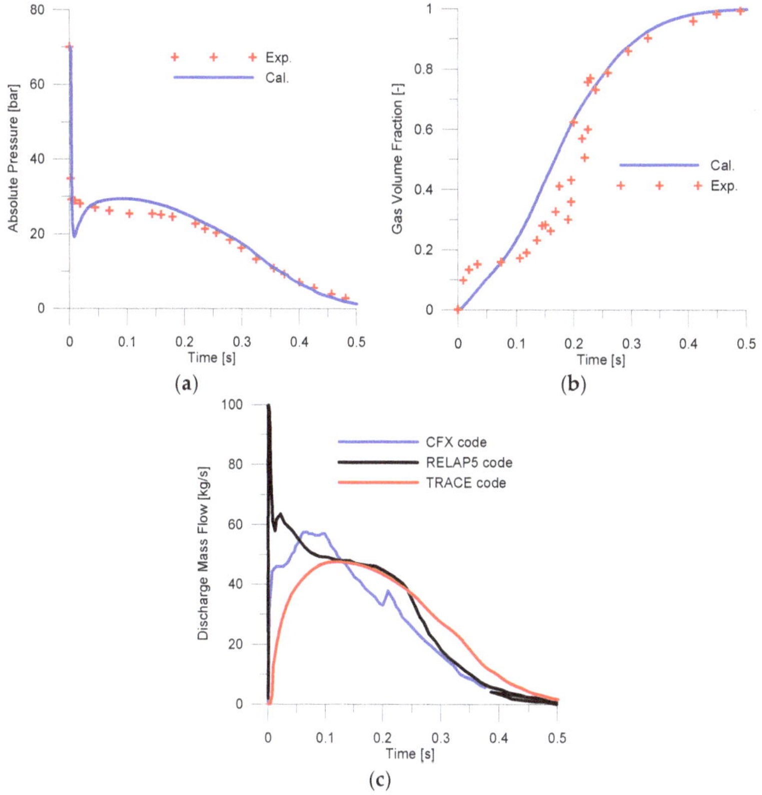

Figure 2. Evolution of flow parameters at x = 1.469 m after the disc burst: (**a**) absolute pressure; (**b**) void fraction; (**c**) discharged mass flow rate change with the time.

One can see that the water inside the pipe vaporizes completely within a half second after the burst of the glass disc. The agreement between simulation and measurement is acceptable, especially at the later stage. In contrast, considerable deviations are present at the early stage of depressurization. The lowest pressure in the simulation, which corresponds to the maximum superheat degree and is required to trigger the onset of flashing, is obviously lower than that in the measurement. Furthermore, the rate of vapor production is under-predicted at the beginning, while in the period from 0.1 s to 0.2 s, it is over-predicted. It is acknowledged that the simplifying assumption of a constant mean bubble size is responsible for the deviations. It is inconsistent with the physical picture of the process, although it is widely used. The size of pre-existing nuclei in the sub-cooled liquid is substantially smaller than the prescribed value of 1 mm. This indicates a significant under-prediction of interfacial area density available for the initiation of flashing. On the other hand, steam bubbles grow rapidly during the flashing process, leading to a mean size much larger than 1 mm in a short time. Nevertheless, the effect of this uncertainty is weakened with the increase of the void fraction. The effect of prescribed bubble size was investigated in [27].

The discharged mass flow from the blowdown pipe is shown in Figure 2c below. Due to lack of experimental information, the results are compared with those obtained by two classical system codes, i.e., RELAP5 and TRACE, from the recent work [28]. One can see that the maximum value is reached shortly after the burst, where the difference between the codes is largest. According to RELAP5, the mass flow increases rapidly from 0 to 100 kg/s at the moment of the burst, while TRACE and CFX give a significantly lower maximum value with a delay around 0.1 s. In addition, the change of mass the flow rate with time delivered by CFX is much more gentle and smooth in comparison to that by RELAP5 and CFX.

4.2. Flashing-Induced Instability Problem in Natural Circulation

4.2.1. Description of the Problem

Since the middle of the 1980s, it has been recognized that the application of passive safety systems can contribute to improving the economics and reliability of nuclear power plants. In some advanced designs, natural circulation is used as a means for residual heat removal. For example, in a German BWR (boiling water reactor) concept, namely, the KERENA™ reactor [29], the containment cooling condenser (CCC) is the key component of a natural cooling circuit; see the red shaded region in Figure 3. In the case of overpressure, surplus steam in the containment condenses on the outside wall of CCC. Heat released by condensation is transferred to the cooling water inside the CCC tubes and finally to the shielding/storage pool vessel (SSPV) through natural circulation.

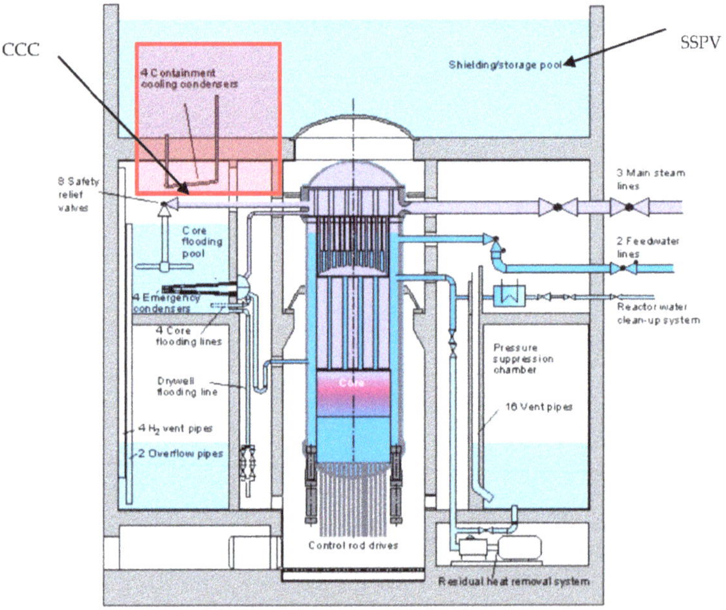

Figure 3. Section picture of KERENA™ containment [30]. CCC: containment cooling condenser; SSPV: shielding/storage pool vessel.

In the experimental investigation performed by AREVA [30], a severe water hammer was experienced. The flow instability is believed to come from the rapid formation and subsequent destruction of steam bubbles inside the circuit. Natural circulation systems are characterized by a downcomer and a riser, which connect to a heated and a cooled section at the ends, respectively. Nearly saturated warm circulating water can flash to vapor inside the riser if no boiling occurs in the heated section. Liquid temperature remains constant till flashing begins if the heat loss through the riser is negligible (adiabatic), while saturation temperature decreases with increased altitude.

The production of vapor in a flashing process taking place in natural circulation has a periodic oscillation character [31]. The circulation flow rate increases as a result of vapor production, which further leads to a low liquid temperature entering the riser. It can be lower than the saturation temperature at the exit of the riser. As a consequence, the flashing is suppressed, which leads to a decrease in flow rate and subsequently an increase in liquid temperature. Therefore, flashing can take place again in the adiabatic riser and generates a self-sustained oscillating flow. The feature of high frequency oscillation and rapid phase change requires high resolution and can cause numerical instability and convergence problems in high-resolved CFD simulations.

4.2.2. Simulation Results

For the simulation, a 3D pipe section is constructed on the basis of the riser in the AREVA test facility; see Figure 4. Two inlets are connected to two CCC condensers. For details on the experiments please, refer to [27,29]. Boundary conditions, such as inlet temperature, mass flow rate and outlet pressure level are provided by the experimental data. Furthermore, the single-phase condition at the inlet is assured by choosing a proper time segment according to the measurement. The start time for simulation is chosen as 4600 s. Furthermore, a mono-disperse approach is utilized, which distinguishes itself from the last case by the prescription of bubble number density instead of bubble size. A constant value of $5 \times 10^4\ \mathrm{m}^{-3}$ is assumed in the following simulation. The settings are believed to be closer to the physical process of flashing since the bubbles are allowed to grow [4].

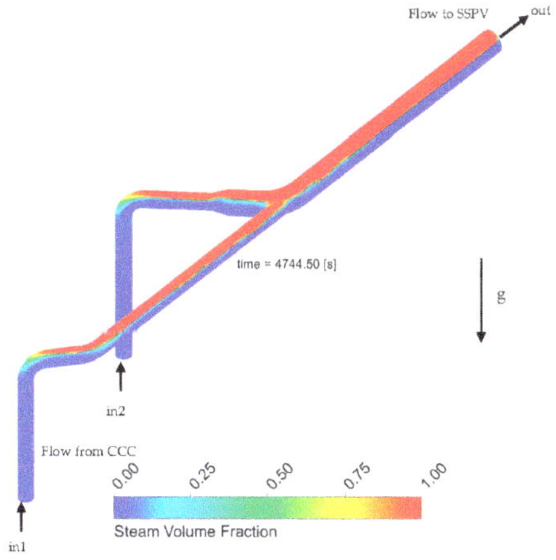

Figure 4. Segment of the riser pipe applied in the simulation colored by steam volume fraction at t = 4744.50 s.

Figure 4 shows the distribution of steam inside the domain at the moment of 4744.50 s. The red color symbolizes steam, while the blue is water. As discussed above, the liquid at the exit becomes first superheated and initiates the flashing process. This gives rise to a high steam volume fraction at the upper part, while single phase flow exists still in the two straight legs near the inlets. Since the gravitational force is considered, steam is accumulated at the top side of the inclined pipe leading to a stratified flow pattern. A broad range of flow patterns and regime transition represents challenges for current two-phase CFD technology.

The temporal course of the inlet pressure and liquid temperature (the measured pressure at the outlet and temperature at the inlet are given as boundary conditions) is depicted in Figure 5. During the time segment from 4600 s to 5200 s, the production of vapor goes periodically through onset, intensification and ceasing, and thermo-hydraulic parameters oscillate correspondingly. The maximum amplitude of pressure and temperature oscillation can reach 0.8 bar and 50 K, respectively. The predicted period agrees well with the measured one, and the amplitude of temperature profile is also satisfying. Nevertheless, clear deviation is observed at pressure valleys, where the onset of flashing occurs. Similar to the limitation of the last method with constant bubble size, the disagreement results from the prescription of a constant bubble number density and neglecting bubble dynamics, such as coalescence and break-up. The sensitivity study on the prescribed values for bubble number density was performed in the previous work of [32].

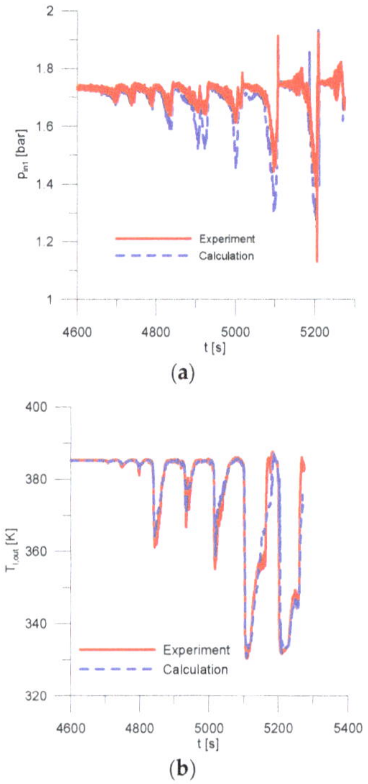

Figure 5. Comparison between calculation and experiment: (**a**) inlet pressure; (**b**) outlet temperature.

4.3. *Abuaf [33] Nozzle Flashing Flow Test*

4.3.1. Description of the Test

In nuclear safety analysis, the flashing of initially sub-cooled liquid through nozzles, orifices or other restrictions has been studied intensively for several decades due to the design basic accident of LOCA (loss-of-coolant accident). Much attention was paid to the critical flow problem, while a deep insight into the transient development of two-phase flow structure is still missing. In comparison with the single phase, the complexity of two-phase critical flow arises from both thermal and mechanical non-equilibrium effects at the interface, which prevent reliable prediction of the critical flow rate from using empirical relations.

The experimental investigation of water flashing through a converging-diverging nozzle presented in [33] has been widely used as a reference for numerical study or model development, e.g., [7,34]. In the frame of this work, several tests with different boundary conditions, such as inlet pressure, temperature and outlet pressure, are reproduced by means of CFD technology. The geometrical sketch of the circular nozzle is shown in Figure 6. Dynamic pressure decreases with the reduction of channel area in the converging part leading to a decrease in the saturation temperature. If it falls below the liquid temperature before reaching the neck position, the onset of flashing will occur near the neck and vaporization continues in the diverging part.

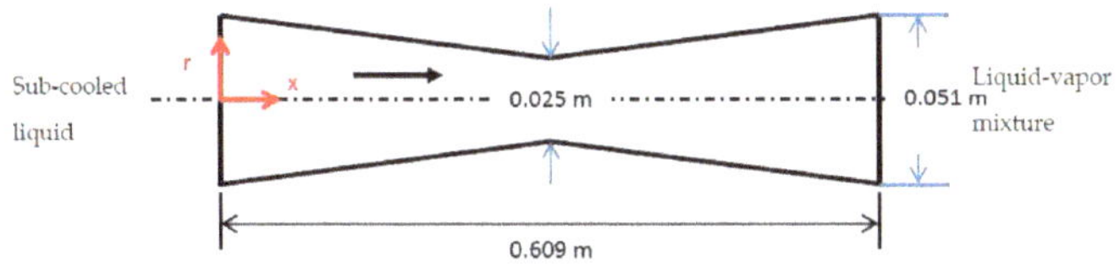

Figure 6. Circular converging-diverging nozzle used in the tests of [33].

4.3.2. Simulation Results

For this case, the generation and transportation of bubbles is traced by solving an additional transport equation for bubble number density. Similar setups were adopted in [6,7]; see Table 1. The formation of bubbles in superheated liquid is called "nucleation", which is deemed to take place under two different mechanisms, namely homogeneous and heterogeneous nucleation. Homogeneous nucleation occurs uniformly in the bulk caused by the thermodynamic state fluctuation of liquid molecules. In contrast, heterogeneous nucleation takes place at foreign "nuclei", which can be microscopic cavities on solid walls or dissolved gasses and impurities pre-existing in the sub-cooled liquid. The latter mechanism has been recognized to be dominant in the real processes, which is therefore taken into account in the current work. Since it allows both bubble size and number to vary, this method is obviously stepwise superior to the previous ones with prescribed bubble size or number density. However, it requires one more closure model, namely the nucleation model, which is still one weak spot in the numerical study of the boiling process. For investigations on the effect of nucleation models, please refer to [9].

From the spatial distribution of steam shown in Figure 7, one can see that the onset of vaporization occurs near the nozzle neck, and bubbles are formed overwhelmingly at the walls and then migrate to the center region. In other words, the mechanism of wall nucleation is dominant in comparison to bulk heterogeneous nucleation. As a result, the radial profile with a clear wall peak is built. The example presented below Figures 8–10 has the boundary conditions of inlet pressure 555.9 kPa, outlet pressure 402.5 kPa and inlet temperature 422.25 K.

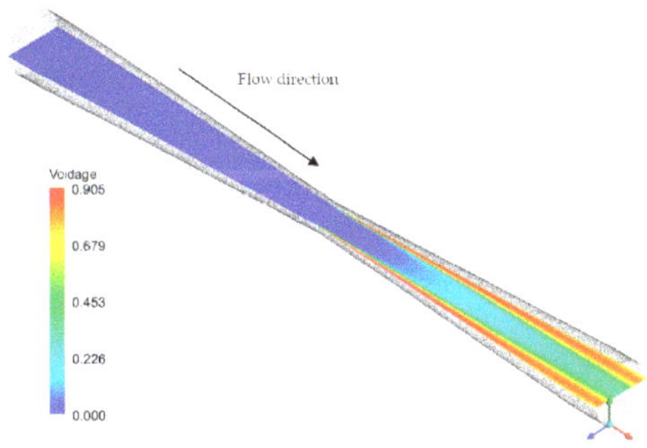

Figure 7. Distribution of the steam volume fraction inside the nozzle.

Furthermore, the nucleation process is found to take place in a narrow region (a few centimeters); see the red shaded region in Figure 8a. The cross-section averaged bubble number density increases steeply in the nucleation region and then remains almost constant in the stable two-phase region. The models described in [16,21] are chosen for the computation of the heterogeneous nucleation rate at the walls and in the bulk. The contribution of bulk nucleation is found to be trivial. In addition, the assumed small initial value is proven to have no influence on the final results [9].

The radial profile of bubble distribution in Figure 8b shows that the majority of steam bubbles remain in the neighborhood of nucleation spots (here, wall cavities), and the concentration in the center region is low and flat. The rate of bubbles migrating from the near-wall region to the nozzle center increases with the bubble size.

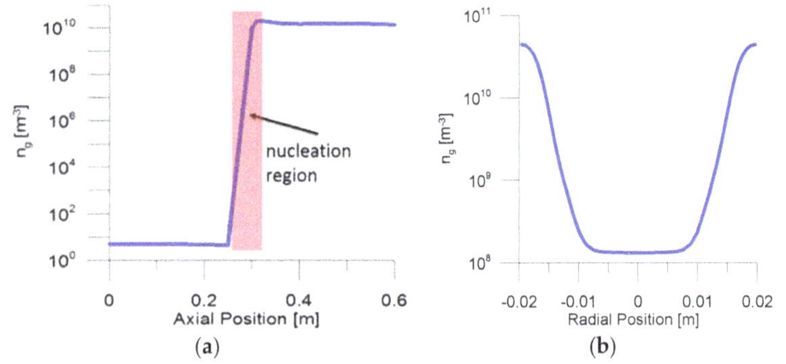

Figure 8. Distribution of bubble number density in (**a**) the axial direction and (**b**) the radial direction at $x = 0.458$ m.

In Figure 9, the simulation results are compared with the data. On the left side is the axial profile of absolute pressure (left y-axis) and steam volume fraction (right y-axis), and on the right side is the radial profile of the steam volume fraction at the position $x = 0.458$ m. Symbols and lines represent measurement and calculation, respectively.

The cross-section averaged steam volume fraction increases in the latter part of flow path, which is in an acceptable agreement with the data. On the other hand, the pressure profile exhibits obvious deviations in the diverging part of the nozzle. The predicted value is lower than the measured one. It suggests that too high a superheat (or pressure undershoot) is required for triggering the onset of flashing in comparison to the experiment. In other words, with the same superheat degree as observed in the experiment, the predicted inter-phase heat transfer rate is insufficient to overcome the latent heat and keep bubbles stable. The interfacial area density or inter-phase heat transfer coefficient is under-predicted at the moment of flashing onset. After triggering the flashing process, the pressure increases under the constraint of the outlet pressure, which is given as the boundary condition. Furthermore, the assumption of pressure equilibrium across the interface might introduce large errors in high pressure undershoot cases.

The transverse distribution of bubbles in the continuous medium is determined by non-drag forces, which depend sensitively on the mean bubble size. It is determined by the bubble number density and nucleation rate, since bubble coalescence and break-up are not considered. As shown in Figures 8b and 9b, bubbles are formed and accumulated in the near wall region and create a void fraction profile with a high wall-peak while a low value in the center region. A similar profile is obtained in the measurement. However, it displays a more gradual transition from the wall peak to the center valley. That means that more bubbles have migrated to the center due to the effect of non-drag forces, i.e., lift and turbulence dispersion. This may imply that bubble growth is under-estimated, and coalescence effects should be taken into account, since lift force pushes large bubbles to the center, while small bubbles to the wall. However, to refine the results, reliable data of local bubble size distributions are indispensable, which are unfortunately missing in most experimental investigations. Furthermore, sufficient turbulent dispersion would also help to improve the agreement. According to the FAD (Favre averaged drag) model, the dispersion force is proportional to eddy viscosity, which is probably under-predicted by the SST model with BIT source terms.

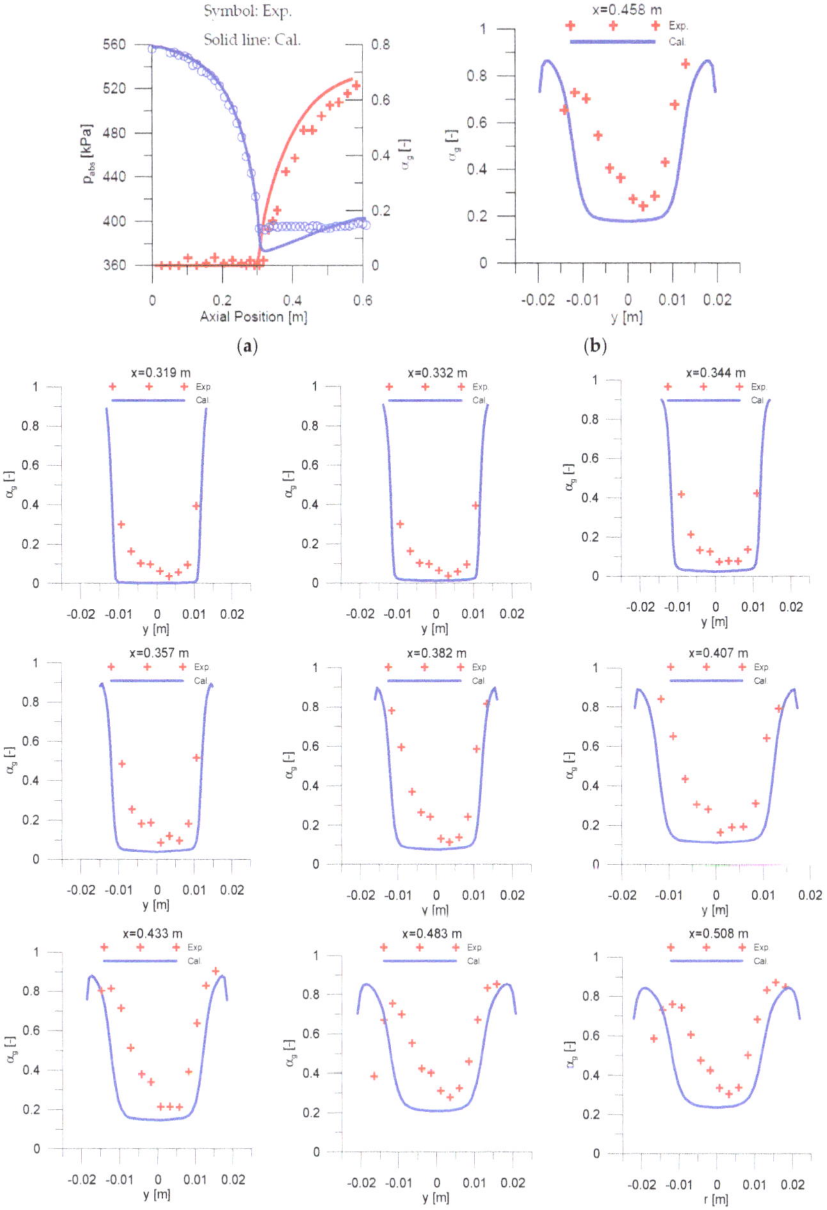

Figure 9. *Cont.*

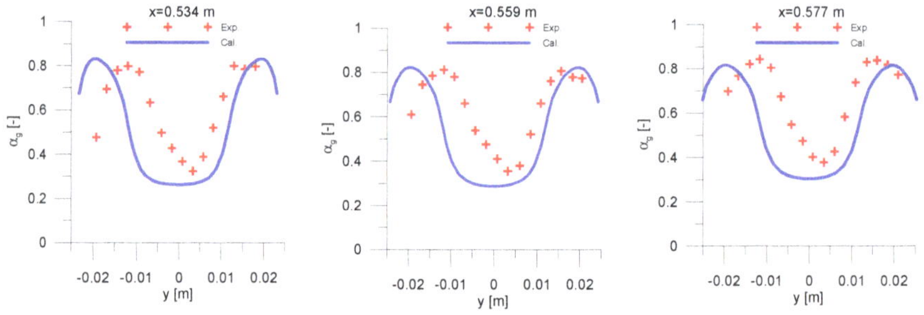

(c)

Figure 9. Comparison between calculation and experiment: (**a**) axial profile of the pressure and void fraction; (**b**) radial profile of the void fraction; (**c**) radial profile of the void fraction at different axial positions.

To give more details about the flow structure, the radial profile of the void fraction is depicted in Figure 9c for several axial positions other than $x = 0.458$ m shown above. The formation of bubbles starts around the throat position ($x = 0.305$ m; see Figure 6) and first leads to a monotonous peak near the wall. It shifts slowly to the nozzle center due to the effect of non-drag forces along the axis, and an "M-shape" profile is developed over the cross-section. The simulated results agree well with the measured ones except for an over-prediction of the wall-peak at the beginning. Another important observation is that in the experiment, the profile is not symmetrical with respect to the axis. According to the experimenter, the asymmetry might be due to the presence of pressure taps on one side of the nozzle [33], which act as nucleation centers.

Several other tests with different inlet and outlet pressures, as well as inlet temperatures are simulated. The critical mass flow rates are summarized in Figure 10 below. The critical rate of flow through the nozzle depends on a variety of parameters, inlet pressure, quality, pressure-undershoot, and so on. In general, increasing the inlet stagnation pressure or decreasing outlet pressure leads to an increase in the critical mass flow rate. Furthermore, the presence of vapor reduces the average density of the two-phase mixture and, thus, favors the deceleration process in the diverging part of the nozzle.

It is shown that the simulated results are in an overall good agreement with the measured ones. The maximum deviation is within 7%. Nevertheless, the simulation tends to under-predict the critical mass flow rate with the increasing of the inlet pressure and decreasing of the outlet pressure.

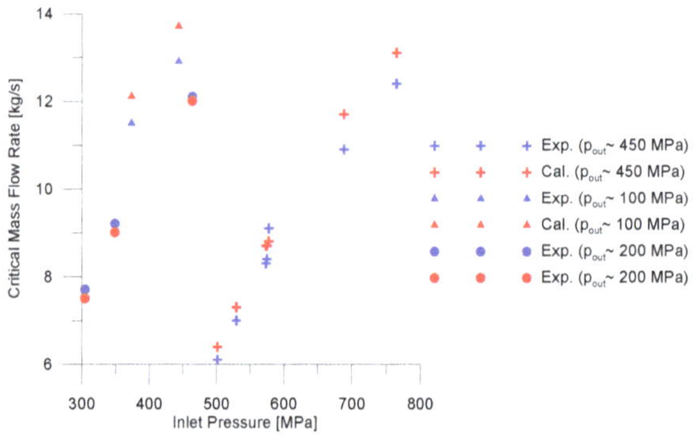

Figure 10. Dependence of the critical mass flow rate on inlet and outlet pressure in the nozzle.

5. Poly-Disperse Simulations

Although it is commonly used, the mono-disperse approach presented above is self-evident limited to situations with a constant bubble size or a narrow distribution. On the other hand, flashing processes are always accompanied with rapid vaporization and a high void fraction. This suggests that a broad spectrum of bubble sizes can be present, which asks for a poly-disperse simulation method. However, to the author's knowledge, no publications in this aspect are so far available for flashing situations, although there is a wide range of poly-disperse simulations for isothermal flows and a few for sub-cooled boiling.

In this work, the first attempts are made to apply the inhomogeneous MUSIG (multiple size group) model [35,36] to controlled pressure release transients, which were carried out on the TOPFLOW facility (see below). MUSIG is a poly-disperse method available in CFX for computing the mean bubble size of a spectrum. It is one kind of class method approximating the size spectrum with several discrete classes; see Figure 11.

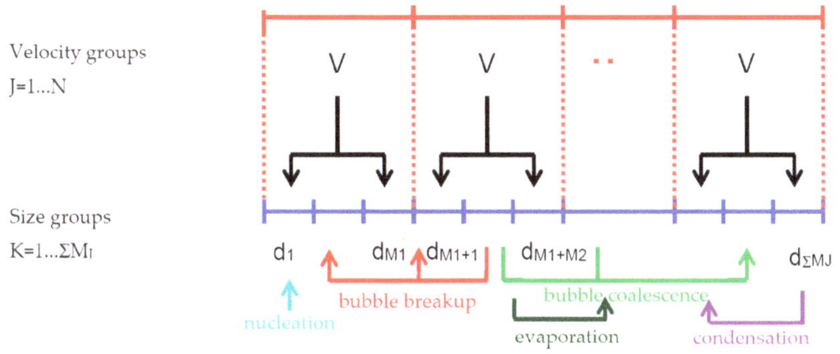

Figure 11. Schema of the inhomogeneous multiple size group (MUSIG) model.

The division of size groups is finer than that of velocity groups due to the limitation of computational speed. Separate momentum equations are solved for each velocity group while all phenomena that change the size distribution, such as nucleation, coalescence and breakup, are considered within the sub-size groups. The exchange between these size groups due to the above-mentioned phenomena is reproduced by solving additional transport equations with corresponding source/sink terms for the fraction of each group f_i. As a result, the mean Sauter diameter of bubbles belonging to a velocity group can be obtained from these size fractions. The MUSIG approach has been expounded in detail elsewhere [24,36].

5.1. TOPFLOW Pressure Release Experiment

One major test section equipped in the TOPFLOW facility is a vertical pipe with a nominal diameter of 200 mm; see Figure 12. The pressure release experiment was carried out at this section. For details about the multipurpose thermal-hydraulic test facility TOPFLOW at Helmholtz-Zentrum Dresden-Rossendorf, please refer to the work of [37].

During the pressure release experiment, water was circulated with a velocity of about 1 m/s and flows upwards through the test section; see Figure 12. The transient pressure course is controlled by the blow-off valve located above the steam drum, where saturation conditions are always guaranteed. As a result, cavitation in the circulation pump below is avoided. At the same time, the maximum evaporation rate in the test section is limited [27].

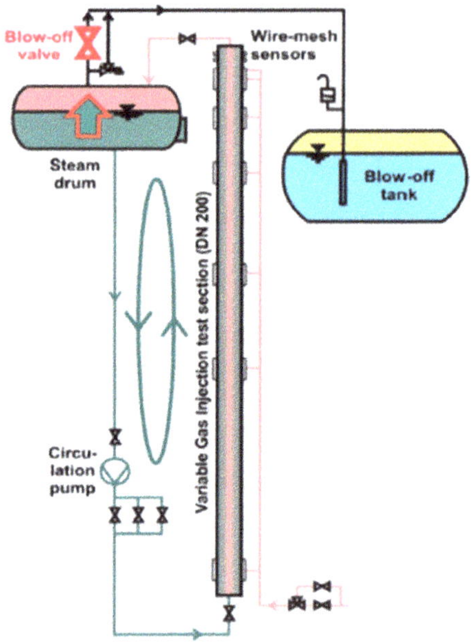

Figure 12. Schema of the experimental procedure.

The blow-off valve was opened to a maximum level and closed again according to the ramp shape as shown in Figure 13.

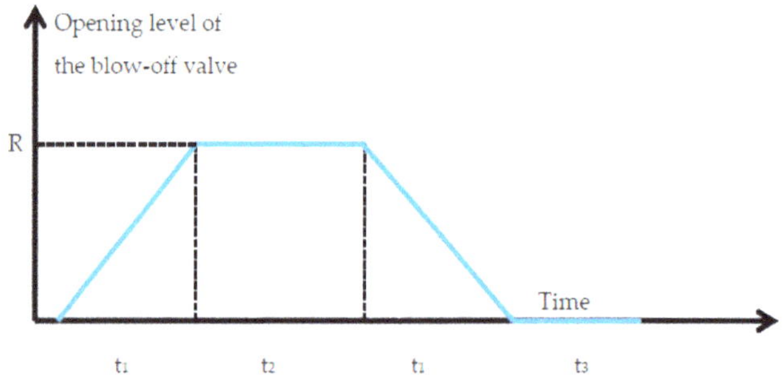

Figure 13. Operation of the blow-off valve.

The operation speed of the valve is controlled by time t_1, while the maximum opening degree and its duration are represented by R and t_2. Correspondingly, the generation and disappearance of steam inside the pipe are determined by the pressure release rate and the operation of the valve. Measurements of the volume fraction, gas velocity and bubble size distribution were realized with the aid of wire-mesh sensors (WMS) at the top; see Figure 12. The highly-resolved data of bubble size distributions and radial void fractions are optimal for the validation of simulation results obtained by the poly-disperse method. Rapid expansion of bubbles and a broad size spectrum were observed in the experiment, and for reliable simulation results, it is necessary to use a poly-disperse method. A detailed introduction to the experimental procedure and discussion on the influence of pressure level on the data were given by Lucas [38].

Two test cases under different pressure levels are investigated in the current work, whose experimental conditions are summarized in Table 2 below.

Table 2. Experimental conditions of the investigated test cases.

Case No.	Pressure (bar)	*R* (%)	t_1 (s)	t_2 (s)	t_3 (s)
1	10	60	21	30	30
2	65	20	7	56	30

Measured inlet temperature, mass flow rate and outlet pressure are used as boundary conditions in the simulation.

5.2. Simulation Results

The aim of poly-disperse simulations is to obtain a realistic mean bubble size by tracing the dynamic change of bubble size distributions, which is important in the case of a broad bubble size spectrum. For this purpose, besides highly-resolved data, reliable closure models for all physical processes that affect the bubble size are indispensable. In this work not only nucleation and phase change, but also bubble coalescence and break-up have been taken into account (see Figure 11). For heterogeneous nucleation, the same models as for the nozzle flow discussed above in Section 4.3.2 were employed, while the models presented in [24] were used for bubble coalescence and break-up. The change resulting from these phenomena is implemented as source or sink in the MUSIG size fraction transport equations mentioned above.

The evolution of the normalized bubble size distribution is displayed in Figures 14 and 15 for the two cases listed in Table 2, respectively. In the vertical axis label, $\Delta\alpha_i$ represents the void fraction portion belonging to the size group *i*, while $\sum\Delta\alpha i$ is the total void fraction. The red line represents experimental data, while the blue one simulation results.

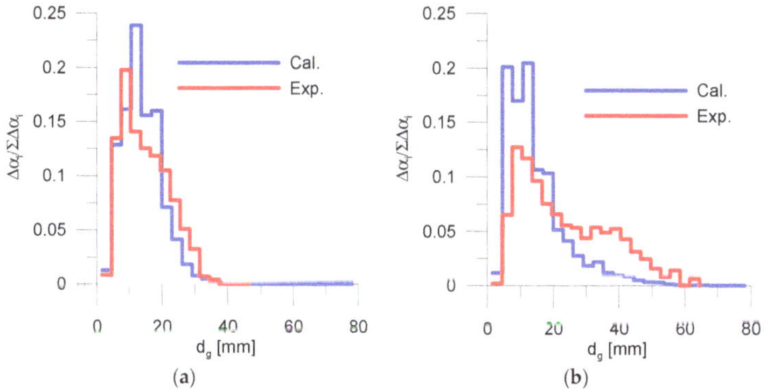

Figure 14. Evolution of normalized bubble size distribution at the wire-mesh sensor (WMS) plane (Case 1): (**a**) t = 39 s; (**b**) t = 45 s.

In Case 1, whose pressure level is 10 bar, bubble growth within the period from t = 39 s to 45 s is significant. The agreement between calculation and experiment at t = 39 s is satisfying, whereas six seconds later, the fraction of large bubbles is obviously under-predicted. As a result, the mean bubble size is smaller than the measured one, although both of them increase. One possible reason for this discrepancy is that the mechanism for inter-phase mass transfer is not completely reproduced by the "thermal phase change model". Furthermore, the break-up rate of large bubbles may be overestimated corresponding to the coalescence rate of small bubbles. The predictability of coalescence and break-up models depends on a number of input parameters, such as turbulence intensity and interfacial shear stress. The evaluation of these parameters in such complex two-phase flows is often difficult. In addition, the superposition of various mechanisms, as well as of coalescence and break-up effects increases the difficulty in further development and improvement of these closures.

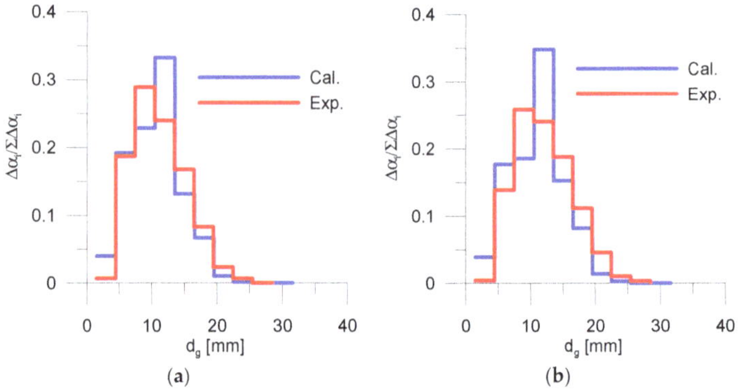

Figure 15. Evolution of normalized bubble size distribution at the WMS plane (Case 2): (**a**) t = 29 s; (**b**) t = 49 s.

The growth of bubbles is substantially slowed down with the increase of pressure level. As shown in Figure 15, in Case 2, the change of bubble size distribution within 20 s is trivial. Consequently, the agreement at both time points is acceptable.

The different performance of the applied model setup at two pressure levels suggests that the contribution of pressure non-equilibrium at the interface to inter-phase mass transfer can be significant in Case 1. An approach that takes both thermal and mechanical non-equilibrium into account is necessary to ensure the reliability of predictions in both low and high pressures.

The pressure and velocity field shown in Figures 16 and 17 below is intended to provide further insight into the flashing flow. In the simulation, the pressure boundary condition was specified for the outlet (top), and the measured pressure was imposed on the boundary. The calculated pressure at the inlet (bottom) was found in good agreement with the data, as shown in Figure 16. This implies that the pressure field in the whole domain is well reproduced. Water flashes into steam during the depressurization, whose rate is controlled by the opening of the blow-off valve (see Figure 13).

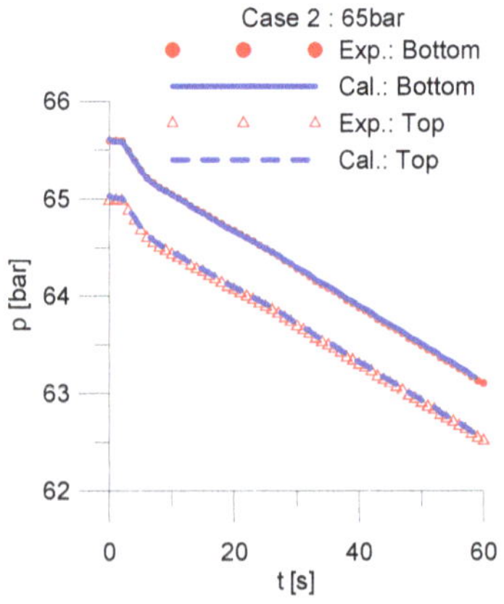

Figure 16. Evolution of the cross-section averaged absolute pressure at the top and bottom of the test section (Case 2).

The comparison of the gas vertical velocity component at the WMS plane, where data are available, is shown in Figure 17. Instead of a core peak as observed in the measurement, the calculated radial profiles show a transit peak. The velocity profile is similar to the void fraction ones, which depends on the magnitude of non-drag forces. Since the bubble size is well captured (see Figure 15), the discrepancy is caused either by an under-prediction of turbulence dispersion force or an over-prediction of lift force. Under the role of lift force, large bubbles move away from the wall and accumulate in the center, while turbulence dispersion tends to counteract this effect. However, for further evaluation of these models, reliable reference information, e.g., turbulence, is still missing.

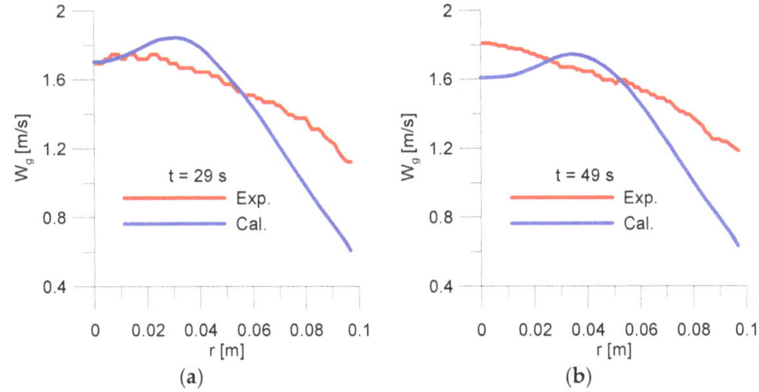

Figure 17. Evolution of radial gas velocity profiles at the WMS plane (Case 2): (**a**) t = 29 s; (**b**) t = 49 s.

Improving the two-phase turbulence, as well as dispersion and lift force modelling will be one of the emphases of future work. Related theoretical and experimental work has been planned and begun.

6. Conclusions

Flashing of liquid to vapor due to pressure drop represents highly complex two-phase situations. Due to its relevance to nuclear safety analysis, the flashing phenomenon has been extensively studied for several decades. Nevertheless, there exists a need to update the analytical approach to a sophistication level that matches available computational technologies, e.g., from system codes to CFD codes. CFD simulations with the simplified mono-disperse approaches, which are commonly used and often deliver satisfying results, are however limited for cases with nearly constant bubble size or number density. Since these conditions are normally not fulfilled, they are shown to have difficulty in capturing the onset of flashing, which consequently affects the agreement in the subsequent phase change stage. In addition, the results are sensitively dependent on the prescribed values and the spatial distribution of phases on the mean bubble size. The poly-disperse approach is promising for removing these restrictions, since it is free of any open parameters and reproduces the realistic change of bubble size distributions. However, as suggested in [12], more elaborate models require more empirical correlations or assumptions for phase interactions which are so far insufficiently tested. As a result, the prediction accuracy of the sophisticated method is greatly affected.

Therefore, before we can exploit the benefits of CFD simulations for flashing flows, we have to understand the physical sub-phenomena, such as nucleation characteristics and inter-phase transfer laws sufficiently, and be able to specify them precisely with closure models. For this purpose, highly-resolved and comprehensive data achieved by experiments or direct numerical simulations are indispensable, especially local bubble size, phase distribution, turbulence, velocity and pressure fields, which are often unavailable, e.g., in Tests 4.1, 4.2 and 4.3. The wire-mesh senor technique applied in the TOPFLOW pressure release experiment can provide the above measurements. Nevertheless, the necessary information on liquid velocity and turbulence parameters is still missing. The development of new measurement techniques is undergoing [39].

The presented results show that the Eulerian CFD technology reproduces global parameters, such as pressure and flow rates, reliably, even in complex practical situations. Nevertheless, the prediction of local phenomena, e.g., phase distribution and velocity fields, is still insufficient. More efforts are required in the assessment and improvement of closures.

Acknowledgments: This work was carried out at Helmholtz-Zentrum Dresden-Rossendorf in the frame of a research project funded by EON Kernkraft GmbH.

Author Contributions: Dirk Lucas and Yixiang Liao conceived of the simulations and evaluated the results. Yixiang Liao performed the simulations and wrote the paper.

Conflicts of Interest: The authors declare that there is no conflict of interest regarding the publication of this paper. The founding sponsors had no role in the design of the study; in the collection, analyses, or interpretation of data; in the writing of the manuscript, and in the decision to publish the results.

Abbreviations

ANSYS	an American computer-aided engineering software developer headquartered south of Pittsburgh in Cecil Township, Pennsylvania, United States
AREVA	a French multinational group specializing in nuclear power and renewable energy headquartered in Paris La Défense
CCC	containment cooling condenser, a passive nuclear safety system
CFD	computational fluid dynamics
CFX	a commercial CFD code developed by ANSYS company
CSNI	committee on the Safety of Nuclear Installations, a committee of the OECD/NEA
FLUENT	a commercial CFD code developed by ANSYS company
IAPWS-IF97	International Association for the Properties of Water and Steam – Industrial Formulation 1997
KERENA^{TM}	a mid-power boiling water reactor developed jointly by the AREVA company and the German energy supply company E.ON
NEA	Nuclear Energy Agency, an intergovernmental agency organized under OECD
NEPTUNE-CFD	a French code created by EDF (Électricité de France) and CEA (Commissariat à I'Energie Atomique) for nuclear reactor thermal-hydraulics simulation and analyses
OECD	Organization for Economic Co-operation and Development
RELAP5	abbreviation of "Reactor Excursion and Leak Analysis Program", A component-oriented reactor systems thermal-hydraulics analysis code developed at Idaho National Laboratory
RPI	Rensselaer Polytechnic Institute
SSPV	Shielding/Storage Pool Vessel in KENERA reactor
TRACE	TRAC/RELAP Advanced Computational Engine, a component-oriented reactor systems thermal-hydraulics analysis code
TRAC	transient reactor analysis code, another reactor system code
WMS	wire mesh sensor, a measurement technique

References

1. Giese, T.; Laurien, E. A thermal based model for cavitation in saturated liquids. *Z. Angew. Math. Mech.* **2001**, *81*, 957–958.
2. Giese, T.; Laurien, E. Experimental and numerical investigation of gravity-driven pipe flow with cavitation. In Proceedings of the 10th International Conference on Nuclear Engineering (ICONE10), Arlington, VA, USA, 14–18 April 2002.
3. Laurien, E.; Giese, T. Exploration of the two fluid model of two-phase flow towards boiling, cavitation and stratification. In Proceedings of the 3rd International Conference on Computational Heat and Mass Transfer, Banff, AB, Canada, 26–30 May 2003.

4. Laurien, E. Influence of the model bubble diameter on three-dimensional numerical simulations of thermal cavitation in pipe elbows. In Proceedings of the 3rd International Symposium on Two-Phase Flow Modelling and Experimentation, Pisa, Italy, 22–24 September 2004.
5. Frank, T. Simulation of flashing and steam condensation in subcooled liquid using ANSYS CFX. In Proceedings of the 5th Joint FZR & ANSYS Workshop "Multiphase Flows: Simulation, Experiment and Application", Dresden, Germany, 26–27 April 2007.
6. Maksic, S.; Mewes, D. CFD-Calculation of the flashing flow in pipes and nozzles. In Proceedings of the Joint U.S.—European Fluids Engineering Division Conference (ASME FEDSM'02), Montreal, QC, Canada, 14–18 July 2002.
7. Marsh, C.A.; O'Mahony, A.P. Three-dimensional modelling of industrial flashing flows. In Proceedings of the Computational Fluid Dynamics (CFD 2008), Trondheim, Norway, 10–12 June 2008.
8. Mimouni, S.; Boucker, M.; Laviéville, J.; Guelfi, A.; Bestion, D. Modelling and computation of cavitation and boiling bubbly flows with the NEPTUNE_CFD code. *Nucl. Eng. Des.* **2008**, *238*, 680–692. [CrossRef]
9. Janet, J.P.; Liao, Y.; Lucas, D. Heterogeneous nucleation in CFD simulation of flashing flows in converging-diverging nozzles. *Int. J. Multiph. Flow* **2015**, *74*, 106–117. [CrossRef]
10. Edwards, A.R.; O'Brien, T.P. Studies of phenomena connected with the depressurization of water reactors. *J. Br. Nucl. Enery Soc.* **1970**, *9*, 125–135.
11. Liao, Y.; Lucas, D.; Krepper, E.; Rzehak, R. Assessment of CFD predictive capacity for flash boiling. In Proceedings of the Computational Fluid Dynamics for Nuclear Reactor Safety-5 (CFD4NRS-5), Zurich, Switzerland, 9–11 September 2014.
12. Wallis, G.B. Critical two-phase flow. *Int. J. Multiph. Flow* **1980**, *6*, 97–112. [CrossRef]
13. Jones, O.C., Jr.; Zuber, N. Bubble growth in variable pressure fields. *J. Heat Transf.* **1978**, *100*, 453–459. [CrossRef]
14. Shin, T.S.; Jones, O.C., Jr. An active cavity model for flashing. *Nucl. Eng. Des.* **1986**, *95*, 185–196. [CrossRef]
15. Shin, T.S.; Jones, O.C. Nucleation and flashing in nozzles—1. *Int. J. Multiph. Flow* **1993**, *19*, 943–964. [CrossRef]
16. Blinkov, V.N.; Jones, O.C.; Nigmatulin, B.I. Nucleation and flashing in nozzles—2. *Int. J. Multiph. Flow* **1993**, *19*, 965–986. [CrossRef]
17. Blander, M.; Katz, J.L. Bubble nucleation in liquids. *AIChE J.* **1975**, *21*, 833–848. [CrossRef]
18. *Release 14.5. ANSYS CFX-Solver Theory Guide*; ANSYS: Canonsburg, PA, USA, 2012.
19. Riznic, J.; Ishii, M. Bubble number density in vapor generation and flashing flow. *Int. J. Heat Mass Transf.* **1989**, *32*, 1821–1833. [CrossRef]
20. Ranz, W.E.; Marshall, W.R. Evaporation from drops—Part I. *Chem. Eng. Prog.* **1952**, *48*, 141–146.
21. Ranz, W.E.; Marshall, W.R. Evaporation from drops—Part II. *Chem. Eng. Prog.* **1952**, *48*, 173–180.
22. Rohatgi, U.; Reshotko, E. Non-equilibrium one-dimensional two-phase flow in variable area channels. In *Non-Equilibrium Two-Phase Flows, Proceedings of the Winter Annual Meeting, Houston, TX, USA, 30 November–5 December 1975*; Meeting Sponsored by the American Society of Mechanical Engineers; American Society of Mechanical Engineers: New York, NY, USA, 1975, Volume 1, pp. 47–54.
23. Rzehak, R.; Krepper, E. CFD modeling of bubble-induced turbulence. *Int. J. Multiph. Flow* **2013**, *55*, 138–155. [CrossRef]
24. Liao, Y.; Rzehak, R.; Lucas, D.; Krepper, E. Baseline closure model for dispersed bubbly flow: Bubble coalescence and breakup. *Chem. Eng. Sci.* **2015**, *122*, 336–349. [CrossRef]
25. Garner, R.W. *Comparative Analyses of Standard Problems, Standard Problem 1 (Straight Pipe Depressurization Experiments)*; Interim Report I-212-74-5.1; Aerojet Nuclear Company: Rancho Cordova, CA, USA, 1973.
26. Stello, V. *Summary of All Participants Results in Comparison to the Experimental Data for the CSNI Standard Problem 1*; For the CSNI-Ad Hoc Working Group on Emergency Core Cooling; OECD/NEA: Paris, Franch, 1975.
27. Liao, Y.; Lucas, D.; Krepper, E.; Rzehak, R. Flashing evaporation under different pressure levels. *Nucl. Eng. Des.* **2013**, *265*, 801–813. [CrossRef]
28. Gurgacz, S.; Pawluczyk, M.; Niewiński, G.; Mazgaj, P. Simulation and analysis of pipe and vessel blowdown phenomena using RELAP5 and TRACE. *J. Power Technol.* **2014**, *94*, 61–68.
29. Dardy, A. *The AREVA Reactor Range—The Nuclear Safety Reference for Europe*; AREVA Day: Warsaw, Poland, 2010.
30. Leyer, S.; Wich, M. The Integral test facility Karlstein. *Sci. Technol. Nucl. Install.* **2012**, *2012*, 439374. [CrossRef]

31. Manera, A. Experimental and Analytical Investigations on Flashing-Induced Instabilities in Natural Circulation Two-Phase Systems. Ph.D. Thesis, Delft University of Technology, Delft, The Netherlands, 2003.
32. Liao, Y.; Lucas, D. 3D CFD simulation of flashing flows in a converging-diverging nozzle. *Nucl. Eng. Des.* **2015**, *292*, 149–163. [CrossRef]
33. Abuaf, N.; Wu, B.J.C.; Zimmer, G.A.; Saha, P. *A Study of Nonequilibrium Flashing of Water in a Converging-Diverging Nozzle: Volume 1—Experimental*; U.S. Nuclear Regulatory Commission: Washington, DC, USA, 1981.
34. Riznic, J.; Ishii, M.; Afgan, N. Mechanistic model for void distribution in flashing flow. In Proceedings of the Transient Phenomena in Multiphase Flow (ICHMT Seminar), Dubrovnik, Yugoslavia, 24–30 May 1987.
35. Krepper, E.; Lucas, D.; Frank, T.; Prasser, H.-M.; Zwart, P.J. The inhomogeneous MUSIG model for the simulation of polydispersed flows. *Nucl. Eng. Des.* **2008**, *238*, 1690–1702. [CrossRef]
36. Liao, Y.; Lucas, D. Poly-disperse simulation of condensing steam-water flow inside a large vertical pipe. *Int. J. Therm. Sci.* **2016**, *104*, 194–207. [CrossRef]
37. Schaffrath, A.; Kruessenberg, A.K.; Weiss, F.P.; Beyer, M.; Carl, H.; Prasser, H.M.; Schuster, J.; Schuetz, P.; Tamme, M.; Zimmermann, W. TOPFLOW—A new multipurpose thermalhydraulic test facility for the investigation of steady state and transient two-phase flow phenomena. *Kerntechnik* **2001**, *66*, 209–212.
38. Lucas, D.; Beyer, M.; Szalinski, L. Experiments on evaporating pipe flow. In Proceedings of the 14th International Topical Meeting on Nuclear Reactor Thermalhydraulics (NURETH-14), Toronto, ON, Canada, 25–30 September 2011.
39. Ziegenhein, T.; Zalucky, J.; Rzehak, R.; Lucas, D. On the hydrodynamics of airlift reactors, Part I: Experiments. *Chem. Eng. Sci.* **2016**, *150*, 54–65. [CrossRef]

energies

MDPI

Article

Computational Study of the Noise Radiation in a Centrifugal Pump When Flow Rate Changes

Ming Gao [1,*], Peixin Dong [1,2], Shenghui Lei [3] and Ali Turan [4]

1 School of Energy and Power Engineering, Shandong University, Jinan 250061, China; paytonsdu@hotmail.com
2 School of Mechanical and Mining Engineering, University of Queensland, Brisbane 4067, Australia
3 Thermal Management Research Group, Efficient Energy Transfer (ηET) Department, Bell Labs Ireland, Nokia, Dublin D15 Y6NT, Ireland; shenghui.lei@nokia-bell-labs.com
4 School of Mechanical, Aerospace and Civil Engineering, University of Manchester, Manchester M60 1QD, UK; a.turan@manchester.ac.uk
* Correspondence: gm@sdu.edu.cn; Tel.: +86-531-8839-9008

Academic Editor: Bjørn H. Hjertager
Received: 1 December 2016; Accepted: 8 February 2017; Published: 14 February 2017

Abstract: Noise radiation is of importance for the performance of centrifugal pumps. Aiming at exploring noise radiation patterns of a typical centrifugal pump at different flow rates, a three-dimensional unsteady hydro/aero acoustic model with large eddy simulation (LES) closure is developed. Specifically, the Ffowcs Williams-Hawkings model (FW-H) is employed to predict noise generation by the impeller and volute. The simulated flow fields reveal that the interactions of the blades with the volute induce root mean square (RMS) pressure and further lead to noise radiation. Moreover, it is found that the profiles of total sound pressure level (*TSPL*) regarding the directivity field for the impeller-generated noise demonstrate a typical dipole characteristic behavior, whereas strictly the volute-generated noise exhibits an apparently asymmetric behavior. Additionally, the design operation (Here, 1 Q represents the design operation) generates the lowest *TSPL* vis-a-vis the off-design operations for all the flow rates studied. In general, as the flow rates decrease from 1 Q to 0.25 Q, *TSPL* initially increases significantly before 0.75 Q and then levels off afterwards. A similar trend appears for cases having the larger flow rates (1–1.25 Q). The *TSPL* deviates with the radiation directivity and the maximum is about 50%. It is also found that *TSPL* by the volute and the blades can reach ~87 dB and ~70 dB at most, respectively. The study may offer a priori guidance for the experimental set up and the actual design layout.

Keywords: centrifugal pump; 3D flow field; varying flow rate; impeller and volute radiation noise; total sound pressure level (*TSPL*)

1. Introduction

Centrifugal pumps as turbomachinery components have been extensively used to transport fluids by converting the rotational kinetic energy consumed to the hydrodynamic energy of the flow [1]. In general, the centrifugal pump routinely operates under various conditions imposed via different operational requirements; hence, a wide frequency range regarding the noise radiation typically occurs and needs to be managed. It has been realized that the noise generation within a centrifugal pump can significantly degrade the pump performance, shift the working point away from the optimal design, reduce the pump efficiency, and further consume extra external energy [2]. In addition, it may give rise to some issues such as safety and reliability. Therefore, it is extremely important to have a better understanding of the hydro/aero acoustics behavior of centrifugal pumps to incorporate in comprehensive pump designs.

In the past, considerable research has been carried out both experimentally and numerically in terms of noise generation in centrifugal pumps. It is generally assumed that the noise sources in the pump mainly comprise of mechanical, (structure borne) noise and flow-induced noise [3]. Choi et al. [4] reported that the noise generation is caused by the large scale flow field instability in a pump impeller. Recently, computational fluid dynamics (CFD) were widely used to elucidate and predict the hydro/aero acoustic generation [5,6]. Chu et al. [7,8] numerically studied a centrifugal pump and reported that the vibration and radiation of noise is mainly determined by the unsteady flow characteristics in centrifugal pumps, which made a great contribution to settle flow-induced noise by CFD methods. More recently, Langthjem et al. [9,10] employed Lighthill's acoustic analogy to predict the radiation of noise for a centrifugal pump based on the 2D flow field. Kato et al. [11,12] reported pressure fluctuation levels on the volute wall using large eddy simulation (LES) in a finite element formulation. Huang [13] also performed numerical simulations on a centrifugal pump via LES and the Ffowcs Williams-Hawkings (FW-H) acoustic model to study the influence of different blade shapes on the efficiency of noise radiation in a centrifugal pump.

Notwithstanding, all the above previous studies concerned radiation of noise in centrifugal pumps, most authors regarded the centrifugal pump as an all-encompassing composite source of sound. However, in this paper, attention is first focused specifically on noise generation by the impeller and/or the volute to map out individual/separate contributions and interactions via each source. In addition, it is rather practical and common for pumps to work with varying flow rates, thus, further emphasis has been put on elucidating the changing rules of radiation noise from both impeller and volute with varying flow rates. To that end a numerical model incorporating LES in conjunction with the FW-H approach in a finite volume environment (FLUENT) is developed. Detailed comparisons with experimental/analytical results on a 2D/3D case are performed to validate the model. The instantaneous flow fields on the blade and volute surfaces are discussed in detail to highlight the nature of the field profiles for the total sound pressure level (*TSPL*, defined using Eqation (8)) obtained to quantitatively reveal the noise distribution induced by different sound sources at varying flow rates.

2. Materials and Methods

2.1. Flow Field Solver

The turbulent flow field within a centrifugal pump is very complex and, in particular, it is challenging to adequately capture the pressure fluctuations accompanying the flow required in the acoustic analysis. To carry out hydro/aero acoustics predictions for a centrifugal pump, three-dimensional transient Navier-Stokes equations are solved for an incompressible fluid using FLUENT [14]. Large eddy simulation is employed to close the governing equations. In the study, a widely accepted sub-grid scale, SGS, model is used to compute the SGS Reynolds stress field via:

$$\tau_{ij} = \frac{1}{3}\delta_{ij}\tau_{kk} - 2\nu_{SGS}\overline{S_{ij}}, \tag{1}$$

where:

$$v_{SGS} = (C_S\Delta)^2|\overline{S}|, \quad |\overline{S}| = \sqrt{2\overline{S_{ij}}\overline{S_{ij}}}, \quad \overline{S_{ij}} = \frac{1}{2}(\frac{\partial \overline{u_i}}{\partial x_j} + \frac{\partial \overline{u_j}}{\partial x_i}), \tag{2}$$

where C_s is the Smagorinsky constant, and Δ is the size of the grid filter.

The governing equations with the LES formulation are discretized using a second-order scheme for the spatial terms and a second-order implicit scheme for the temporal terms. The pressure implicit with splitting of operator (PISO) algorithm is used to solve the pressure-velocity coupling equation [15].

2.2. The Acoustic Submodel

To further capture the noise radiation in a centrifugal pump induced by the inner flow field encompassing the impellers and/or volute, the FW-H acoustic model is employed, in line with the built in assumptions that the flow speed is low, noise is primarily produced by the unsteady pressure field and the chosen noise sources sit in an open region when calculating the level of radiation noise which means the interaction between solid and fluid is neglected. For this model, it is inherently required that all the receivers are located far away from the primary sources of sound. Based on the transient flow simulation discussed above, the FW-H model can be expressed as [16]:

$$\frac{1}{a_0^2}\frac{\partial^2 p'}{\partial t^2} - \nabla^2 p' = \frac{\partial}{\partial t}\{[\rho_0 v_n + \rho(u_n - v_n)]\delta(f)\} - \frac{\partial}{\partial x_i}\{[p_{ij}n_j + \rho u_i(u_n - v_n)]\delta(f)\} + \frac{\partial^2}{\partial x_i \partial x_j}\{T_{ij}H(f)\}, \quad (3)$$

where p' is the far-field acoustic pressure, a_0 is the speed of sound, ρ_0 is the reference density of the fluid, f is the surface, u_i and v_i are the flow velocity component and the surface velocity component in the x_i direction respectively, u_n and v_n is the flow velocity component and the surface velocity normal to the surface f respectively, n_j is the unit normal vector, $\delta(f)$ is Dirac delta function and $H(f)$ is Heaviside function. Here T_{ij} is the Lighthill stress tensor given by:

$$T_{ij} = \rho u_i u_j + p_{ij} - a_0^2(\rho - \rho_0)\delta_{ij}, \quad (4)$$

where p_{ij} is the stress tensor expressed as:

$$p_{ij} = p\delta_{ij} - \mu\left[\frac{\partial u_i}{\partial x_j} + \frac{\partial u_j}{\partial x_i} - \frac{2}{3}\frac{\partial u_k}{x_k}\delta_{ij}\right], \quad (5)$$

Physically, the three RHS terms appearing in Equation (3) correspond to monopole, dipole, and quadrupole acoustic sources respectively. In this study, the quadrupole noise source is ignored [17] in view of the flow having low speed compared with the local acoustic velocity. In addition, for a well-designed centrifugal pump, the monopole noise source is relatively weak and also neglected. Thus, the dipole noise induced by the wall surface encompassing the transient flow RMS pressure is the primary concern in this paper.

Furthermore, according to the GB/T29529-2013 standard [18], the volute and/or impeller wall surfaces chosen as monitoring stations are employed to study the impeller and/or volute noise radiation. As shown in Figure 1, the monitoring surface is circular, having a radius of 1000 mm; 24 receivers denoted by P_1–P_{24} are evenly arranged along the surface and, in particular, P_1 is placed strictly towards the volute tongue. A Fourier transformation involving the acoustic signals obtained at these receivers is applied to extract the relevant acoustic spectra. The sound pressure level (*SPL*) is calculated by:

$$SPL = 20\ \log_{10}\frac{P_e}{P_{ref}}, \mathrm{dB}, \quad (6)$$

where P_{ref} is the reference sound pressure (=2 × 10^{-5} Pa in air), and P_e is the effective value of acoustic pressure defined as:

$$P_e = \sqrt{\frac{1}{T}\int_0^T p'^2 dt}, \quad (7)$$

To reveal the intensity of noise radiation, it is necessary to derive a temporal intensity profile involving a superposition of acoustic pressures at each Fourier frequency. In this regard, the *TSPL* is introduced and expressed as:

$$TSPL = 10\lg\sum_{i=1}^{n} 10^{SPL_i/10}, \quad (8)$$

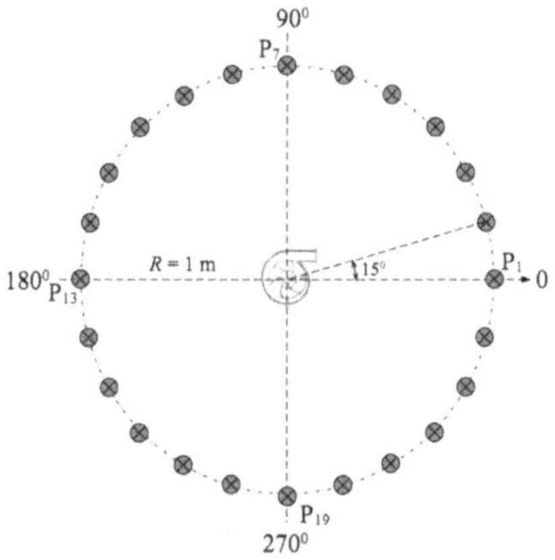

Figure 1. Arrangement of receivers outside the centrifugal pump.

2.3. Numerical Issue Concerning the Set-Up

In this study, a single-stage single-suction centrifugal pump (media, water) having a medium specific speed of 112 is selected. Table 1 lists all the geometric and performance parameters. The 3D computational domain for the simulations comprises of the inlet pipe, impeller, and the volute, as shown in Figure 2. It should be noted that the practical disk friction losses and leakage losses are neglected, but an account is taken of the potential backflow at the inlet and outlet. A hybrid mesh system is generated using GAMBIT (2.4.6, Fluent, Inc., New York, NY, USA) [19], as shown in Figure 3. Considering the complexity of the geometrical layout and the volute tongue, an unstructured grid is used for these domains and an appropriate mesh refinement is further applied, whereas, a structured mesh is used for the remaining regions. Here, particular attention is paid to the near wall regions and the unstructured mesh is automatically refined to have appropriate y^+ values ($y^+ \approx 1$) and thus to ensure the requirement of wall-resolved LES for the unsteady calculation [20]. It is observed that different levels of mesh refinement may lead to different average y^+ values, and when the grid number reaches almost 2.5 million (2.5 M) cells, the average y^+ values decrease to around 1, as shown in Figure 4. With the increase of mesh refinement, the average y^+ values gradually reduce. With a combining consideration of both the mesh accuracy and the limited computing resources, the grid of 2.5 M cells is chosen for the calculation. What is more, it is not difficult to find that the average y^+ values are correctly consistent with the sizes of the first boundary layer. Therefore, different grid systems shown are tested for the required grid-dependency based on the convergence pattern for the pump hydraulic head to the designed value (~28 m), as shown in Table 2. To improve the prediction accuracy while reducing the attendant computational costs, Grid-C is chosen in the study.

Table 1. Geometric and performance parameters used for a simulated centrifugal pump (water).

Parameter	Value	Unit
Impeller outlet diameter D_2	270	mm
Impeller outlet width D	30	mm
Design flow rate Q	80	m^3/h
Design head H	28	m
Design rotational speed n	1450	r/min
Blade number Z	6	1
Axis pass frequency f_a (APF)	24.2	Hz
Blade pass frequency f_b (BPF)	145	Hz

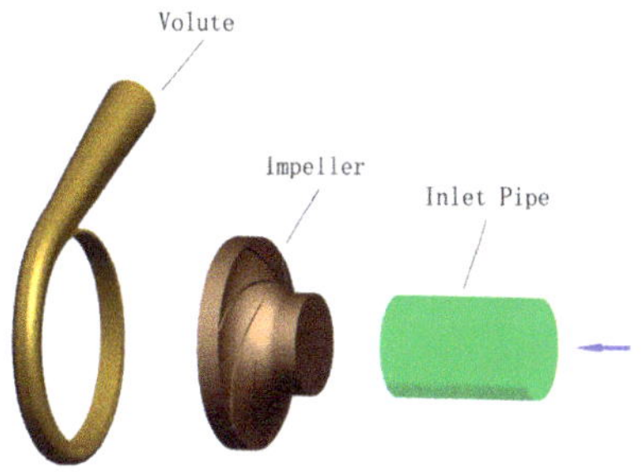

Figure 2. Computational geometry of the centrifugal pump for each part.

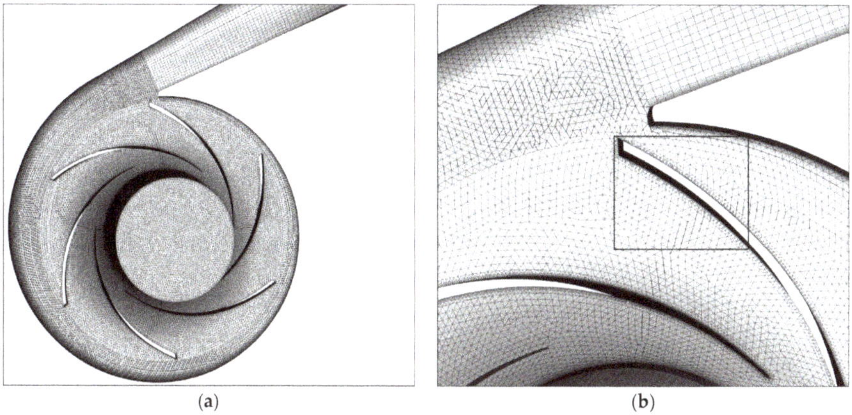

Figure 3. Grid System of the centrifugal pump. (**a**) Overall view; and (**b**) magnified view of the volute tongue.

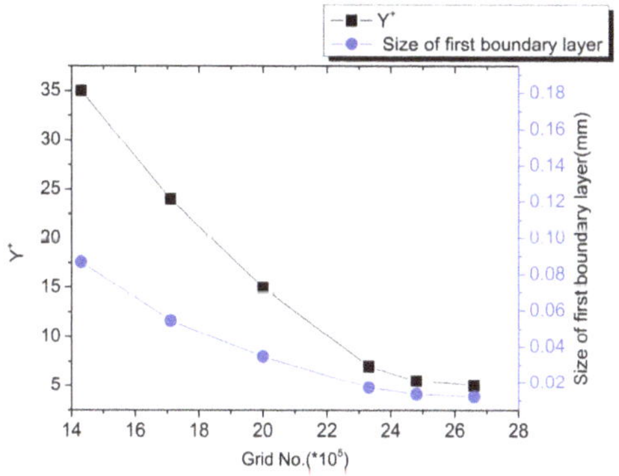

Figure 4. Y^{+} values and corresponding sizes of first boundary layer.

Table 2. Grid independent Analysis.

Grid System	Grid Number	Head (m)
Grid-A	2,012,658	28.82
Grid-B	2,324,567	28.70
Grid-C	2,496,543	28.63
Grid-D	2,676,328	28.62

For the boundary conditions, a uniform axial velocity profile is imposed at the inlet while 'outlet' boundary conditions are adopted at the outlet. All the walls are treated as non-slip boundaries. The pre-converged steady flow field obtained via the k-ε is adopted as the initial condition for the subsequent transient wall-resolved LES simulations with a Subgrid-Scale Model of Smagorinsky-Lilly. The time step is specified by:

$$\Delta t = \frac{60}{nK}, \tag{9}$$

where K is the step number in a rotational period of the impeller (=360 in the study), and n is the rotational speed. With the designed values given in Table 1, Δt is 1.1494×10^{-4} s in the study.

3. Results and Discussion

3.1. Numerical Validaton

To verify the FW-H model used, a 2D cylinder case was first studied and its computational domain and the boundary conditions are shown in Figure 5. It should be pointed out that the receiver is placed at a distance of 1.3 m from the center of the cylinder to accord with the experimental set-up [21]. A constant velocity of 40 m/s is specified at all the inlets, whereas an ambient pressure of 1 atm is used at the outlet. A structured mesh system of about 90,000 cells is used and the finest cell has a minimum edge length of 0.005 mm. Figure 6a shows the predicted and measured sound pressure spectra. As can be seen, the prediction matches rather well with the measurements at the receiver. Furthermore, the two peak frequencies at 800 Hz and 1500 Hz are adequately captured. Figure 6b displays the pressure coefficients around the cylinder surface. The simulation displays the overall trend regarding the pressure coefficient obtained in the experiment with a maximum discrepancy less than 10%.

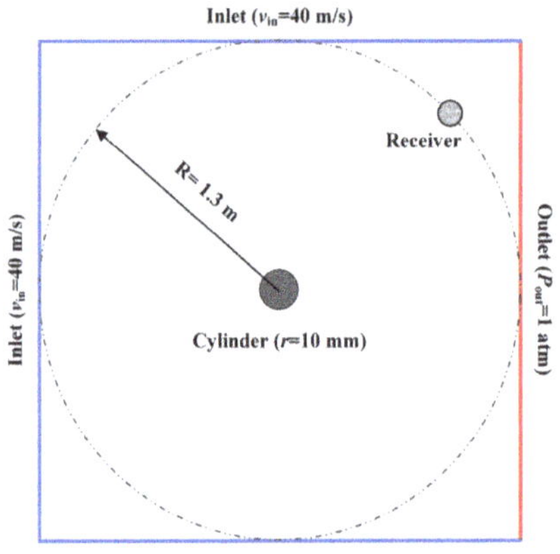

Figure 5. The validation case of a 2D cylinder (not to scale).

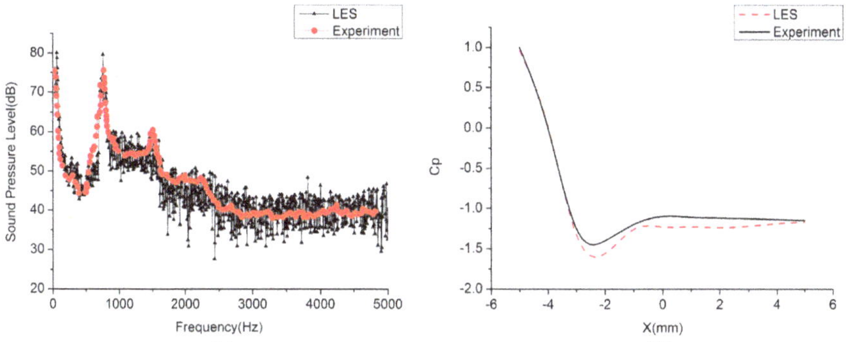

(a) (b)

Figure 6. Comparison of numerical predictions with experimental measurements. (**a**) Sound pressure spectra; and (**b**) pressure coefficient (C_p).

Additionally, the numerical method adopted should also be verified. In terms of the 3D case shown in Figure 2, the hydraulic head predicted by the numerical model is compared with the nominal design head for the centrifugal pump. The simulation yields a head of 28.63 m. This prediction agrees well with the designed valued (28 m) and the relative error of the calculation is less than 2.3%. The discrepancy is essentially due to the fact that the disk friction losses and leakage losses appearing in practice are neglected in the simulations. The sound spectra for the receiver 1 in terms of the design flow rate are also shown in Figure 7. The predicted rotational frequency (APF) and blade pass frequency (BPF) are 24.1 Hz and 142.3 Hz respectively, very close to the design values of 24.2 Hz and 145 Hz marked in dashed lines. Clearly, the numerical model presented in the paper is able to yield reasonably accurate predictions on the flow field as well as the noise.

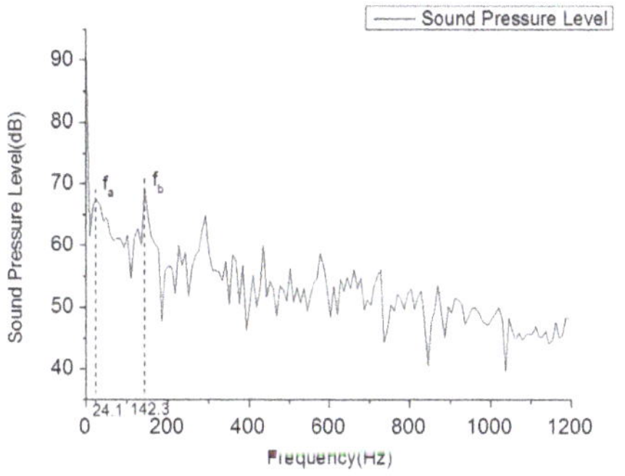

Figure 7. The sound spectra of receiver 1.

3.2. Flow Field

Throttle governing is a common control mode routinely employed in the operational envelope of the centrifugal pump system. By such means installation of the throttle employing valves or baffles and adjusting the opening of the throttle can change the local loss and achieve the desired goal of throttle governing. In this study, 13 flow rate conditions are studied to analyze the noise radiation, including 20 m^3/h, 30 m^3/h, 40 m^3/h, 50 m^3/h, 60 m^3/h, 70 m^3/h, 78 m^3/h, 80 m^3/h (which is the design flow rate), 82 m^3/h, 90 m^3/h, and 100 m^3/h, respectively.

To elucidate the noise generation by the flow field, an index employing a value of RMS pressure field is introduced to provide information for the pressure variations with time step on the wall of the impeller and volute, P_{rms} defined by:

$$P_{rms} = \sqrt{\frac{1}{N}\sum_{i=0}^{N-1}\left[P(t) - \overline{P}\right]^2}, \tag{10}$$

where $P(t)$ represents the time-dependent static pressure, $\overline{P}$ is the time-averaged static pressure in a circulation period (N = 360 in this study).

Figure 8 describes the distribution of RMS values on the surfaces of impeller and volute. When the flow rate is 0.75 Q, the RMS pressure field primarily resides on the side of inlet, displaying values much stronger vis-a-vis the opposite side, as is shown in Figure 8a. It can be inferred from such considerations that while the flow rate is less than the design value, the velocity field distribution on the impeller is substantially non-uniform, providing for extremely unsteady flow near the inlet side vis-a-vis the outlet. Thus, the primary reason for the impeller noise is the unsteady flow field at 0.75 Q. When the flow rate reaches 1 Q, the average RMS is a minimum and the flow field is also smoothly distributed, especially near the inlet side, yielding the impeller noise under the design flow rate condition to be substantially lower than other flow rate conditions. Additionally, the RMS pressure field displays extreme values near both the inlet and the outlet regions under the 1.25 Q flow rate condition to yield a maximum noise field generated.

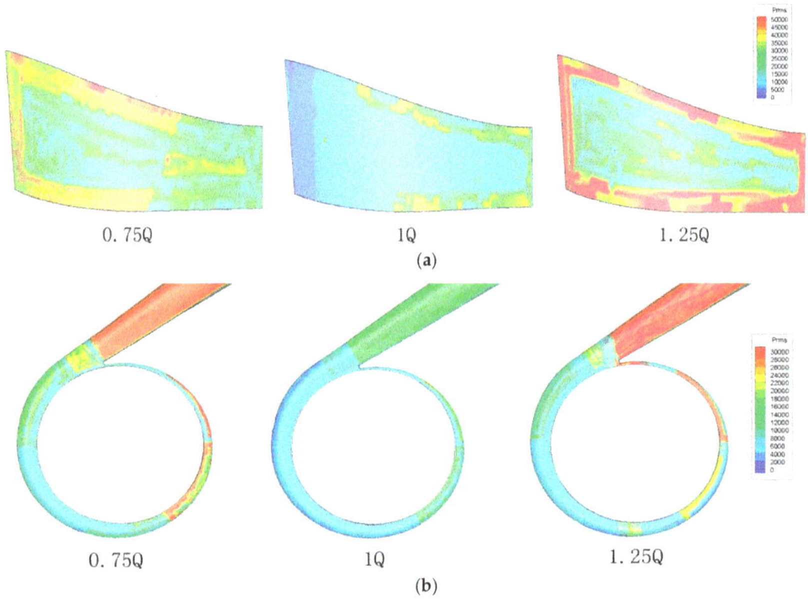

Figure 8. Distribution of root mean square (RMS) value on both impeller and volute under three separate flow rates, including 0.75 Q, 1 Q and 1.25 Q. (**a**) Distribution of RMS value on impeller; and (**b**) distribution of RMS value on volute.

It can also be inferred from a study of the RMS pressure field occurring in the volute at different flow rates that the resulting acoustic behavior is rather similar to that of the impeller (seen in Figure 8b). The overall distribution of RMS at the wall of the volute under the three flow rate conditions is that the RMS values in the diffuser are unmistakably the highest and, near the volute tongue, the fluctuation field is rather "stiff", particularly when the fluid is just transported past the tongue to flow into the volute, implying that the volute tongue is the primary source of the RMS pressure and even the noise. Although the RMS at 1 Q is smaller than those at other flow rates, the distribution resembles that under 0.75 Q and 1.25 Q flow rate conditions.

3.3. Noise Analysis

3.3.1. The Impeller Radiation Noise

The noise generated by the blades is originally studied by defining only the surfaces of the blades as the sources of sound. The distribution of impeller radiation noise for different flow rates calculated using the FW-H model is shown in Figure 9. The profiles of *TSPL* of directivity field shown in Figure 9a apparently demonstrate the dipole characteristic behavior. It is also found that two *TSPL* valleys appear at θ = 75° and 270°, but the weak asymmetric distribution occurs along these two points. Figure 9b reveals that the design operation (1 Q) generates the smallest *TSPL* than off-design operations. As the flow rates decrease from 1 Q to 0.25 Q, the *TSPL* initially increases significantly before 0.75 Q and then level off afterwards. A similar trend occurs for the cases for the larger flow rates (1–1.25 Q), but the *TSPL* increases very quickly. For example, the 1.25 Q case produces much more noise in comparison with 0.75 Q, even though they have the same off-design deviation. In general, the noise level deviates about 50% in different directions as demonstrated by the deviation bar in Figure 9b. Nevertheless, the *TSPL* is eventually saturated at ~70 dB with the increase/decrease of the flow rate.

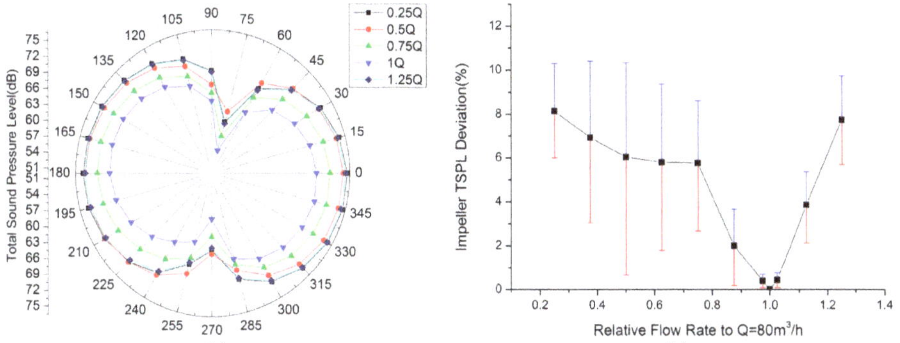

Figure 9. Impeller radiation noise. (**a**) Total sound pressure level (*TSPL*) of directivity field from the impellers; and (**b**) the relationship of *TSPL* deviation vs. the flow rate.

3.3.2. The Volute Radiation Noise

Similarly, the noise generated by the volute is further studied by defining the volute surface as the source of sound. Figure 10a shows the *TSPL* pattern for the volute radiation noise. Interestingly, the radiation field displays very different behavior in comparison with the blade noise discussed above. The *TSPL* profiles exhibit an apparent asymmetric behavior and *TSPL* on the right hand side is much larger than the one on the left. This may be explained by the fact that the volute tongue is the key noise source and the non-uniform distance between the receivers and the tongue leads to the asymmetric distribution. Figure 10b reveals a similar trend shown in Figure 9b. It is interesting to find that the mean *TSPL* levels off in the range of 0.5–0.75 Q. With the increase/decrease of the flow rate vis-a-vis the design value, the *TSPL* deviation increases and the maximum is about 50%. It should be noted that the *TSPL* can reach ~87 dB, which informs us that the noise contribution by the volute is substantial.

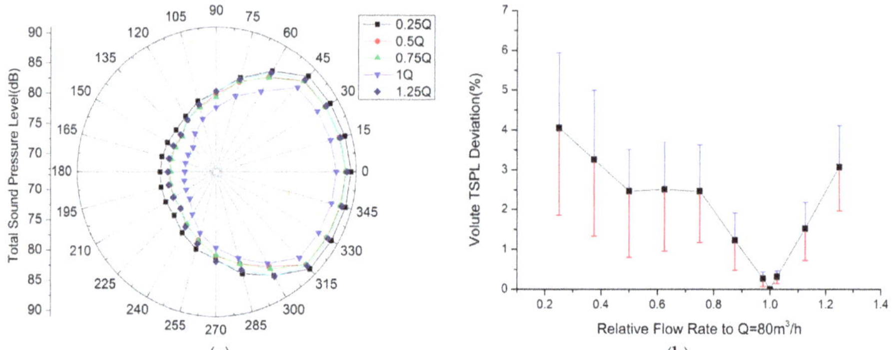

(a) (b)

Figure 10. Volute radiation noise. (**a**) Total sound pressure level of directivity field from the volute; and (**b**) the relationship of *TSPL* deviation vs. the flow rate.

4. Conclusions

In terms of a single-stage singe-suction centrifugal pump, a 3D numerical model with the LES closure is developed for elucidation and reliable prediction of the pertinent hydro/aero acoustic behavior. In particular, the FW-H model was employed to predict the noise generation by the blades as well as the volute. Comparisons of numerical predictions with the measured/analytical results reveal that the model can yield good results on the noise and the flow field. The simulations further show that the interaction of the blades with the volute induces high RMS pressure and further leads to additional noise radiation. Moreover, the sources of noise generated by the blade surfaces and the volute surface at different flow rates were studied. It is found that the profiles of *TSPL* of directivity field for the impeller-generated noise demonstrate a typical dipole characteristic behavior, whereas the ones for the volute-generated noise exhibit an apparent asymmetric behavior. In addition, within the flow rate range studied, the design operation (1 Q) generates the smallest *TSPL* than off-design operations. As the flow rates decrease from 1 Q to 0.25 Q, *TSPL* initially increases significantly before 0.75 Q and then levels off afterwards. A similar trend occurs for the cases in the larger flow rates (1–1.25 Q). The *TSPL* deviates with the radiation directivity and the maximum is about 50%. It is also found that *TSPL* by the volute and the blades can reach ~87 dB and ~70 dB at most, respectively.

Acknowledgments: Shenghui Lei would appreciate the financial support by the Industrial Development Agency (IDA) Ireland.

Author Contributions: Ming Gao conceived the initial idea of this research and guided the work; Peixin Dong carried out the simulations and collected and analyzed the data; Shenghui Lei analyzed the data and guided the simulations; Ali Turan contributed valuable scientific discussion; all the authors commented and wrote the paper.

Conflicts of Interest: The authors declare no conflict of interest. The founding sponsors had no role in the design of the study; in the collection, analyses, or interpretation of data; in the writing of the manuscript, and in the decision to publish the results.

Nomenclature

C_s	Smagorinsky constant
Δ	Size of the grid filter
p'	Far-field acoustic pressure
a_0	Speed of sound
ρ_0	Reference density of fluid
$\delta(f)$	Dirac delta function
$H(f)$	Heaviside function
P_{ref}	Reference sound pressure

K	Step number in a rotational period
n	Rotational speed
$P(t)$	Transient static pressure
$\overline{P}$	Time-averaged static pressure

References

1. Johann, F.G. *Centrifugal Pumps*; Springer: Berlin/Heidelberg, Germany; New York, NY, USA, 2008.
2. Brennen, C.E. A review of the dynamics of cavitating pumps. *J. Fluids Eng.* **2012**, *135*. [CrossRef]
3. Wang, J.; Feng, T.; Liu, K.; Zhou, Q.J. Experimental research on the relationship between the flow-induced noise and the hyaraulic parameters in centrifugal pump. *Fluid Mach.* **2007**, *6*, 1–13.
4. Choi, J.S.; Mclaughlin, D.K.; Thompson, D.E. Experiments on the unsteady flow field and noise generation in a centrifugal pump impeller. *J. Sound Vib.* **2003**, *263*, 493–514. [CrossRef]
5. Parrondo, J.; Pérez, J.; Barrio, R.; González, J. A simple acoustic model to characterize the internal low frequency sound field in centrifugal pumps. *Appl. Acoust.* **2011**, *72*, 59–64. [CrossRef]
6. Liu, H.L.; Dai, H.W.; Ding, J. Numerical and experimental studies of hydraulic noise induced by surface dipole sources in a centrifugal pump. *J. Hydrodyn. Ser. B* **2016**, *28*, 43–51. [CrossRef]
7. Chu, S.; Dong, R.; Katz, J. Relationship between unsteady flow, pressure fluctuations, and noise in a centrifugal pump—Part A: Use of PDV data to compute the pressure field. *J. Fluids Eng.* **1995**, *117*, 24–29. [CrossRef]
8. Chu, S.; Dong, R.; Katz, J. Relationship between unsteady flow, pressure fluctuations, and noise in a centrifugal pump—Part B: Effects of blade-tongue interactions. *J. Fluids Eng.* **1995**, *117*, 30–35. [CrossRef]
9. Langthjem, M.A.; Olhoff, N. A numerical study of flow-induced noise in a two-dimensional centrifugal pump. Part I. Hydrodynamics. *J. Fluids Struct.* **2004**, *19*, 349–368. [CrossRef]
10. Langthjem, M.A.; Olhoff, N. A numerical study of flow-induced noise in a two-dimensional centrifugal pump. Part II. Hydroacoustics. *J. Fluids Struct.* **2004**, *19*, 369–386. [CrossRef]
11. Kato, C.; Yamade, Y.; Wang, H.; Guo, Y.; Miyazawa, M.; Takaishi, T.; Takano, Y. Numerical prediction of sound generated from flows with a low Mach number. *Comput. Fluids* **2007**, *36*, 53–68. [CrossRef]
12. Kato, C.; Kaiho, M.; Manabe, A. An overset finite-element large-eddy simulation method with applications to turbomachinery and aeroacoustics. *Trans. Am. Soc. Mech. Eng. J. Appl. Mech.* **2003**, *70*, 32–43. [CrossRef]
13. Huang, J. Comparison of noise characteristics in centrifugal pumps with different types of impellers. *Acta Acust.* **2010**, *35*, 113–118.
14. Li, K.; Yu, J.; Shi, F.; Huang, A.X. Dimension splitting method for the three dimensional rotating Navier-Stokes equations. *Acta Math. Appl. Sin. Engl. Ser.* **2012**, *28*, 417–442. [CrossRef]
15. Seif, M.S.; Asnaghi, A.; Jahanbakhsh, E. Implementation of PISO algorithm for simulating unsteady cavitating flows. *Ocean Eng.* **2010**, *37*, 1321–1336. [CrossRef]
16. Williams, J.F.; Hawkings, D.L. Sound generated by turbulence and surfaces in arbitrary motion. *Philos. Trans. R. Soc. A Math. Phys. Eng. Sci.* **2010**, *264*, 321–342. [CrossRef]
17. Ghasemian, M.; Nejat, A. Aerodynamic noise prediction of a horizontal axis wind turbine using improved delayed detached eddy simulation and acoustic analogy. *Energy Convers. Manag.* **2015**, *99*, 210–220. [CrossRef]
18. Tao, J.; Lu, X.; Wang, L.; Wang, L.; Li, J. *Methods of Measuring and Evaluating Noise of Pumps*; Chinese Quality Supervision Bureau: Beijing, China, 2013; pp. 4–6.
19. Fluent Inc. *Gambit 2.4 User's Guide*; Fluent Inc.: New York, NY, USA, 2007; pp. 1–17.
20. Choi, H.; Moin, P. Grid-point requirements for large eddy simulation: Chapman's estimates revisited. *Phys. Fluids* **2012**, *24*. [CrossRef]
21. Tsai, C.H.; Fu, L.M.; Tai, C.H.; Huang, Y.L.; Leong, J.C. Computational aero-acoustic analysis of a passenger car with a rear spoiler. *Appl. Math. Model.* **2009**, *33*, 3661–3673. [CrossRef]

energies

MDPI

Article

Study the Flow behind a Semi-Circular Step Cylinder (Laser Doppler Velocimetry (LDV) and Computational Fluid Dynamics (CFD))

S. M. Sayeed-Bin-Asad *, Tord Staffan Lundström and Anders Gustav Andersson

Division of Fluid and Experimental Mechanics, Luleå University of Technology, SE-971 87 Luleå, Sweden; staffan.lundstrom@ltu.se (T.S.L.); anders.g.andersson@ltu.se (A.G.A.)

* Correspondence: sayeed.asad@ltu.se; Tel.: +46-920-492-886

Academic Editor: Bjørn H. Hjertager

Received: 1 February 2017; Accepted: 6 March 2017; Published: 9 March 2017

Abstract: Laser Doppler Velocimetry (LDV) measurements, flow visualizations and unsteady Reynolds-Averaged Navier-Stokes (RANS) Computational Fluid Dynamics (CFD) simulations have been carried out to study the turbulent wake that is formed behind a semi-circular step cylinder at a constant flow rate. The semi-circular cylinder has two diameters, a so-called step cylinder. The results from the LDV measurements indicate that wake length and vortex shedding frequency varies with the cylinder diameter. This implies that a step cylinder can be used to attract fish of different size. By visualizations of the formation of a recirculation region and the well-known von Kármán vortex street behind the cylinder are disclosed. The simulation results predict the wake length and shedding frequency well for the flow behind the large cylinder but fail to capture the dynamics of the flow near the step in diameter to some extent and the flow behind the small cylinder to a larger extent when compared with measurements.

Keywords: laser doppler velocimetry (LDV); computational fluid dynamics (CFD); fish migration; step cylinder; shedding; vortex street

1. Introduction

Hydro and nuclear power are the major sources of electricity in Sweden. Swedish Energy [1] has recently published some statistics regarding power production in Sweden showing that hydro and nuclear power stand for 47% and 34% of the total electricity in Sweden, respectively. In order to increase the percentage renewable energy overall, to keep the target of greenhouse gas emission [2], systems are constantly developed to harvest wind, wave, tidal, and solar energy. These sources of energy are typically irregular and continuous power production is required to cover for this irregularity. The availability, storage capacity, and flexibility of hydropower make it a perfect contender for the power regulation role. However, hydropower has some local impacts on the environment, such as creating dams in rivers and changing water flow directions, which lead to some problems for fauna passage, in general, and migrating fishes, in particular. Fish migration issues are mostly studied from a biological point of view and more detailed studies from a fundamental fluid mechanics point of view could lead to new insights and solutions.

Some hydropower dams have fishways or fish ladders to allow fish to migrate past dams, as well as past rotating turbines and, during swimming or passing this fishway or fish ladder, fish have to tackle different flow obstructions or disturbances, like turbine intakes, stones, and concrete structures. Fluid flow characteristics in fish ladders or fishways during fish migration are crucial when designing effective fishways so that fishes migrate in an efficient way. Studies [3–7] have shown that cylindrical obstructions can be used to create favorable conditions for upstream migrating fishes of appropriate size. The concept is to create periodic vortices that the fish uses in an energy efficient way (less energy than in a free stream) to maintain a position in the flow. Liao [8] describes the structure of the flow around a half cylinder and the positions of fish swimming within the wake behind and around the cylinder. It was found that the fishes exploit the flow conditions created by the semicircular cylinder when swimming upstream around it. Apart from this, Liao discovered the altered shedding frequencies and wakes for various cylinder diameters. Different sizes of fish exploit vortices with different shedding frequencies. Therefore, it is important to have proper shedding for appropriate sizes of fish to swim past obstructions.

There are many applications for step full cylinders, including airport control towers, street lamp posts, antenna members and radio communication towers. Hence, there are several driving forces for engineers to examine this type of flow phenomena, both numerically and experimentally.

Dunn et al. [9] carried out velocity measurement with laser doppler velocimetry (LDV) and flow visualization around a step cylinder. One result is that the recirculation near a step has significant effects on the shedding frequencies as compared to locations away from the step. Similarly, another study from Dunn et al. [10] also found significant effects of free-stream shearing on vortex-shedding characteristics from a step-cylinder. Chris Morton et al. [11–16] carried out several studies using both numerical and experimental with 2D and 3D particle image velocimetry (PIV) investigations on step cylinders, such as single step dual step, etc. Detailed studies of the wake vortex dynamics behind step-cylinder are provided. He [17] has recently used tomographic-PIV (3D-PIV) behind a step cylinder for three-dimensional (3D) information of vortex shedding. Rafati [18] conducted a series of experimental studies with various diameters of step cylinders using tomographic-PIV measurement. One result is that for high flow (higher velocity than free stream velocity), streamwise vortices are introduced into the wake that complicates the vortex dynamic, and vortex interactions. Rafati [18] also produced a table of studies on step cylinders available (Table 1).

Table 1. Previous studies on flow behind step cylinders. Reproduced from [18].

Study	Re_D	D/d
N.Ko et al. [19]	8×10^4	2
Yagita et al. [20]	8×10^2–10^4	1.25–5
Lewis et al. [21]	67–200	1.14–1.76
Dunn et al. [10]	62–1230	2
Norberg [22]	3×10^3–13×10^3	1.25–2
N.Ko et al. (1984, 1990, 1992) [23–25]	8×10^4	1.33–2.78
Chua et al. [26]	4.72×10^3	3
Morton et al. [11,27,28]	150–2100	2
Dunn et al. [9] (span-wise sheared flow)	152–764	1.92

Despite having some investigations of flow behavior around stepped cylinders, little or no scientific works, especially experimental, on half cylinders with steps has been performed. Therefore, in the current study the flow around such an object is studied with LDV and computational fluid dynamics (CFD). The flow dynamics of the step and diameter change is interesting from a fluid mechanics point of view but as the half cylinders also have shown to be interesting for fish migration, the difference in wake formation and shedding of vortices behind the different cylinder diameters could also be designed to fit different sizes and species of fish.

2. Experimental Arrangement

Experiments were carried out in a water flume at the John Field Laboratory at Luleå University of Technology, Sweden. The flume is 7.5 m long with a cross-section of 0.295 m × 0.310 m. The semi-circular step cylinder employed in this investigation is made of acetal (POM) and the diameter of the small (*d*) and large (*D*) cylinders are 50 mm and 100 mm, respectively. Figure 1 describes the configuration and measuring zones of the cylinder centerline.

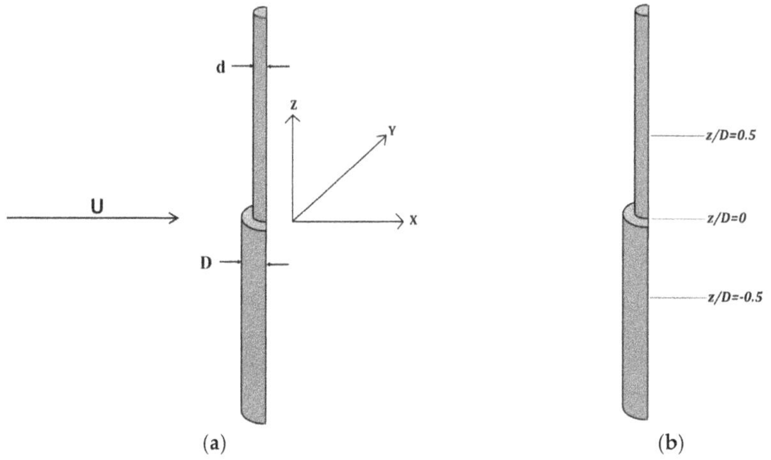

Figure 1. Cylinder configuration; (**a**) flow direction and coordinate description; and (**b**) measuring zones.

A submersible pump was used to pump the water through the channel from a rectangle water reservoir tank with 2 m^3 storage capacity. The water reservoir is placed at the outlet of the channel and the temperature of the water was controlled using an automatic cooling arrangement. This is essential since variation in water temperature changes its viscosity and refractive index properties. Changes in water temperature leads to laser beam displacements from the desired point of measurements. A Danfoss MassFlo Corolios flow meter was used to control the water flow rate.

To obtain a uniform velocity distribution through the channel there is a steel net and a honeycomb at the inlet of the channel. The thickness of the honeycomb is 75 mm and the diameter of its wholes is 7.6 mm. The steel net geometry consists of 2.5 mm × 2.5 mm square holes and it is 0.8 mm thick. A schematic arrangement of the experimental set-up is presented Figure 2.

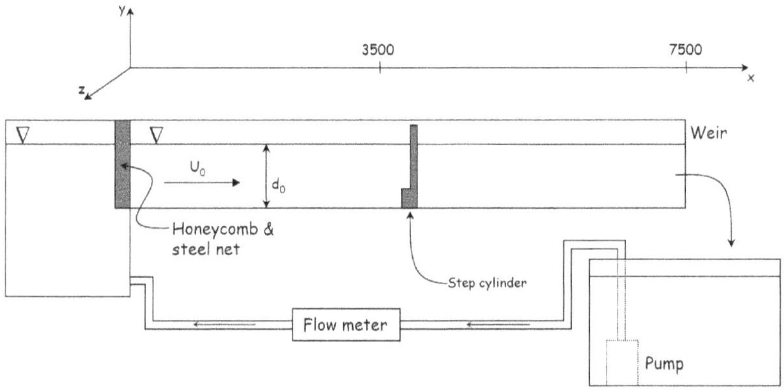

Figure 2. Schematic presentation of water circulation in the flume.

This LDV system, used in this investigation, is a two component configuration with an 83 mm diameter optical fiber probe and a front lens with 360 mm focal length which consists of a 300 mW diode-pumped solid-state laser, and transmitting optics, including a beam splitter Bragg-cell, photo detector, and signal processor. The system is used in backscatter mode so that the probe is both transmitting and receiving signals. The laser has two channels with a wavelength of 532 and 561 nm, respectively. The laser power is 300 mW for each channel. The smallest dimensions of the measurement volume are approximately $0.123 \times 0.117\ mm^2$ for both colors, and the system is operated in coincidence mode, which ensures correlated velocity components. The water was seeded with polyamide particles with a diameter of 5 μm (Dantec's PSP-5). TSI's *FlowSizer 64* software (Shoreview, MN, USA) was used for the data acquisition and data acquisition rates were tried to maintain greater than 50 Hz. The 2D-LDV probe was mounted on a 3D traversing mechanism (can travel maximum 600 mm in each direction) controlled by the software. The sampling time was maintained as 121 seconds at each measurement point, corresponding to at least 10,000 samples, which are essential for statistical precision. All of the measurements were carried out in coincidence mode [29] to obtain an equal number of velocity counts for each direction, and the free stream velocity (U_∞) was determined from repeated measurements at a position 155 mm upstream of the cylinder at $z/D = 0$ as Norberg [30] determined. Free stream (U_∞) is also calculated theoretically using the flow rate ($0.0077\ m^3 \cdot s^{-1}$) and it is found to be $U_\infty = 0.1451\ m \cdot s^{-1}$. Repeatability tests are completed to estimate the first-order variable uncertainty for the experiments [31]. Four different measurements for all experiments were completed to estimate the random error introduced by the experimental facility and changes in experimental conditions, such as the water temperature, position of the cylinder, etc.

Flow visualization was also conducted to explore the wake structure using a digital single-lens reflex (DSLR) camera (Nikon D90) and two tungsten light heads to illuminate the flow area. Two food dyes (red and green) were injected from both sides of the cylinder using a programmable syringe pump running at constant speed. Videos were recorded from both the side and top of the flume and the recordings were converted into images in equal time step using MATLAB (The MathWorks, Natick, MA, USA).

Except otherwise mentioned for scaling the results, all positions and distances are scaled with D and velocities with the free stream velocity U_∞.

3. Numerical Setup

The simulations were performed with the commercial software Ansys CFX 16 (ANSYS, Inc, Canonsburg, PA, USA). A structured grid with 10.1 million nodes and an average wall resolution of $y^+ = 0.7$ and maximum $y^+ = 2.8$ was created in ANSYS IcemCFD (275 Technology Drive, Canonsburg, PA, USA). The volume of fluids (VOF) method was used to account for both the water and air phase. The numerical domain did not include the full flume, the inlet was placed 1 m upstream the cylinder and the outlet was placed 1.5 m downstream to reduce computational time. To generate a fully developed inlet profile an additional simulation model was created for the upstream part of the flume with a constant velocity inlet assumed at the honeycomb and the velocity profile from the outlet of that simulation was used as inflow for the domain with the cylinder. A second-order accurate advection scheme was used to solve the flow equations and a second-order backward Euler transient scheme was used for the temporal discretization. A timestep of 0.025 s was selected which corresponds to a RMS Courant number of ~5 and the relevant transient results were saved every 10 time steps. The streamwise velocity component was monitored at 16 points in the central plane on three different heights corresponding to the measurement positions from the LDV. The shear stress transport (SST) turbulence model was selected [32] as it utilizes the near-wall capabilities of the k-ω model and at the same time uses the bulk flow from the k-ε model where the k-ω is weaker. The flow was initialized with a constant velocity corresponding to the measured mass flow ($0.0077\ m^3 \cdot s^{-1}$). After an initial settling period of ~1000 iterations the flow became more periodic and a developed vortex shedding was assumed. The simulation data was then saved for 2500 iterations or 62.5 s resulting in a total computational time of ~4 weeks on 64 CPU cores.

4. Results and Discussion

Migrating fishes swimming around cylindrical obstructions face vortices and, interestingly, these vortices provide extra energy to fishes to improve their swimming abilities. These vortices are influenced by wake and shedding frequency, but these wake and vortex shedding frequencies are greatly dependent on the cylinder diameter [8]. A cylinder with different diameters, such as a step cylinder, could generate an interesting flow behind it and variable wake and shedding frequencies. This warrants this study of wake and vortex shedding behind a semicircular step cylinder with LDV and CFD.

4.1. Wake

Time-averaged measurements of centerline velocity distribution behind the step cylinder at $z/D = -0.5$, $z/D = 0.0$, and $z/D = 0.5$ are shown in Figure 3a–c and notice that for all positions the negative values captured indicate a wake. A main result is that behind the step ($z/D = 0.0$), the wake is longer compared to the situation behind both the large and small cylinders (see Figure 3a–c and Table 2). The simulations, on the other hand, yield that the length of the wake behind the large cylinder ($z/D = -0.5$) is shorter than for the other cases and the time averaged local velocity $\bar{u}$ swiftly goes back toward free stream velocity, U_∞. There is, therefore, good agreement between the simulation and experiments (see Figure 3a). The agreement is also fairly good between simulations and measurements at the step while the simulations severely over predict the length of the wake behind the small cylinder ($z/D = 0.5$) and the simulated results never approach the free stream velocity in the interval of measurement (see Figure 3c) and Table 2 also yields that although simulations compare well to experiments behind the large cylinder and at the step, the simulations always overpredict the length of the wake.

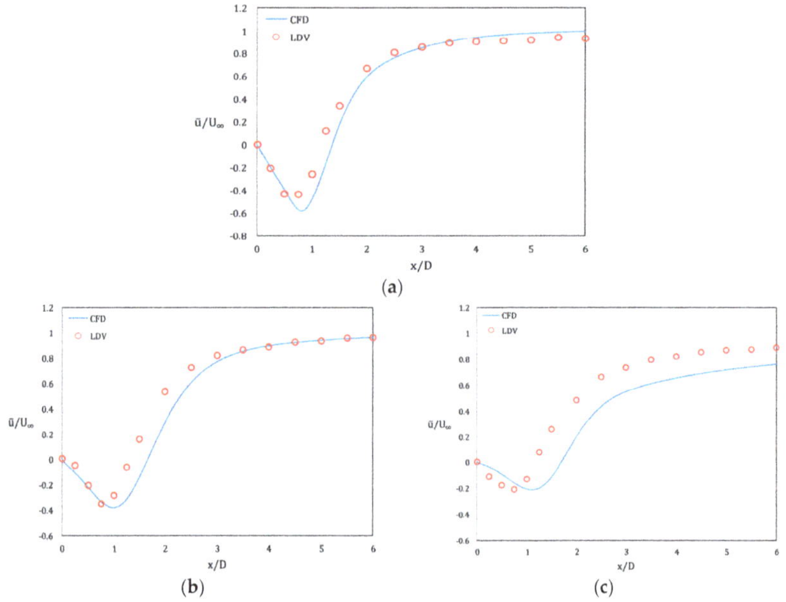

Figure 3. Wake centerline velocity at (**a**) $z/D = -0.5$; (**b**) $z/D = 0.0$; and (**c**) $z/D = 0.5$.

Table 2. Wake length behind the step cylinder.

Description	Position (*z*/*D*)	Case	Position (*x*/*D*)	Wake Length (mm)
Downward 50 mm	−0.5	CFD	1.4	139
		LDV	1.25	125
Along step	0	CFD	1.7	169
		LDV	1.5	150
Upward 50 mm	0.5	CFD	1.76	176
		LDV	1.25	125

The wake behind the cylinder can be qualitatively visualized by the average of streamwise velocity as derived in the simulations, as seen in Figure 4. The wake is symmetrical in the horizontal plane (Figure 4a) which suggests that the velocity field was averaged for a sufficiently long time; the bow wake [7] in front of the cylinder is also captured. When scrutinizing the vertical plane (Figure 4b) the wake is longer down-stream the smaller cylinder. It can also be seen that within the wake behind the step cylinder a recirculation region develops immediately after the cylinder where reverse flow ($\bar{u} < 0$) is observed (Figures 3 and 4), which extends roughly one diameter downstream of the cylinder. Mean velocity, $\bar{u}$ rapidly approaches the free stream velocity U_∞ after the recirculation region. The velocity after the wake is not capable of recovering completely as the drag-induced momentum deficit must remain constant behind the cylinder (see Figure 3). The continuum is, instead, reserved by a slight increase in the level of the surface.

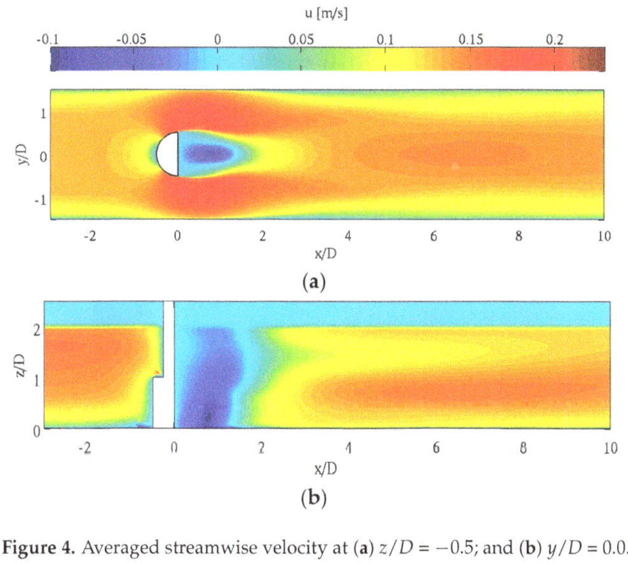

Figure 4. Averaged streamwise velocity at (**a**) $z/D = -0.5$; and (**b**) $y/D = 0.0$.

The wake structures have also been visually observed from the side of a semi-circular cylinder with a step. For instance, the structures of a wake were found behind the step cylinder at $z/D = -0.5$ with a 1 s interval (t, $t + 1$ and $t + 1 + 1$) are presented in Figure 5a–c, respectively. Dyes (red and green) were injected from the both sides of the step cylinder. It is seen that the dyes were swept into the well-known von-Karman vortex street right after downstream of the cylinder. Additionally, three-dimensional (3D) flow structures were also observed behind the cylinder, and a mainly downward flow was noticed.

(a)

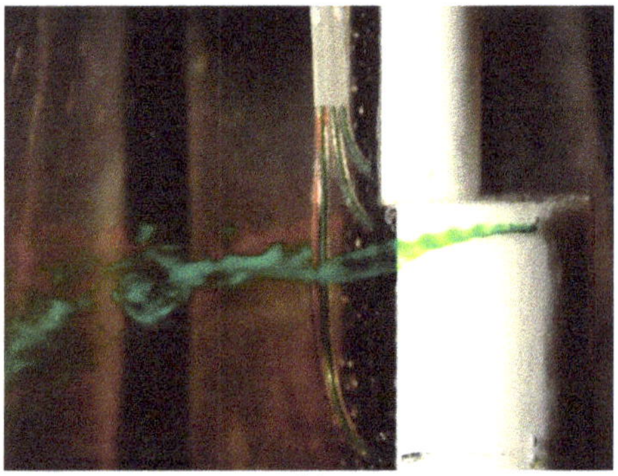

(b)

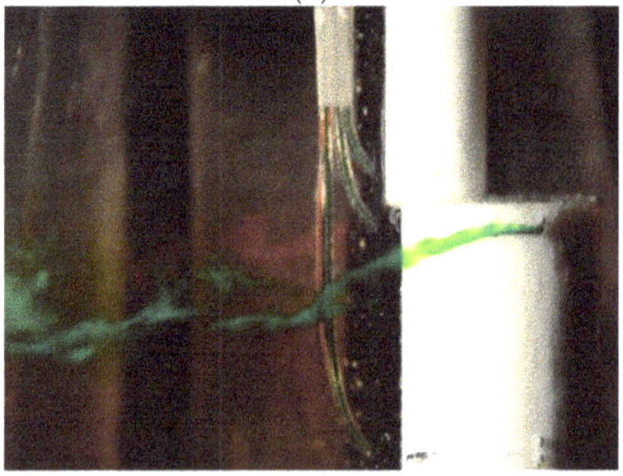

(c)

Figure 5. Wake structures around step cylinder at $z/D = -0.5$; (**a**) $[t = 0]$; (**b**) $[t + 1]$; (**c**) $[t + 1 + 1]$.

4.2. Vortex Shedding

The vortex shedding behind the cylinder cannot be described with averaged data. Therefore, Figure 6 shows an instant velocity field captured at the final time step of the simulation (t = 87.5 s) for the horizontal planes in the center of the large cylinder, small cylinder, and at the cylinder step, respectively. As seen there is a clear shedding for all planes. The impacts of channel blockage, i.e., the ratio of the cylinder diameter (D or d) to channel width (W) should be taken into consideration. The blockage ratio, D/W and d/W, are 0.34 and 0.17 for large and small cylinders, respectively, in this study.

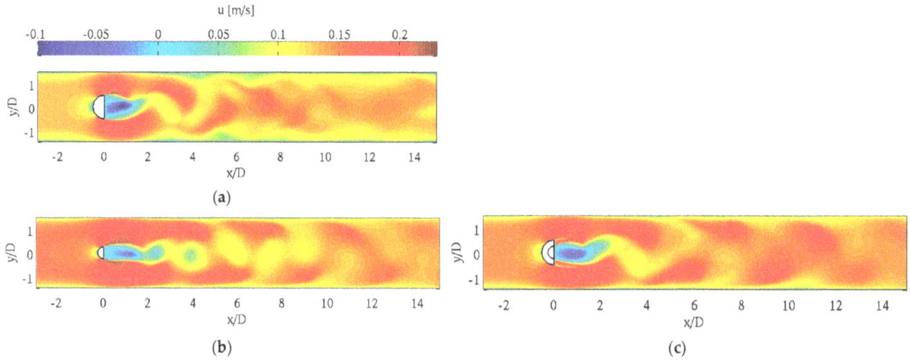

Figure 6. Streamwise velocity of the final time step (t = 87.5 s) at (**a**) $z/D = -0.5$; (**b**) $z/D = 0.5$; and (**c**) $z/D = 0.0$.

The shedding can also be studied in the vorticity field, for instance, Figure 7 shows the vertical vorticity component at one of the last time steps (t = 86.5 s). Counter-rotating vortices can be seen for all three planes and in the plane at the step there seems to be some interaction between the vortices shed from the small and large cylinder. Visualization experiments were also carried out for $z/D = -0.5$ and $z/D = 0.5$ to verify the vorticity obtained from CFD. Figure 7 depicts the comparison of these with CFD at a similar stage of vortex shedding and it is shown that there is a qualitative agreement between experiments and simulations.

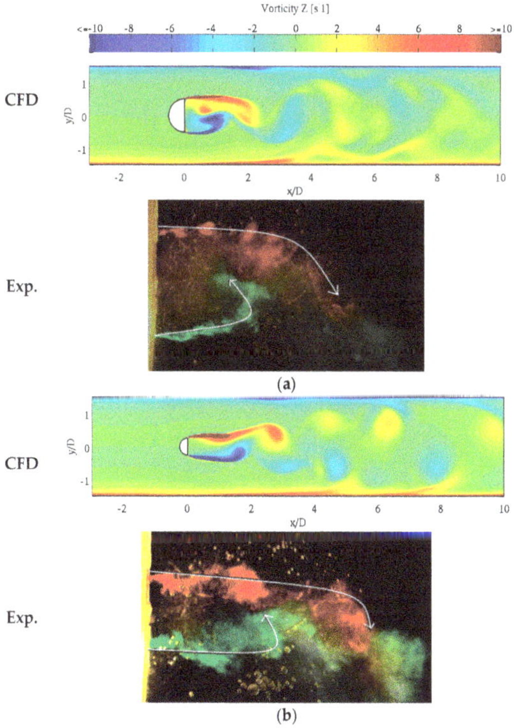

Figure 7. *Cont.*

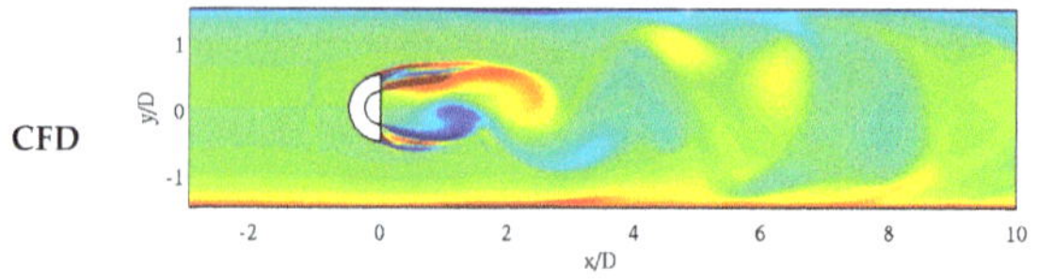

(c)

Figure 7. Vertical vorticity component (t = 86.5 s) at (**a**) $z/D = -0.5$; (**b**) $z/D = 0.5$; and (**c**) $z/D = 0.0$ along with the experimental visualization for $z/D = -0.5$ and $z/D = 0.5$.

Another interesting variable is the fluid pressure. Figure 8 shows the pressure in the final time step at $z/D = -0.5$ and $z/D = 0.5$. Zones where the pressure is lower than the surrounding pressure can be found throughout the vortex street, which could be of some importance when predicting and describing fish movement in this type of flow.

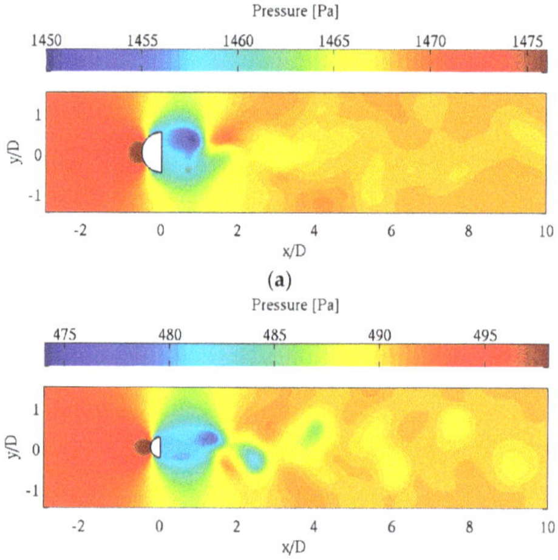

(b)

Figure 8. Fluid pressure at the final time step (t = 87.5 s) at (**a**) $z/D = -0.5$; and (**b**) $z/D = 0.5$.

The shedding frequency (f) was determined from the velocity spectra using the fast Fourier transform (FFT). The concept of LDV is that measurements are carried out if a particle travels through the volume of measurement, meaning that often there is, no signal available [33–35]. Thus, sampling data from LDV are random offering challenges for the estimation of the power spectrum. Nobach [29,31,36] described the details of up-to-date available methods to analyze the frequency spectrum from arbitrarily sampled data acquired by LDV measurements.

The estimation of power spectra presented in this paper is the interpolation of the arbitrarily or randomly sampled LDV data to obtain a uniform velocity over time, which is then re-sampled in equal steps with a known sampling frequency close to the average sampling frequency. The mean velocity for each signal was then subtracted as the fluctuations around the mean u' describing the periodicity. The dominant peak in each spectrum is then assumed to correspond to the vortex shedding frequency (f).

The shedding frequency for the large cylinder and step are evaluated at $x/D = 2.5$, $y/D = 0.75$ and the shedding frequency from the small cylinder at $x/d = 2.5$ and $y/d = 0.75$, see Figure 9. For the shedding behind the large cylinder the dominating peak gives f = 0.34 Hz for the experiments and f = 0.38 Hz for the simulations corresponding to Strouhal numbers $St_{LDV} = fD/U_\infty \approx 0.234$ and

$St_{CFD} \approx 0.262$ which can be considered to be in good agreement. For the shedding behind the smaller cylinder the simulation gives the same frequency as behind the large cylinder whereas the LDV gives a peak at $f = 0.6$ Hz which is more consistent with other studies [37–39] behind fully-circular stepped cylinders. For the shedding behind the step, the experiments yield two dominant peaks, one at $f = 0.36$ Hz, which should correspond to the shedding from the large cylinder and $f = 0.6$ Hz which corresponds to the shedding from the small one, whereas the simulation once again gives the same frequency as for the large cylinder, not capturing the dynamics of the small cylinder or the step. This may be traced to the SST turbulence model's inability to capture the shedding correctly; other possible contributors could be that the time-step is too large during the experiments so that some spatial or temporal details are neglected or that the flow at the smaller cylinder is not yet fully developed even after 3500 iterations. In addition, the effect of blockage on vortex shedding for the flow past a step cylinder can also be significant. Some previous studies [37,38,40–43] have shown that the effect of channel blockage on vortex shedding for the flow past a step cylinder can also be significant as the channel blockage affects f, consequently influencing St. The increase of shedding frequency with the increase of the blockage ratios is evidently observed in these spectral results (Figure 6). Therefore, channel blockage could also be the reason on varying the shedding frequency behind step cylinder (large and small cylinders).

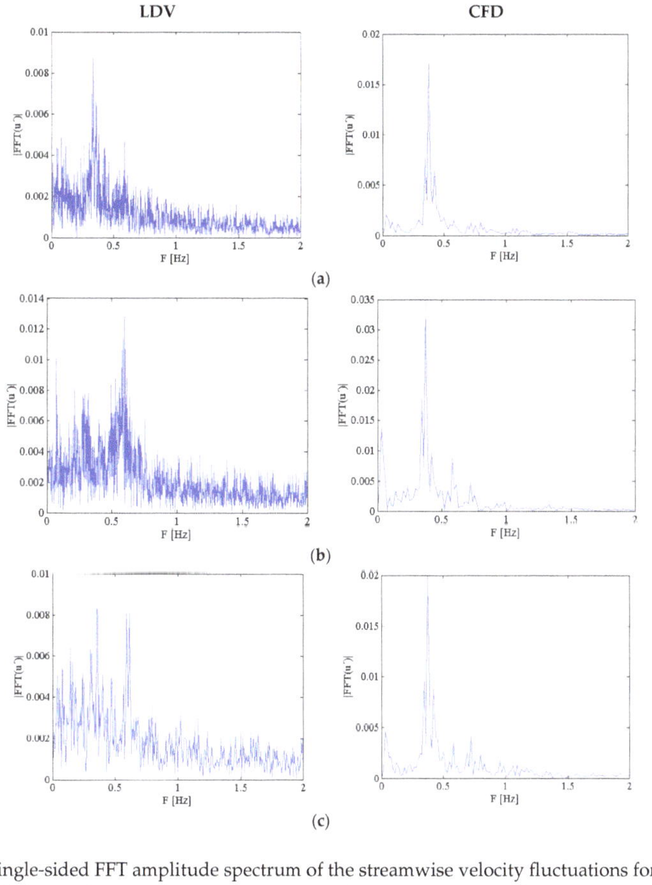

Figure 9. Single-sided FFT amplitude spectrum of the streamwise velocity fluctuations for LDV and CFD at (**a**) the large cylinder, $x/D = 2.5$, $z/D = -0.5$, $y/D = 0.75$ (**b**) small cylinder, $x/d = 2.5$, $z/D = 0.5$, $y/d = 0.75$; and (**c**) diameter step, $x/D = 2.5$, $z/D = 0.0$, $y/D = 0.75$.

5. Concluding Remarks

An investigation with LDV measurements and CFD simulations behind a semi-circular cylinder with a step has been conducted. Additional flow visualization was also carried out to explore the wake structure. A summary of key features of the wake flows obtained in this study is given below:

1. The main result is that the measured vortex shedding frequency is, $f = 0.34$ Hz behind the large cylinder and $f = 0.6$ Hz behind the small cylinder, while it has two dominating frequencies across the step, $f = 0.36$ Hz and $f = 0.6$ Hz. This implies that a step cylinder can be used to attract fish of different sizes and facilitate their motion up-stream the fishway.
2. Experiments yield that the wake length is longer along the step ($z/D = 0.0$) and it reduces for the large cylinder ($z/D = -0.5$), as well as the small cylinder ($z/D = 0.5$). The simulation over-predicts all of these wake lengths.
3. Flow visualization finds the development of three-dimensional (3D) flow structures along with well-known Von Karman vortex street.
4. Simulation results show acceptable agreement with the experimental measurements behind the large cylinder, but fail to capture the dynamics of the small cylinder diameter.

Future studies with tomographic-PIV would enable the detailed vortex dynamics of the flow behind the stepped half-cylinders to be investigated. Additional simulations with more advanced models, such as the Reynolds stress models and large eddy simulations, could also provide additional insights into this type of problem.

For the application of fish migration, it would also be interesting to test these configurations on fish in laboratory studies, as well as in the field, to see how different species and sizes of fish respond to different geometries and flow conditions.

Acknowledgments: This work has been funded by the collaboration initiative *StandUp for Energy*. The research program is a part of the Swedish government's commitment to high quality research in areas of strategic importance.

Author Contributions: S. M. Sayeed-Bin-Asad, Tord Staffan Lundström and Anders Gustav Andersson conceived and designed the experiments; S. M. Sayeed-Bin-Asad performed the experiments and analyzed the data; Anders Gustav Andersson performed the simulation; S. M. Sayeed-Bin-Asad wrote the paper; All authors read and approved the final manuscript.

Conflicts of Interest: The authors declare no conflict of interest.

References

1. Swedenergy. Svensk Energi—Swedenergy—AB. Available online: http://www.svenskenergi.se/ (accessed on 12 October 2016).
2. Kaveh, A. Experimental Investigation of a Kaplan Runner under Steady-State and Transient Operations. Ph.D. Thesis, Luleå University of Technology, Luleå, Sweden, 2016.
3. Liao, J.C.; Beal, D.N.; Lauder, G.V.; Triantafyllou, M.S. Fish exploiting vortices decrease muscle activity. *Science* **2003**, *302*, 1566–1569. [CrossRef] [PubMed]
4. Liao, J.C. A review of fish swimming mechanics and behaviour in altered flows. *Philos. Trans. R. Soc. B Biol. Sci.* **2007**, *362*, 1973–1993. [CrossRef] [PubMed]
5. Stewart, W.J.; Tian, F.B.; Akanyeti, O.; Walker, C.J.; Liao, J.C. Refuging rainbow trout selectively exploit flows behind tandem cylinders. *J. Exp. Biol.* **2016**, *219*, 2182–2191. [CrossRef] [PubMed]
6. Akanyeti, O.; Thornycroft, P.J.M.; Lauder, G.V.; Yanagitsuru, Y.R.; Peterson, A.N.; Liao, J.C. Fish optimize sensing and respiration during undulatory swimming. *Nat. Commun.* **2016**, *7*, 11044. [CrossRef] [PubMed]
7. Sayeed-Bin-Asad, S. Velocity Distribution Measurements in A Fishway Like Open Channel by Laser Doppler Anemometry (LDA). *EPJ Web. Conf.* **2016**. [CrossRef]
8. Liao, J.C.; Beal, D.N.; Lauder, G.V.; Triantafyllou, M.S. The Kármán gait: Novel body kinematics of rainbow trout swimming in a vortex street. *J. Exp. Biol.* **2003**, *206*, 1059–1073. [CrossRef] [PubMed]

9. Dunn, W.; Tavoularis, S. Vortex shedding from a step-cylinder in spanwise sheared flow. *Phys. Fluids* **2011**, *23*, 035109. [CrossRef]
10. Dunn, W.; Tavoularis, S. Experimental studies of vortices shed from cylinders with a step-change in diameter. *J. Fluid Mech.* **2006**, *555*, 409–437. [CrossRef]
11. Morton, C.; Yarusevych, S. Vortex shedding in the wake of a step cylinder. *Phys. Fluids* **2010**, *22*, 083602. [CrossRef]
12. Morton, C.; Yarusevych, S. Three-dimensional flow and surface visualization using hydrogen bubble technique. *J. Vis.* **2015**, *18*, 47–58.
13. Morton, C.; Yarusevych, S. On vortex shedding from low aspect ratio dual step cylinders. *J. Fluids Struct.* **2014**, *44*, 251–269. [CrossRef]
14. Morton, C.; Yarusevych, S. Vortex dynamics in the turbulent wake of a single step cylinder. *J. Fluids Eng.* **2014**, *136*, 031204. [CrossRef]
15. McClure, J.; Morton, C.; Yarusevych, S. Flow development and structural loading on dual step cylinders in laminar shedding regime. *Phys. Fluids* **2015**, *27*, 063602. [CrossRef]
16. Morton, C.R.; Yarusevych, S. A combined experimental and numerical study of flow past a single step cylinder. In Proceedings of the ASME 2010 3rd Joint US-European Fluids Engineering Summer Meeting Collocated with 8th International Conference on Nanochannels, Microchannels, and Minichannels, Montreal, QC, Canada, 1–5 August 2010; American Society of Mechanical Engineers: New York, NY, USA, 2010.
17. Morton, C.; Yarusevych, S.; Scarano, F. A tomographic particle image velocimetry investigation of the flow development over dual step cylinders. *Phys. Fluids* **2016**, *28*, 025104. [CrossRef]
18. Rafati, S. *Investigating Step Cylinder Wake Topology with Tomographic PIV*; Delft University of Technology: Delft, The Netherlands, 2014.
19. Ko, N.; Leung, W.; Au, H. Flow behind two coaxial circular cylinders. *J. Fluids Eng.* **1982**, *104*, 223–227. [CrossRef]
20. Yagita, M.; Kojima, Y.; Matsuzaki, K. On vortex shedding from circular cylinder with step. *Bull. JSME* **1984**, *27*, 426–431. [CrossRef]
21. Lewis, C.; Gharib, M. An exploration of the wake three dimensionalities caused by a local discontinuity in cylinder diameter. *Phys. Fluids A Fluid Dyn.* **1992**, *4*, 104–117. [CrossRef]
22. Norberg, C. An experimental study of the flow around cylinders joined with a step in diameter. In Proceedings of the 11th Australasian Fluid Mechanics Conference, Hobart, Australia, 14–18 December 1992.
23. Ko, W.-M.; Chan, S.-K. *Pressure Distributions on Circular Cylinders with Stepwise Change of the Diameter*; American Society of Mechanical Engineers: New York, NY, USA, 1984.
24. Ko, N.; Chan, A. In the intermixing region behind circular cylinders with stepwise change of the diameter. *Exp. Fluids* **1990**, *9*, 213–221. [CrossRef]
25. Ko, N.; Chan, A. Wakes behind circular cylinders with stepwise change of diameter. *Exp. Therm. Fluid Sci.* **1992**, *5*, 182–187. [CrossRef]
26. Chua, L.; Liu, C.; Chan, W. Measurements of a step cylinder. *Int. Commun. Heat Mass Transf.* **1998**, *25*, 205–215. [CrossRef]
27. Morton, C.; Yarusevych, S. Modeling Flow over a Circular Cylinder with a Stepwise Discontinuity. In Proceedings of the 39th AIAA Fluid Dynamics Conference, San Antonio, TX, USA, 22–25 June 2009.
28. Morton, C.; Yarusevych, S. Cross flow over cylinders with two stepwise discontinuities in diameter. In Proceedings of the Turbulence and Shear Flow Phenomena Conference, Ottawa, ON, Canada, 28–30 July 2011.
29. Nobach, H. Present Methods to Estimate the Cross-Correlation and Cross-Spectral Density for Two-Channel Laser Doppler Anemometry. In Proceedings of the 18th International Symposium on the Application of Laser and Imaging Techniques to Fluid Mechanics, Lisbon, Portugal, 4–7 July 2016.
30. Norberg, C. LDV-Measurements in the Near Wake of a Circular Cylinder. Available online: http://www.ht.energy.lth.se/fileadmin/ht/Norberg-FEDSM98-5202.pdf (accessed on 12 October 2016).
31. Hirt, C.W.; Nichols, B.D. Volume of fluid (VOF) method for the dynamics of free boundaries. *J. Comput. Phys.* **1981**, *39*, 201–225. [CrossRef]
32. Menter, F.R. Review of the shear-stress transport turbulence model experience from an industrial perspective. *Int. J. Comput. Fluid Dyn.* **2009**, *23*, 305–316. [CrossRef]

33. Morton, C.; Yarusevych, S. An experimental investigation of flow past a dual step cylinder. *Exp. Fluids* **2012**, *52*, 69–83. [CrossRef]
34. Velte, C.M.; George, W.K.; Buchhave, P. Estimation of burst-mode LDA power spectra. *Exp. Fluids* **2014**, *55*, 1–20. [CrossRef]
35. Müller, E.; Nobach, H.; Tropea, C. A refined reconstruction-based correlation estimator for two-channel, non-coincidence laser Doppler anemometry. *Meas. Sci. Technol.* **1998**, *9*, 442. [CrossRef]
36. Nobach, H. Laser Doppler Data Processing. Available online: http://ldvproc.nambis.de/programs/pyLDV.html (accessed on 10 October 2016).
37. Turki, S.; Abbassi, H.; Nasrallah, S.B. Effect of the blockage ratio on the flow in a channel with a built-in square cylinder. *Comput. Mech.* **2003**, *33*, 22–29. [CrossRef]
38. Zong, L.; Nepf, H. Vortex development behind a finite porous obstruction in a channel. *J. Fluid Mech.* **2012**, *691*, 368–391. [CrossRef]
39. Morton, C.R. Experimental and Numerical Investigations of the Flow Development over Circular Cylinders with Stepwise Discontinuities in Diameter. Master's Thesis, University of Waterloo, Waterloo, ON, Canada, 2010.
40. Sahin, M.; Owens, R.G. A numerical investigation of wall effects up to high blockage ratios on two-dimensional flow past a confined circular cylinder. *Phys. Fluids (1994–Present)* **2004**, *16*, 1305–1320. [CrossRef]
41. Mettu, S.; Verma, N.; Chhabra, R. Momentum and heat transfer from an asymmetrically confined circular cylinder in a plane channel. *Heat Mass Transf.* **2006**, *42*, 1037–1048. [CrossRef]
42. Okajima, A.; Yi, D.; Sakuda, A.; Nakano, T. Numerical study of blockage effects on aerodynamic characteristics of an oscillating rectangular cylinder. *J. Wind Eng. Ind. Aerodyn.* **1997**, *67*, 91–102. [CrossRef]
43. Park, C.-W.; Lee, S.-J. Free end effects on the near wake flow structure behind a finite circular cylinder. *J. Wind Eng. Ind. Aerodyn.* **2000**, *88*, 231–246. [CrossRef]

energies

Article

Numerical Investigation of Periodic Fluctuations in Energy Efficiency in Centrifugal Pumps at Different Working Points

Hehui Zhang [1], Shengxiang Deng [1,*] and Yingjie Qu [2]

[1] School of Energy Science and Engineering, Central South University, Changsha 410083, China; hehuizhang@csu.edu.cn

[2] Hunan M &W Energy Saving Tech. Co., Ltd., Changsha 410208, China; pump118@163.com

* Correspondence: csdsx@163.com; Tel.: +86-731-8887-9863

Academic Editor: Bjørn H. Hjertager

Received: 10 December 2016; Accepted: 2 March 2017; Published: 10 March 2017

Abstract: In order to simulate the energy efficiency fluctuation behavior of an industrial centrifugal pump with a six-blade impeller, a full-scale three-dimensional (3D) an unsteady state computational fluid dynamics (CFD) model was used. Five operational points with different flow fluxes were numerically investigated by using the Navier–Stokes code with shear-stress transport (SST) k-ω turbulence model. The predicted performance curves agreed well with the test data. A sine function was fitted to the transient calculation results and the results show that the efficiency fluctuates mainly on the blade passing frequency, while the fluctuation level varies with flow rate. Furthermore, high efficiency is not necessarily associated with low fluctuation level. The efficiency fluctuation level is high at part-load points, and becomes relatively low when flow rate exceeds the design value. The effect of change in torque is greater than that of the head lift with respect to fluctuations of efficiency. Based upon the analysis of velocity vector distribution of different impeller phase positions, a hypothesis which considers both the effect of pump's structural shape and flow fluxes was proposed to explain the above behavior by analyzing the impeller–tongue interaction. This work enriches the theoretical system of flow parameters fluctuation of centrifugal pump, and provides useful insight for the optimal design of centrifugal pumps.

Keywords: centrifugal pump; energy efficiency; computational fluid dynamics; fluctuation

1. Introduction

The centrifugal pump is a kind of energy conversion device widely used in various process industries. However, it consumes large amounts of energy for fluid transportation and pressure enhancement, making energy efficiency one of its most important performance indicators [1]. A great deal of research has been done on centrifugal pumps for improving their efficiency under rated conditions, and for expanding the range of the high-efficiency region. Research has mainly focused on the optimization of inner flow field in centrifugal pumps. Generally, computational fluid dynamics (CFD) methods are used for flow field analysis and prediction of performance of centrifugal pumps due to their low cost, quick development speed and convenience for comparing different design schemes [2–4]. According to a number of published reports, flow field structures and performance curves predicted by CFD methods are the predominant basis for optimization of pump product design to obtain higher efficiencies [5–8]. For example, Shi et al. [8] numerically investigated the quantitative relationship between efficiency and impeller design parameters of a forward-extended double-blade sewage pump, and obtained a response surface of the highest efficiency under different design factors through regression analysis. In their work, the change in efficiency predicted by CFD calculations under all factors and levels set by orthogonal table was less than 5% [8].

Generally, most CFD research is aimed at improving the efficiency of the centrifugal pump and is based on steady state models, in which the impeller region with a fixed phase position is set in rotating reference frame to simulate its rotation. In this case, the possibility of a change in efficiency under different impeller phase positions is completely neglected. In fact, the inside flow field fluctuation of centrifugal pump is obvious due to the interaction between impeller blade and volute tongue, which has been popular field of research in recent years. Flow field information such as secondary flow structure changes and pressure fluctuations are obtained through unsteady CFD calculations, while sliding mesh technique for the impeller region is used to reflect the real dynamic effect of the rotating impeller. As early as 2002, González et al. [9] systematically studied the dynamic effects of the impeller–volute interaction of a centrifugal pump on the inside flow by using 3D transient flow numerical simulations, which was focused on the change in flow pattern with different impeller circumferential positions on both design and off-design points. On each operational point, the pressure inside the volute fluctuated mainly at the blade passing frequency, whereas the fluctuation amplitude and irregularity were larger on the off-design point than those on the design point, which led to corresponding force changes on the main shaft of the pump. Also in 2002, Treutz [10] performed transient flow analysis in centrifugal pumps with commercial CFD code CFX, and validated the computation results with experimental measurements including pump characteristics and local pressure fluctuation in the volute casing. Treutz's work focused more on the development of a transient simulation method and the detailed analysis of the secondary flow field itself, but seldom involved the energy efficiency fluctuation which is likely to be caused by the interaction between volute casing and the blades within it. In 2003, Guo et al. [11] experimentally studied the fluctuation behavior of fluid forces on shaft, and reported that there was no necessary correlation for fluctuation amplitude between pressure and fluid forces. Later, around the inside flow field fluctuation caused by the impeller–volute interaction, more experimental and numerical studies were undertaken to describe the transient flow phenomena and explain the formation mechanism behind it (e.g., Cheah et al. [12] in 2008, Lucius et al. [13] in 2010, Huang et al. [14] and Dai et al. [15] in 2013). The unsteady state analysis on fluctuation of flow parameters, such as pressure and hydraulic forces, can provide an essential reference for blade design [16,17], flow rate setup [18,19] and impact assessment under special working conditions [20].

The hydraulic efficiency of centrifugal pump merely depends on the values of shaft torque and head lift under constant flow rate and rotation speed, while these parameters are controlled by fluid forces and pressure. Since there is obvious fluctuation of fluid forces and pressure, it is obvious that the efficiency will also show some periodic trends of fluctuation. However, compared with the considerable research on the fluctuation of fluid forces and pressure, few reports are found on the fluctuation of efficiency. Thus, the major objective in this article is to fill the gap in research on the efficiency fluctuation behavior of centrifugal pumps. A full-scale 3D numerical simulation of unsteady flow was carried out to study an industrial centrifugal pump with six blades, and the time-varying flow parameters were recorded to investigate the efficiency fluctuation behavior at five different operational points.

2. Numerical Procedure

2.1. Pump Geometry and Design Parameters

An industrial horizontal double-entry split centrifugal pump with six blades (manufactured by Hunan M & W Energy Saving Tech. Co. Ltd., Changsha, China) was studied in this work. The pump was designed to work with clean water with the highest temperature of 65 °C. The inner hydraulic region of pump was modeled to be a set of three different parts: a suction chamber, an extruding chamber and the impeller between the above two chambers. The 3D geometry of the hydraulic model is shown in Figure 1, while the main structure and performance parameters of the pump are listed in Table 1. Both the suction and exit centers are located at the same horizontal level.

Figure 1. 3D geometry of inner fluid region of the pump.

Table 1. Important design parameters of the pump studied in this work.

Structural Parameter	**Value**
Suction diameter	$D_s = 0.35$ m
Exit diameter	$D_e = 0.35$ m
Impeller inlet diameter	$d_1 = 0.08$ m
Impeller outlet diameter	$d_2 = 0.29$ m
Impeller outlet width	$b_2 = 0.136$ m
Rated flow	$Q_{des} = 0.35\ m^3/s$
Rated head	$H_{des} = 17$ m
Rotation speed	$n = 1480$ rpm
NPSHr	$\Delta h_r = 3.2$ m

To better understand the inner structure of the volute and impeller, a middle section of the volute fluid region and impeller are illustrated in Figure 2. The plane shown in Figure 2 is perpendicular to the main shaft of the pump. An X-Y rectangular coordinate system is given on the plane, where the original point of the coordinate system and the center point of the impeller coincide with each other. Both the volute tongue and exit face are in the fourth quadrant of the coordinate system. As shown in Figure 2, the impeller rotates anticlockwise, and the rotary phase is set to be the zero phase position when the impeller center, one blade tip and the volute tongue vertex are located on the same straight line.

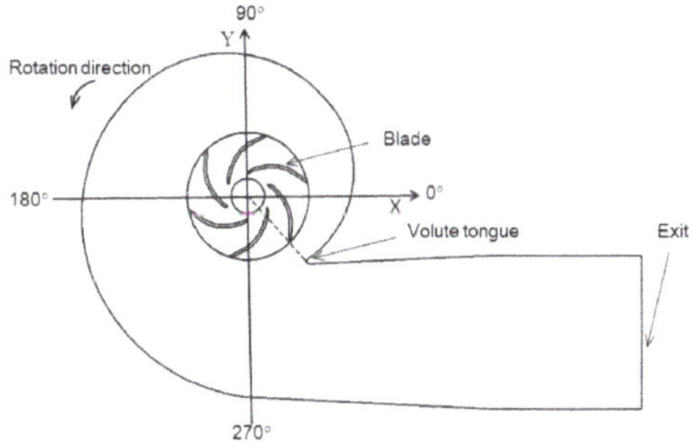

Figure 2. Mid-plane view of the inner fluid region of the pump.

2.2. Numerical Model and Sliding Technology

The commercial CFD software Fluent 14.0 was used to perform the calculations for 3D full-scale model by solving Reynolds averaged incompressible Navier–Stokes equations based on finite volume method. The mass and momentum conservation equations are given as:

$$\frac{\partial \rho}{\partial t} + \nabla \cdot (\rho \vec{u}) = 0 \tag{1}$$

$$\frac{\partial}{\partial t}(\rho\vec{u}) + \nabla\cdot(\rho\vec{u}\vec{u}) = -\nabla p + \nabla\cdot[\mu(\nabla\vec{u} + (\nabla\vec{u})^{T})] + \rho\vec{g} + \vec{f} \tag{2}$$

where $\vec{u}$ is the velocity vector ($m\cdot s^{-1}$), t represents time (s), ρ is density ($kg\cdot m^{-3}$), p is static pressure (Pa), $\vec{f}$ is external body force term ($N\cdot m^{-3}$), $\vec{g}$ is the gravity constant ($m\cdot s^{-2}$) and μ is the effective viscosity ($kg\cdot m^{-1}\cdot s^{-1}$).

The full model was made up of two different kinds of zones: the stationary zone for the suction and extruding chambers, and the motion zone for the impeller region. In the stationary zone, for a general scalar, ϕ, in an arbitrary control volume, V (m^3), the convection–diffusion-reaction (CDR) equation in its integral form can be written as Equation (3):

$$\frac{d}{dt}\int_{V}\rho\phi dV + \int_{\partial V}\rho\phi\vec{u}\cdot d\vec{A} = \int_{\partial V}\Gamma\nabla\phi\cdot d\vec{A} + \int_{V}S_{\phi}dV \tag{3}$$

where $\vec{A}$ is the area vector of the faces on the control volume (m^2), ϕ is the generic variable for solving the equation, and Γ and S_ϕ are the diffusion coefficient and source term of ϕ, respectively. In the equation of continuity, ϕ is set as 1 and Γ is 0. However, in the momentum equation, ϕ is the velocity component and Γ is the dynamic viscosity ($kg\cdot m^{-1}\cdot s^{-1}$). From left to right, the four items in Equation (3) represent the mass variation over time, the convection term, the diffusion term and the source term, respectively.

For the motion zone, Fluent 14.0 provided a dynamic mesh model with sliding technology [21] to better reflect the real fluid flow under the impeller–tongue interaction. With this technology, all the nodes in impeller region rotate rigidly around the center, and the dynamic mesh zone gets connected to the stationary zone of extruding chamber through non-conformal interfaces. When the mesh rotation is updated each time, the non-conformal interfaces are also likewise updated, so the new phase difference between the impeller and tongue can be truly reflected. The general transport equation in Equation (3) also applies to the dynamic mesh model, except for the convection term, which is left out due to the consideration of the mesh's motion. Therefore, the conservation equation for the above general scalar, ϕ, in the motion zone can be represented by Equation (4) [21].

$$\frac{d}{dt}\int_{V}\rho\phi dV + \int_{\partial V}\rho\phi(\vec{u} - \vec{u}_{g})\cdot d\vec{A} = \int_{\partial V}\Gamma\nabla\phi\cdot d\vec{A} + \int_{V}S_{\phi}dV \tag{4}$$

where $\vec{u}_g$ is the mesh velocity vector of the moving mesh ($m\cdot s^{-1}$).

As there is no shape change for all of the mesh cells during the rotation of impeller, the control volume V remains constant, and therefore the time derivative term in Equation (4) can be simplified to Equation (5) by using first-order backward difference.

$$\frac{d}{dt}\int_{V}\rho\phi dV = \frac{(\rho\phi V)_{n+1} - (\rho\phi V)_{n}}{\Delta t} \tag{5}$$

where n denotes the respective value at the nth time level, and $n + 1$ represents its next time level.

For the turbulence model, the k-ω-based shear-stress transport (SST) model developed by Menter [22] was used to solve the mass and momentum conservation equations. The k-ω-based SST model is especially well-suited for the turbulence problems occurring in high rotational speed turbomachinery due to its reasonable balance between the computational cost and accuracy. In the far field region with free shear flow, the standard k-ε model is used to speed up the calculations, while in the near-wall region with intense separation flow induced by the multi-curvature revolving blade walls, the turbulence model is gradually transformed into k-ω model to ensure the resolution of the analysis. Details of the coefficients of turbulence model can be found in a previously published report [22].

The calculation procedure was separated into two stages. The first stage was for a steady model, while the second was for an unsteady model. There were five different cases to be computed in this study, while each one corresponded to a certain working condition. The flow rates for the five cases were $0.6Q_{des}$, $0.8Q_{des}$, $1.0Q_{des}$, $1.2Q_{des}$ and $1.4Q_{des}$, respectively. The steady model was computed in advance to validate the calculation results with test data and then provide the initial condition for subsequent transient numerical simulation, where the impeller was located on the zero phase position. In the steady model, the full fluid region consists of two different reference frames. These reference frames were the stationary frame for the suction and extruding chambers, and the rotating frame with a rotating speed of 1480 $r \cdot min^{-1}$ for the impeller zone. A specified flow velocity was set on the suction face as the inlet boundary condition with flow direction perpendicular to the face, and the turbulence intensity was set at 5%. An outflow condition was given on the exit face of the extruding chamber to capture the fully developed pressure and velocity distribution. All interior faces of the fluid region consisted of no-slip rough walls, and according to the specifications of the processing technology, the roughness of the blades and other faces were set to be 5×10^{-5} m and 10^{-4} m, respectively. The internal leakage of pump was ignored for simplification. Physical property parameters of pure water, including density and dynamic viscosity at 25 °C were given as the fluid properties. The convergence criterion was less than 10^{-4} for mean residues of both mass and momentum. The whole fluid region was meshed with hexahedral structure elements to ensure convergence and computational speed. To run the grid sensitivity test, seven computational cases with different mesh sizes but fixed flow rate $1.0Q_{des}$ were studied to obtain the changes in head and efficiency values. Both the predicted head and efficiency values become very stable after the fourth point (see Figure 3), whose rangeability becomes less than 0.3% and 0.2% respectively, so the mesh of that point was selected for computation in this work in order to save computation time. The mesh information including elements and nodes number was listed in Table 2, and the element quality, skewness factor and orthogonal quality of the mesh was 0.73, 0.76 and 0.81 respectively. The maximum non-dimensional wall distance y+ < 60 was obtained in the full flow field of the final computation results by performing the Yplus mesh adaption refinement function [21] during numerical computations.

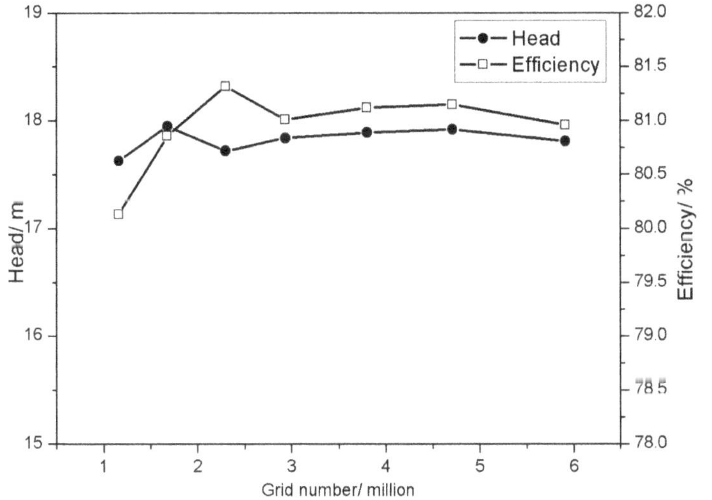

Figure 3. Predicted head values with different grid numbers.

Table 2. Mesh information obtained in the current work.

Fluid Zone	Number of Elements	Number of Nodes
Suction chamber	832,871	2,966,467
Extruding chamber	1,105,902	4,331,943
Impeller	999,772	3,474,289
Total	2,938,544	10,772,699

2.3. Transient Calculations

After steady state computation, each case was transformed into a transient one with the steady result as its initial condition. There were mainly two differences between the unsteady and steady state models. First of all, the rotating reference frame for the impeller fluid was discarded, and a sliding mesh technique was used with a mesh motion speed of 1480 $r \cdot min^{-1}$ to simulate the real fluid effect of the impeller's rotation. For the solution setup, the total calculation time was 0.2432 s, corresponding to six shaft rotation cycles. The time step Δt was defined according to the Equation (6).

$$\Delta t = \frac{T}{360} \tag{6}$$

where T is the rotation period of the main shaft and had the value of 0.04054 s. This means that the impeller fluid region meshes rotate 1° anticlockwise for each time step. The unsteady flow calculation results of the last shaft rotation cycle were chosen for analysis when the transient iteration was fully convergent and the fluctuation in flow parameters was regular. The head lift and fluid torque on impeller were monitored during the whole computational process, and the energy efficiency for each time step was calculated using Equation (7).

$$\eta = \frac{955 \rho g Q H}{n N} \tag{7}$$

where g is the gravitational acceleration (9.8 $N \cdot kg^{-1}$), Q is the volume flow rate ($m^3 \cdot s^{-1}$), H is the head of the pump (m), n is the shaft's rotational speed ($rev \cdot min^{-1}$), and N is the torque ($N \cdot m$) on the impeller. The pump head H can be obtained using Equation (8).

$$H = \frac{P_o - P_i}{\rho g} \tag{8}$$

where P_o and P_i represent the total pressure of the exit and entrance faces of the pump, respectively.

3. Numerical Data Processing and Validation

3.1. Regression Analysis

Regression analysis was performed to quantitatively research the efficiency of the fluctuation behavior. For each working condition with a fixed flow rate, it is obvious from Equation (5) that all influencing factors on efficiency are constants except for the torque N and pump head H. According to the published experimental and simulation reports on the flow field parameters [7,8], a sine function was chosen for regression analysis of the fluctuation tendency of the torque, head and finally the efficiency in the transient calculation results. The sine function for fitting the above three flow parameters can be expressed using Equation (9).

$$f(t) = B \sin(\omega t + \varphi) + C \tag{9}$$

where B and C are peak value and displacement distance on vertical coordinate, respectively, and share the same units with the fitting object. Furthermore, ω is the angular rotational rate of the shaft ($rad \cdot s^{-1}$), t is the time (s), and φ is the initial phase of sine function (rad).

3.2. Curve Comparison after Normalization

The main aim of the regression analysis is to describe the periodic fluctuation characters quantitatively, especially to reflect the amplitude of the fluctuation. As represented by Equation (5), there are three flow parameters, namely the efficiency and its influencing factors torque and head, which will be studied by regression analysis. However, it is impossible to apply comparative study on these three flow parameters because they have different dimensions and value ranges. Thus, the normalization process was carried out to make the value ranges of different parameters less than 1. This way, it will become convenient to compare the fluctuation amplitude of different flow parameters at different working points, and hence, identify the underlying leading factors contributing to the fluctuation in the efficiency.

A method named maximum normalization was chosen to normalize the values. The method makes linear transformation of the original data according to Equation (10).

$$f(t)^* = \frac{f(t)}{f(t)_{\max}} \tag{10}$$

where $f(t)$ is the original sine function for fitting, and $f(t)^*$ is its value after normalization. Furthermore, $f(t)_{\max}$ is the maximum value ($A + B$) of the sine function. In this way, the peak value reflecting fluctuation amplitude is turned into a dimensionless normalized form $A/(A + B)$, which allows comparison among different flow parameters after the regression analysis.

3.3. Test Validation

To ensure the correctness of the CFD results, a test on a full-size pump was made according to Grade 2 of the ISO 9906-2012 standard [23], as shown in Figure 4. The pumped medium was cold clean water. Two pressure gauges with an accuracy class of 0.4 were mounted on the pipe near the pump flange, with one upstream of the entrance face while the other was downstream of the exit face. An electromagnetic flowmeter was installed on the straight pipe, and was used to measure the water flow velocity, which can be multiplied by the pipe sectional area to obtain the volumetric flow rate. A torque-meter and a laser tachometer (each having 0.2 accuracy class) were applied to record the motor torque and rotation speed, respectively.

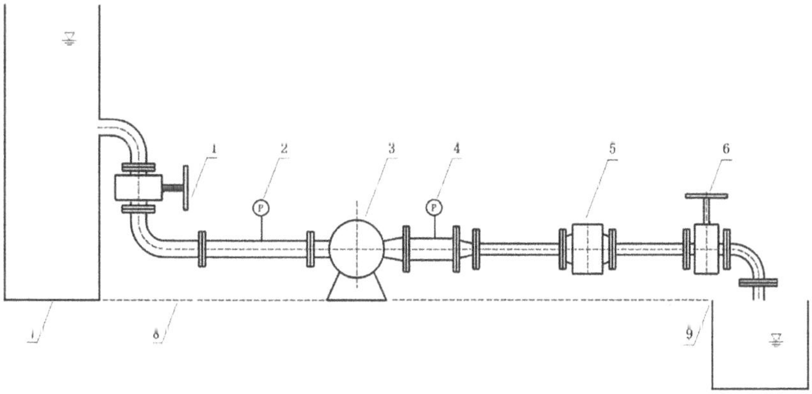

Figure 4. Schematic diagram of the test rig. 1: inlet valve; 2: vacuum gauge; 3: centrifugal pump driven by motor; 4: pressure gauge; 5: electromagnetic flowmeter; 6: outlet valve; 7: upstream tank; 8: mounting datum; 9: downstream tank.

4. Results and Discussion

4.1. Performance Curve Verification

At first, the pump performance curves including flow~head curve (Q~H) and flow~efficiency curve (Q~η) were drawn in Figure 5 to compare the calculated results with the test data. Beside the results from steady state model, the average head and efficiency values during the last shaft rotating cycle in the unsteady state model were also included in the comparison. Generally, the performance curves predicted by the CFD agreed well with the test data, and the small difference between them might have been caused by the computational and test random errors, processing deviations of flow-passing surfaces of the test pump, and the CFD model simplification. Besides, there is also mechanical power loss in experiment due to engineering factors in practice, such as the rotation of coupling, friction within bearings, friction between bearing and shaft, and friction between the sealing ring and shaft. Although the power loss due to internal fluid leakage was ignored in computation model, the CFD calculation efficiency values might still be lower than the test ones when the power loss due to friction exceeds the loss caused by internal leakage. This phenomenon can be found in Figure 5b. On the point with flow flux of $1.0Q_{des}$, the highest energy efficiency is reached, which conforms to the design specification. Calculation results from steady and transient models have the same trend, though the predicted head and efficiency values of the transient model are closer to the experimental results than those obtained from the steady state model for most of the operating points. A preliminary explanation is that the steady state model can only represent the pump performance on a certain impeller phase position, while the mean values of the transient state model reduce the deviation of fluctuation by considering all the impeller phase positions.

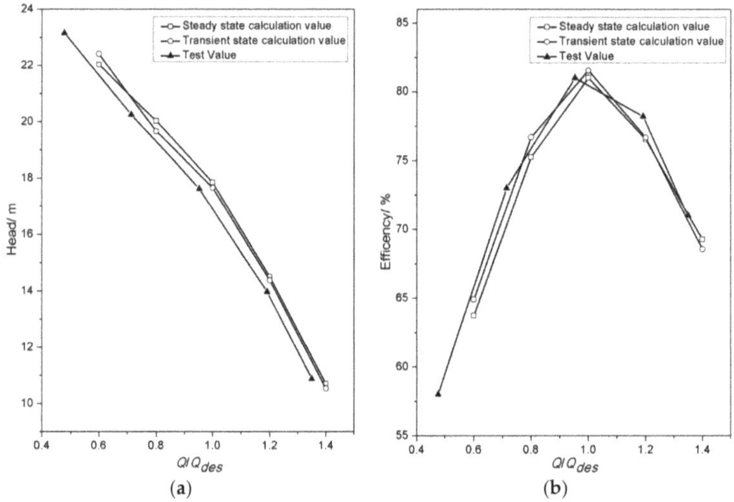

Figure 5. Comparison between the simulated and experimental results: (**a**) flow-head curve (Q~H); (**b**) flow-efficiency curve (Q~η).

4.2. Periodic Fluctuation Behavior

A detailed quantitative analysis can be made based upon the transient CFD results to discover the fluctuation behavior of the flow field. In the last shaft rotation cycle, the time-variation of the flow field parameters, including torque, head and finally efficiency, for different working conditions are shown in Figures 6–8, respectively. The red dots in these figures represent the calculated values for each time step. To better understand the fluctuation trend, the discrete time series values were fitted using the sine function. The fitting parameters are listed in Tables 3–5, and the fitted curves were drawn using black solid lines as shown in Figures 6–8. The fitting optimization index R^2 for all the curves exceeded 0.8, showing a very good fit. Since the efficiency values do not come directly from the CFD results, but from the calculation of the torque and head, the R^2 values are slightly lower for them.

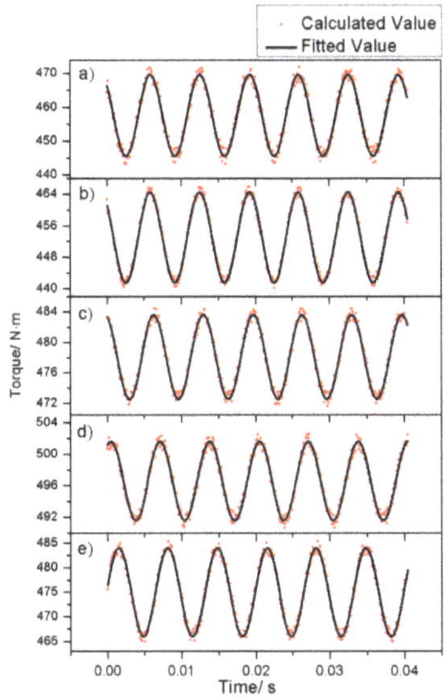

Figure 6. Fluctuation of torque in one shaft rotation cycle on different working points: (**a**) $0.6Q_{des}$; (**b**) $0.8Q_{des}$; (**c**) $1.0Q_{des}$; (**d**) $1.2Q_{des}$; (**e**) $1.4Q_{des}$.

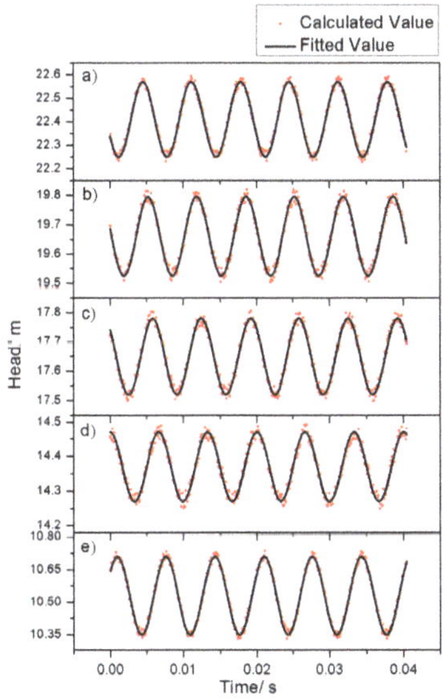

Figure 7. Fluctuation of head in one shaft rotation cycle at different working points: (**a**) $0.6Q_{des}$; (**b**) $0.8Q_{des}$; (**c**) $1.0Q_{des}$; (**d**) $1.2Q_{des}$; (**e**) $1.4Q_{des}$.

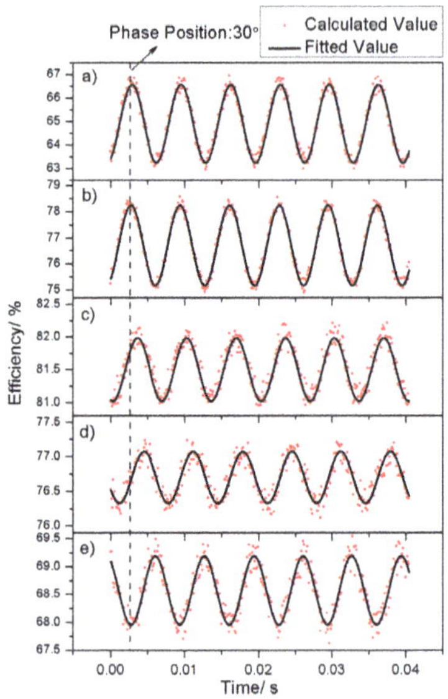

Figure 8. Fluctuation of efficiency in one shaft rotation cycle at different working points: (**a**) $0.6Q_{des}$; (**b**) $0.8Q_{des}$; (**c**) $1.0Q_{des}$; (**d**) $1.2Q_{des}$; (**e**) $1.4Q_{des}$.

Table 3. Fitting parameters for torque fluctuation in different working conditions. B: peak value on the vertical coordinate; C: displacement distance on the vertical coordinate; ω: angular rotational rate of the shaft (rad·s^{-1}); ϕ: initial phase of sine function (rad); R^2: fitting optimization index.

	$0.6Q_{des}$	$0.8Q_{des}$	$1.0Q_{des}$	$1.2Q_{des}$	$1.4Q_{des}$
B	12.10	11.50	5.51	5.04	8.96
ω	940.1	942.0	940.1	940.1	940.1
ϕ	2.33	2.34	1.93	1.15	0.18
C	457.5	453.0	478.0	497.0	475.0
R^2	0.93	0.92	0.90	0.94	0.91

Table 4. Fitting parameters for head fluctuation in different working conditions.

	$0.6Q_{des}$	$0.8Q_{des}$	$1.0Q_{des}$	$1.2Q_{des}$	$1.4Q_{des}$
B	0.159	0.135	0.130	0.101	0.180
ω	942.9	951.5	940.1	940.1	940.1
ϕ	−2.70	2.95	2.37	1.63	0.66
C	23.4	20.1	17.7	15.2	11.5
R^2	0.92	0.95	0.94	0.91	0.92

Table 5. Fitting parameters for efficiency fluctuation in different working conditions.

	$0.6Q_{des}$	$0.8Q_{des}$	$1.0Q_{des}$	$1.2Q_{des}$	$1.4Q_{des}$
B	1.66	1.54	0.484	0.374	0.617
ω	940.1	940.1	940.1	940.1	943.0
ϕ	−1.09	−0.97	−1.81	−2.64	2.16
C	64.9	76.7	81.5	76.7	68.6
R^2	0.91	0.92	0.89	0.84	0.88

Three main findings can be obtained from Figures 6–8. Firstly, the sine function periods of all curves are equal to one-sixth-fold of the shaft rotation cycle. Since it triggers obvious second flow when each blade passes near the tongue, the dominating fluctuation frequency of flow field parameters is usually the blade passing frequency (BPF). This phenomenon has been validated by many researchers [4,7–9]. Although all curves share the same period time, their initial phase values are different. Furthermore, there is no unified law for the initial phase for each flow parameter on each operational point. This is due to the initial phase value of flow parameters, which depends on the detailed flow field distribution including both velocity and pressure at the start of each shaft rotation cycle. Lastly but most importantly, there are huge differences between the peak values of each flow field parameter among different working conditions.

The peak values listed in Table 5 show that for the pump studied in this work, the maximum fluctuation value of efficiency is as high as 1.66%, while the minimum value exceeds 0.37%. The fluctuation level basically has the same magnitude as the reported increased efficiency value based on steady state CFD models (e.g., Shi et al. [8], Kim et al. [24] and Zhao et al. [25]). Therefore, it is preferable to take the unsteady state CFD model to optimize the design of the centrifugal pump which can improve its determinacy.

The peak value can provide the most intuitive information about the fluctuation behavior. The comparison of dimensionless wave peak values on different working conditions for torque, head and efficiency are shown in Figure 9. As listed in Tables 3–5, the original peak values vary considerably for different parameters, however their normalized quantities are in the same order of magnitude, thus providing a common comparison basis. Unlike the simple intuition, low flow parameter fluctuation lever does not necessarily associate with the high energy efficiency. Instead, the minimum normalized wave peak values of all three parameters appear on the operating point with a $1.2Q_{des}$ flow flux. Furthermore, all of these values clearly rise when the flow flux exceeds $1.2Q_{des}$. Generally, the torque and efficiency curves (shown in Figure 9) share similar trends. On the part-load points, the normalized peak values for the above two parameters are very high, and they both reach the maximum values on the point with $0.6Q_{des}$ flow flux.

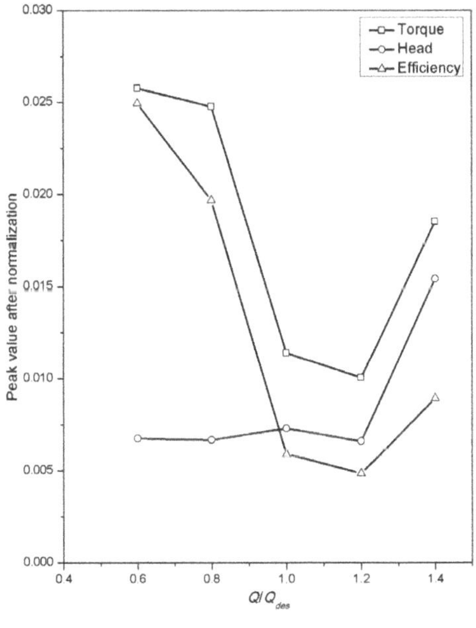

Figure 9. Dimensionless peak values after the process of normalization. Q: volume flow rate ($m^3 \cdot s^{-1}$).

The curve of the normalized peak value of the head has completely different characteristics from the other two curves (as shown in Figure 9). This is because there is little change before the flow flux reaches the value of $1.2Q_{des}$. Thus, it can be speculated that there is a high degree of correlation between the fluctuation of torque and efficiency, especially for the part load working points. Further research is still needed to discover the detailed mechanism of the fluctuation of efficiency. However, the current findings suggest that the toque fluctuation might be the leading factor (among others) which contributes to the fluctuation of efficiency.

4.3. Flow Structures Analysis

It can be said superficially that the fluctuation in the pump efficiency is caused by the periodic behavior of the macroscopic flow parameters, such as torque and head. However, the actual root cause can only be revealed through the structural analysis of the flow field. To do so, the operating points including $0.6Q_{des}$, $1.0Q_{des}$, and $1.4Q_{des}$ were chosen for further study, and the velocity vector distributions on the mid-plane are shown in Figures 10–12, respectively. In each figure, a rectangular area near the tongue with the size of 0.25 m × 0.20 m is picked out and magnified for better observation, which is marked with black lines. For each working point, the flow fields on the phase positions of both 0° and 30° are shown for comparison. Note that the time corresponds to the phase position of 30° is marked as a vertical dash line in Figure 8. Therefore, the efficiency value in that position can be observed clearly. In addition, the initial point with a 0° phase is also shown in the figure.

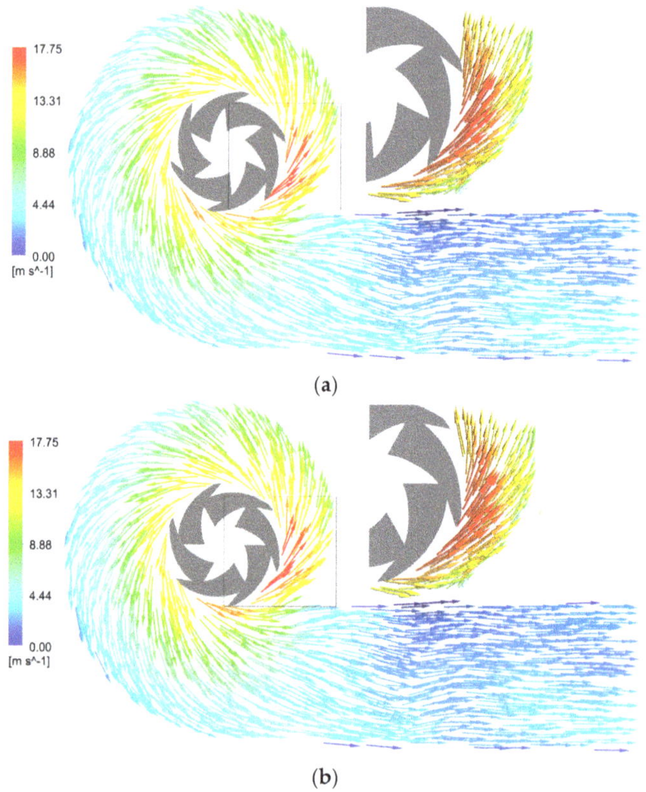

(a)

(b)

Figure 10. Velocity vector distribution at the point of $0.6Q_{des}$ with different phase positions: (**a**) 0°; (**b**) 30°.

(a)

(b)

Figure 11. Velocity vector distribution at the point of $1.0Q_{des}$ with different phase positions: (**a**) 0°; (**b**) 30°.

(a)

Figure 12. *Cont.*

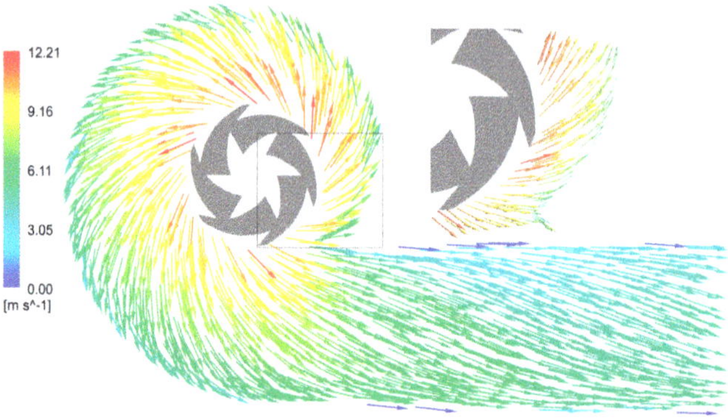

(b)

Figure 12. Velocity vector distribution at the point of $1.4Q_{des}$ with different phase positions: (**a**) 0°; (**b**) 30°.

Primary observation shows that there are mainly two obvious differences in the comparison of flow structures. First of all, the direction of the velocity vectors near the interface between impeller and exit chamber changes a lot along the flow flux. The incidence angles for difference computation cases, which is the intersection angle of the local velocity vector and the circle tangent of the above interface, are listed in Table 6. There are little differences when the phase position changes for the data in Table 6, but both the maximum and averages values of the incidence angle increase dramatically when the flow flux rises. Then, there are some flow structure changes near the tongue along with the rotation of impeller due to the interaction factors, but the specific difference level does not keep the same under each flow flux.

Table 6. Fluid incidence angle on the interface between impeller and exit chamber. Unit: Degree.

Flow Flux	$0.6Q_{des}$		$1.0Q_{des}$		$1.4Q_{des}$	
Phase Position	0°	30°	0°	30°	0°	30°
Maximum Value	29.33	24.57	34.4	36.57	41.81	43.21
Average Value	8.21	8.97	14.84	13.18	20.85	19.22

To investigate the mechanism of the fluctuation in the efficiency at different working points, a hypothesis about the refinement of the widely applicable impeller–volute interaction theory was proposed in this work. The core idea of this hypothesis is that the fluctuation of flow parameters of centrifugal pump depends on both the structural shape of hydraulic model and flow flux. In fact, it can be judged that the affection of flow flux acts through the change of incidence angle, as can be observed in Table 6. First of all, with fixed shaft rotating speed, there is an optimal hydraulic model shape and flow flux combination for each centrifugal pump. Under these circumstances, inner fluid flow can fit well with the impeller rotation, and distribute fluid evenly with the increment of impeller's rotational angle, thus the efficiency fluctuation is relatively low. However, when the flow rate varies and disturbs the balance of the above combination, the fluctuation will increase.

At lower flow rates, the effect of hydraulic modelled shape becomes the dominating factor. As the flow rate further decreases, the impeller–volute interaction becomes stronger. This has been shown in Figure 10. There is an obvious vector distribution difference between the two phase positions. At the 0° position, the velocity in the blade passage near the tongue (see the rectangular dotted box in Figure 10a) is much larger than the other region, and therefore it produces more friction, thus reducing the energy efficiency. Meanwhile, at the 30° position (as shown in the rectangular dotted box of Figure 10b), the region with high flow velocity is divided into two blade passages. Therefore, the vector distribution becomes more uniform and the energy efficiency gets higher. Similar patterns can also be found in Figure 11, but the difference in velocity vector distribution and efficiency value between the above two phase positions becomes smaller because the flow flux rises and gets closer to the optimal rate.

However, the dominant factor changes to the flow flux when the flow rate is larger than the optimal value, and the effect of impeller–volute interaction on flow field becomes weaker. For example, in Figure 12, there is little difference between the velocity vector distributions of 0° and 30° positions. Due to this, the efficiency fluctuation at the $1.4Q_{des}$ point is small.

Since the changes in velocity vector are only limited to the local blade-tongue region and they would not significantly affect the main flow features of the full field, the fluctuation of the torque applied on the impeller is larger than the head lift during the rotation. This explains why the torque fluctuation is the main factor for efficiency wave behaviors rather than the head for this research subject.

5. Conclusions

The present work numerically investigates the fluctuation behavior in the efficiency of a centrifugal pump with detailed and quantitative analysis. Numerical methods can provide a convenient and swift method for quantitative evaluation of the efficiency fluctuation without disturbance from the external environment factors. Some important conclusions as obtained from the current work are as follows.

(1) Centrifugal pumps show efficiency fluctuation with common dominant frequency equaling the blade passing frequency. However, the initial phase of the fluctuation curves is different for each working point.
(2) The efficiency fluctuation level is significant, and its value varies for different working points. It is relatively higher for the part-load point, however high efficiency does not necessarily result in low efficiency fluctuation.
(3) The efficiency fluctuation is caused by the impeller–volute interaction, and its effect is strong on the part-load condition. When the flow flux rises, its effect becomes less obvious and the level of efficiency fluctuation becomes relatively smaller.
(4) Compared with the head, the torque acting on the impeller provides greater influence on the efficiency fluctuation.

Taking into account many research reports on the fluctuation of pressure and fluid forces, more intensive study is still needed to understand efficiency fluctuation in centrifugal pumps. This should especially be done by using direct experimental testing and validation work with more precise measurement instruments.

Acknowledgments: The present work was supported by the Science and Technology Plan of Hunan Province, China (2014GK3150). Anonymous reviewers provided some constructive reviews, which is gratefully acknowledged.

Author Contributions: Hehui Zhang built the math model, carried out simulation computation and analyzed the data; Shengxiang Deng guided the research work and made scientific discussion; Yingjie Qu contributed experiment test; all the authors commented and wrote the article.

Conflicts of Interest: The authors declare no conflict of interest.

References

1. Zhang, H.H.; Deng, S.X.; Qu, Y.J. High working efficiency of rapid custom design. *World Pumps* **2016**, *3*, 34–36. [CrossRef]
2. Shah, S.R.; Jain, S.V.; Patel, R.N.; Lakhera, V.J. CFD for centrifugal pumps: A review of the state-of-the-art. *Proc. Eng.* **2013**, *51*, 715–720. [CrossRef]
3. Lei, T.; Zhu, B.S.; Cao, S.L.; Wang, Y.C.; Wang, B.B. Influence of prewhirl regulation by inlet guide vanes on cavitation performance of a centrifugal pump. *Energies* **2014**, *7*, 1050–1065.
4. Ding, H.; Visser, F.C.; Jiang, Y.; Furmanczyk, M. Demonstration and validation of a 3D CFD simulation tool predicting pump performance and cavitation for industrial applications. *J. Fluids Eng.* **2011**, *133*. [CrossRef]

5. Kean, W.C.; Winoto, S.L.; Cheah, K.W. Numerical study of inlet and impeller flow structures in centrifugal pump at design and off-design points. *Int. J. Fluid Mach. Syst.* **2011**, *4*, 25–32.
6. Sten, M.; Martin, G. Experimental and numerical investigation of centrifugal pumps with asymmetric inflow conditions. *J. Therm. Sci.* **2015**, *24*, 516–525.
7. Wang, C.L.; Zeng, C.; Peng, X.Y.; Peng, H.B.; Liu, D. Numerical simulation of internal flow field and performance prediction of reversible double suction pump. *J. Drain. Irrig. Mach. Eng.* **2015**, *33*, 577–582.
8. Shi, W.D.; Chen, K.Q.; Zhang, D.S.; Xing, J. Numerical optimization and regression analysis of forward-extended double-blade sewage pump. *J. Huazhong Univ. Sci. Technol.* **2015**, *43*, 49–63.
9. González, J.; Fernández, J.; Blanco, E.; Santolaria, C. Numerical simulation of the dynamic effects due to impeller-volute interaction in a centrifugal pump. *J. Fluids Eng.* **2002**, *124*, 348–355. [CrossRef]
10. Treutz, G. Numerische Simulation der instationären Strömung in einer Kreiselpumpe. Ph.D. Thesis, University of Damstadt, Damstadt, Germany, 2002.
11. Guo, S.J.; Okamoto, H. An experimental study on the fluid forces induced by rotor-stator interaction in a centrifugal pump. *Int. J. Rotating Mach.* **2003**, *9*, 135–144. [CrossRef]
12. Cheah, K.W.; Lee, T.S.; Winoto, S.H. Unsteady fluid flow study in a centrifugal pump by CFD Method. In Proceedings of the 7th Asean ANSYS Conference, Biopolos, Singapore, 30–31 October 2008.
13. Lucius, A.; Brenner, G. Unsteady CFD simulations of a pump in part load conditions using scale-adaptive simulation. *Int. J. Heat Fluid Flow* **2010**, *31*, 1113–1118. [CrossRef]
14. Huang, S.; Yang, F.X.; Guo, J. Numerical simulation of 3D unsteady flow in centrifugal pump by dynamic mesh technique. *Proc. Eng.* **2013**, *61*, 270–275.
15. Dai, C.; Kong, F.Y.; Dong, L. Pressure fluctuation and its influencing factors in circulating water pump. *J. Cent. South Univ.* **2013**, *20*, 149–155. [CrossRef]
16. Yuan, S.Q.; Zhou, J.J.; Yuan, J.P.; Zhang, J.F.; Xu, Y.P.; Li, T. Characteristic analysis of pressure fluctuation of unsteady flow in screw-type centrifugal pump with small blade. *Trans. Chin. Soc. Agric. Mach.* **2012**, *43*, 83–87.
17. Zhang, J.F.; Wang, W.J.; Fang, Y.J.; Ye, L.T.; Yuan, S.Q. Influence of splitter blades on unsteady flow and structural dynamic characteristics of a molten salt centrifugal pump. *J. Vib. Shock* **2014**, *33*, 37–41.
18. Barrio, R.; Parrondo, J.; Blanco, E. Numerical analysis of the unsteady flow in the near-tongue region in a volute-type centrifugal pump for different operating points. *Comput. Fluids* **2010**, *39*, 859–870. [CrossRef]
19. Pei, J.; Yuan, S.Q.; Li, X.J.; Yuan, J.P. Numerical prediction of 3-D periodic flow unsteadiness in a centrifugal pump under part-load condition. *J. Hydrodyn.* **2014**, *26*, 257–263. [CrossRef]
20. Liu, H.L.; Liu, D.X.; Wang, Y.; Wu, X.F.; Wang, J.; Du, H. Experimental investigation and numerical analysis of unsteady attached sheet cavitation flows in a centrifugal pump. *J. Hydrodyn.* **2013**, *25*, 370–378. [CrossRef]
21. ANSYS, Inc. *ANSYS Help System*; ANSYS, Inc.: Canonsburg, PA, USA, 2011.
22. Menter, F.R. Two-equation eddy-visocity turbulence models for engineering applications. *AIAA J.* **1994**, *32*, 1598–1605. [CrossRef]
23. The International Standards Organization for Standardization. *ISO 9906–2012. Rotodynamic Pumps—Hydraulic Performance Acceptance Tests—Grades 1, 2 and 3*, 2nd ed.; The International Standards Organization for Standardization: Geneva, Switzerland, 2015.
24. Kim, J.H.; Oh, K.T.; Pyun, K.B.; Kim, C.K.; Choi, Y.S. Design optimization of a centrifugal pump impeller and volute using computational fluid dynamics. In Proceedings of the 26th IAHR Symposium on Hydraulic Machinery and Systems, Beijing, China, 19–23 August 2012.
25. Zhao, W.G.; Sheng, J.P.; Yang, J.H.; Song, Q.C. Optimization design and experiment of centrifugal pump based on CFD. *Trans. Chin. Soc. Agric. Mach.* **2015**, *31*, 125–131.

energies

MDPI

Article

An Investigation of the Restitution Coefficient Impact on Simulating Sand-Char Mixing in a Bubbling Fluidized Bed

Xinjun Zhao, Qitai Eri * and Qiang Wang

School of Energy and Power Engineering, Beihang University, Beijing 100191, China; kunpengzhao@buaa.edu.cn (X.Z.); qwang518@buaa.edu.cn (Q.W.)
* Correspondence: eriqitai@buaa.edu.cn; Tel.: +86-10-8233-9759

Academic Editor: Bjørn H. Hjertager
Received: 17 March 2017; Accepted: 27 April 2017; Published: 3 May 2017

Abstract: In the present work, the effect of the restitution coefficient on the numerical results for a binary mixture system of sand particles and char particles in a bubbling fluidized bed with a huge difference between the particles in terms of density and volume fraction has been studied based on two-fluid model along with the kinetic theory of granular flow. Results show that the effect of restitution coefficient on the flow characteristics varies in different regions of the bed, which is more evident for the top region of the bed. The restitution coefficient can be categorized into two classes. The restitution coefficients of 0.7 and 0.8 can be included into one class, whereas the restitution coefficient of 0.9 and 0.95 can be included into another class. Moreover, four vortices can be found in the time-averaged flow pattern distribution, which is very different from the result obtained for the binary system with the similar values between particles in density and volume fraction.

Keywords: restitution coefficient; segregation; flow pattern; bubbling fluidized bed; binary particles; sand; char

1. Introduction

Bubbling fluidized beds are commonly used in industrial processes [1], such as combustion and gasification, due to their good mixing ability and heat transfer characteristic between the gas and solid phases. In the process of biomass gasification, the first step is pyrolysis. The gas component products after pyrolysis are non-condensable gases and tars, and char is left as a solid residue [2]. Due to the irregular shape and low density of char particles, it is difficult to attain a stable fluidization status. Generally, inert particles such as sand particles are added to the fluidized bed to improve the fluidization status and the heat transfer effect. During the process of fluidization, char particles will accumulate toward the top of the bed, whereas the inert particles will sink towards the bottom of the bed. Hence, segregation is a widespread phenomenon in a binary system of particles in a bubbling fluidized bed [3,4].

The complicated flow characteristics in the bubbling fluidized bed have a strong effect on the reaction process. Therefore understanding the flow behavior is important to design a fluidized bed and optimize the operation conditions. Although the most accurate method is still based on the experimental data, the application of this method has limitations, in terms of the longer time required and high costs. With advances in computational fluid dynamics (CFD), simulation studies have become a useful method to analyze the flow characteristics in bubbling fluidized beds [5,6]. The most widely used approach for simulating dense gas-solid flow is the Eulerian–Eulerian model (two fluid model) along with the kinetic theory of granular flow [7,8].

The main parameters affecting the simulation results include the drag model, particle collision characteristics, and solid phase wall boundary conditions. The particle collision characteristics are defined based on the restitution coefficient, and the wall boundary conditions are defined in at least three ways, which are the traditional no-slip boundary conditions and two partial slip conditions [9,10]. The effects of these parameters on the simulation results have been studied by some researchers. Chao et al. [10] and Bai et al. [11] focused on the effect of the drag model on the mixing and segregation behavior of biomass mixtures in a fluidized bed. Tagliaferri et al. [12] and Mostafazadeh et al. [13] have studied the influence of the restitution coefficient on the flow dynamics of a binary solid mixture. Zhong et al. [14] have investigated the influence of wall boundary conditions on the concentration and velocity distribution of particles. In these studies, the density or volume fraction of binary particles was usually similar. However, in actual biomass gasification, the char has lower density than the sand. Meanwhile, the char generated from the pyrolysis will continue to react with the gases, hence, the volume fraction of the char in the solid phase mixtures is very low [15]. When the simulation conditions do not agree with actual situation, the reasonability and accuracy of simulation results is questionable.

Many studies [16–18] have studied the suitability of different drag models. Generally, the Gidaspow drag model is suitable for simulation in a dense bubbling fluidized bed. Loha et al. [6] reported that the model predictions were sensitive to the specularity coefficient and the simulation results with a higher value of specularity coefficient were in good agreement with the experimental results. According to a comparison of the differences between the specularity coefficients of one wall boundary conditions and traditional no-slip wall boundary conditions, Zhong et al. [19] found that the no-slip wall boundary condition was more suitable for simulating the dynamic segregation process of binary particles. In this study, based on a Gidaspow drag model and no-slip wall boundary conditions, a simulation is carried out to investigate the effect of the restitution coefficient on the segregation and flow characteristics of a binary particles system in a bubbling fluidized bed with a huge difference between the particles in terms of density and volume fraction. The particle system consists of char particles and sand particles with bulk density of 120 and 1590 kg/m^3 and volume fraction of 11% and 89%. According to the comparison of results between computation and experiment, a suitable scope of the restitution coefficient is determined, and meanwhile, the effect of different particles characteristics in the binary mixture on the fluid dynamic behavior is also compared.

2. Model Description

In the two fluid model (TFM), the gas and solid phases are considered as inter-penetrating continua, hence the governing equations for the solid phase are similar to those for the gas phase. For the case of cold fluidization with no chemical reactions, the conservation equations for mass and momentum are represented as follows:

The continuity equation for the gas phase is expressed as follows:

$$\frac{\partial(\alpha_g\rho_g)}{\partial t} + \nabla\cdot(\alpha_g\rho_g\vec{u}_g) = 0 \tag{1}$$

The continuity equation for the sth solid phase is written as:

$$\frac{\partial(\alpha_s\rho_s)}{\partial t} + \nabla\cdot(\alpha_s\rho_s\vec{u}_s) = 0 \tag{2}$$

The momentum balance equation for the gas phase is as follows:

$$\frac{\partial(\alpha_g\rho_g\vec{u}_g)}{\partial t} + \nabla\cdot(\alpha_g\rho_g\vec{u}_g\vec{u}_g) = -\alpha_g\nabla p + \nabla\cdot\overline{\overline{\tau}}_g + \alpha_g\rho_g\vec{g} + \sum_{m=1}^{N} K_{mg}(\vec{u}_m - \vec{u}_g) \tag{3}$$

The momentum balance equation for the sth solid phase can be written as:

$$\frac{\partial(\alpha_s\rho_s\vec{u}_s)}{\partial t} + \nabla\cdot(\alpha_s\rho_s\vec{u}_s\vec{u}_s) = -\alpha_s\nabla p - \nabla p_s + \nabla\cdot\overline{\overline{\tau}}_s + \alpha_s\rho_s\vec{g} + \sum_{n=1}^{N} K_{ns}(\vec{u}_n - \vec{u}_s) \tag{4}$$

The total volume fraction of the phases is equal to 1 and it is expressed in the following way:

$$\alpha_g + \sum\alpha_s = 1 \tag{5}$$

where t is the time. The subscript "g" refers to the gas phase and the subscript "s" refers to the sth solid phase. $\vec{g}$, $\vec{u}$, and $\overline{\overline{\tau}}$ are the acceleration due to the gravity, velocity vector, and stress-strain tensor, respectively. p, α, and ρ are the pressure, volume fraction, and density, respectively. K is the momentum exchange coefficient between phases.

The solid pressure is evaluated using the expression proposed by Lun et al. [20], and it is composed of a kinetic term and a second term due to particle collisions:

$$p_s = \alpha_s\rho_s\Theta_s + 2\rho_s(1+e_{ss})\alpha_s^2 g_{0,ss}\Theta_s \tag{6}$$

where $g_{0,ss}$ and Θ_s are the radial distribution function and granular temperature of the sth solid phase, respectively. e_{ss} is the restitution coefficient for particle collision, which quantifies the elasticity of particle collision.

3. Simulation Setup

In this study, the experimental setup by Park et al. [21] forms the basis for the simulation. The geometry of the rectangular fluidized bed with 200 mm width and 50 mm depth is schematically shown in Figure 1. In the experiments, sand particles and char particles are used. The properties of the particles in this study are summarized in Table 1.

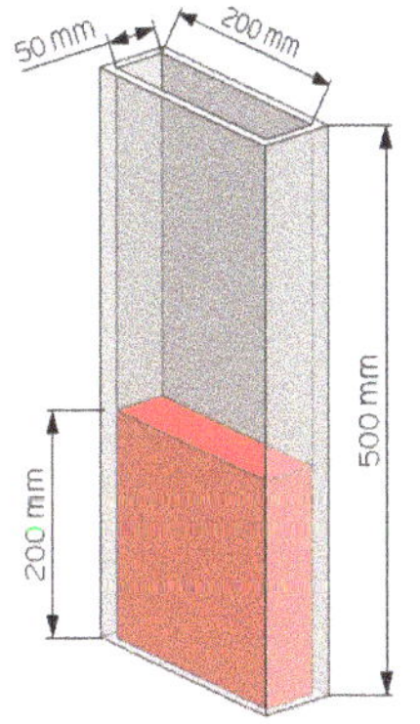

Figure 1. Schematic representation of the fluidized bed.

Table 1. Properties of particles.

Particles	Mean Diameter (μm)	Void Fraction	Bulk Density (kg/m^3)
Sand	387	0.333	1590
Char	957	0.693	120

The 2D simulation is carried out using the TFM method, and the computational domain is accordingly set to be 200 mm width and 500 mm height. The mesh geometry and coordinate system is schematically shown in Figure 2. The Z direction represents the axial direction or height direction, whereas X direction represents the radial direction or lateral direction.

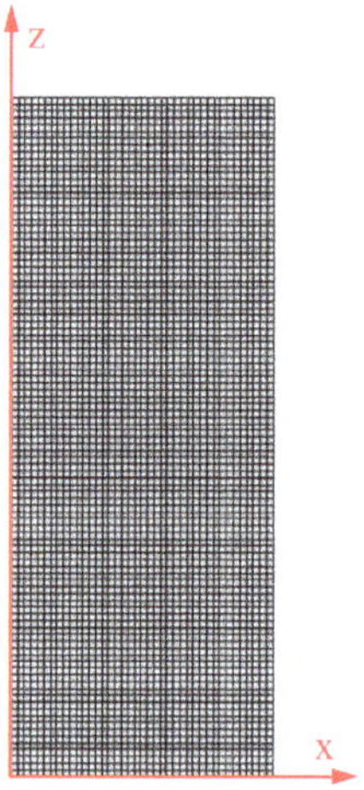

Figure 2. Mesh geometry and coordinate system.

In the simulation, ambient air is used as the fluidizing medium. Meanwhile, the uniform gas velocity is specified at the bottom of the bed and the atmospheric pressure boundary condition is used at outlet of the bed. The detailed parameters including the material properties and operating conditions for simulation are summarized in Table 2.

Table 2. Summary of the simulation parameters.

Parameter	Value or Model
Bed height (m)	0.5
Bed width (m)	0.2
Minimum fluidization velocity (m/s)	0.14
Superficial gas velocity (m/s)	0.19
Total particle weight (g)	Sand: 2000 Char: 40
Particle volume fraction	Sand: 89% Char: 11%
Drag coefficient	Gidaspow
Granular viscosity	Syamlal-O'Brien
Granular bulk viscosity	Lun et al.
Restitution coefficient	e_{ss} = 0.7, 0.8, 0.9, 0.95
Wall boundary condition	No-slip

At the beginning of the computation, the sand particles and char particles are perfectly mixed at the indicated ratio. A time step of 5×10^{-4} s was employed in the simulation using Fluent 15.0 software. When TFM is used for the simulation, the grid size should be much smaller than the physical dimensions of the geometry. Meanwhile, it should also be bigger than the particle diameter, which will ensure that the solid phase can be treated as a continuous flow. Therefore, the grid size has a drastic effect on the flow behavior [22]. A coarse grid could lead to an overprediction of the solid expansion height of the bed [23]. Sande et al. [17] found that the numerical results agree well with the experimental results when the grid size is about 10 times the particle diameter and there is no improvement in capturing homogeneous expansion when the mesh is further refined. Hence, a grid size of 5 mm was chosen in this simulation.

4. Results and Discussion

4.1. Stationary Condition

In the fluidization of the binary mixture, the particles that sink at the bottom of the bed are known as jetsam, while those that accumulate at the top of the bed are known as flotsam. In this study, the sand particles are the jetsam particles, and the char particles are the flotsam particles.

For the case with e_{ss} = 0.9, the time evolution of particle volume fraction in different layers along the height direction is monitored to obtain the statistical steady-state for solution, as shown in Figure 3. As time increases, the flotsam moves upwards while the jetsam moves downwards. Hence, with increase in time, the jetsam volume fraction increases while the flotsam volume fraction decreases at the height of 0.053 m, and the jetsam volume fraction decreases while the flotsam volume fraction increases at the height of 0.153 m. At the beginning of fluidization, the volume fractions vary rapidly, whereas the values change slowly after 30 s, corresponding to complete fluidization. Hence, the time-averaged variables are computed between 30 s and 50 s. Moreover, the result shows that the simulation time corresponding to the stationary state should be increased with increase in restitution coefficient. For the cases with e_{ss} = 0.7 and 0.8, the simulation time of 50 s is long enough. However, for e_{ss} = 0.95, the simulation has not converged to a stationary state until the time exceeds 65 s. Hence, for the case with e_{ss} = 0.95, the time-averaged variables are computed between 65 s and 85 s. When the restitution coefficient increases, it means that there is lesser dissipation of kinetic energy of particle, due to a more significant elastic particle-particle collision. This may be the reason why a longer simulation time is required before the mixing pattern reaches a stationary state when the restitution coefficient increases.

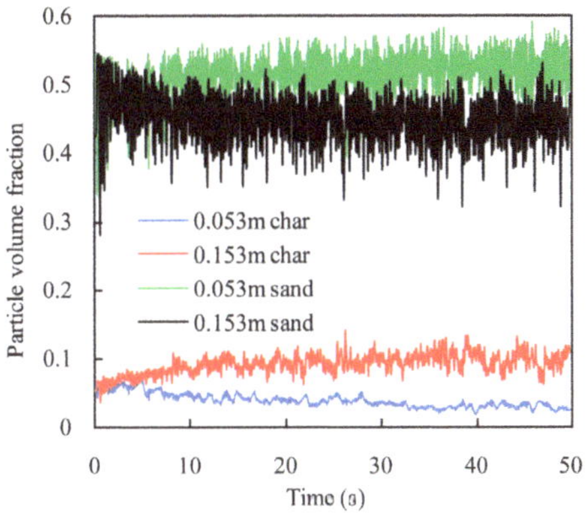

Figure 3. Time profiles of the particle volume fraction in different layers along the height direction.

4.2. Particle Velocity and Flow Pattern

Figure 4 shows the lateral profiles of time-averaged axial velocity of the flotsam at the bed height of 0.153 m. In this figure, the velocity profiles of the flotsam for e_{ss} = 0.7, 0.8, 0.9, and 0.95 are similar. This clearly shows that the axial velocity of the flotsam is directed upward in the center of the bed and the axial velocity at the wall is nearly equal to zero, which represents the characteristic of the no-slip wall boundary condition.

In order to reflect the movement relevance between the jetsam and flotsam, in the case of e_{ss} = 0.9, the predicted flow patterns for the flotsam and jetsam are shown in Figures 5 and 6, respectively. The flow patterns for the flotsam and jetsam are quite similar. Two vortices show at the bottom of the bed and two vortices close to the top of the bed. At the bottom of the bed, the particles generally rise towards the wall, whereas they fall down towards the central region. However, at the top of the bed, the movement direction of particles is reversed. The flow pattern obtained in this study is very different from the result obtained by other researchers [24]. In those studies, only two vortices can be found throughout the entire region of the dense bed layer. However, in this study, four vortices are obtained in the simulation. Moreover, the two vortices at the bottom of the bed are relatively close to the centerline of the bed, whereas the two vortices at the top of the bed are generated near the wall. It can also be seen that the distance between the two vortices at the top of the bed is farther than the value for the vortices at the bottom of the bed. This phenomenon might be due to the relationship between the binary particles in density and volume fraction. When the density and volume fraction between the binary particles changes not much [14,24], the segregation of particles does not play a leading role in fluidization process. Hence, in the top and bottom regions of the bed, uniform patterns can be found. However, in this study, a binary particles system with a huge difference between particles in density and volume fraction is used, which causes a much more notable segregation. Hence, the flow pattern of particles at the top region of the bed is different from the result for the bottom region.

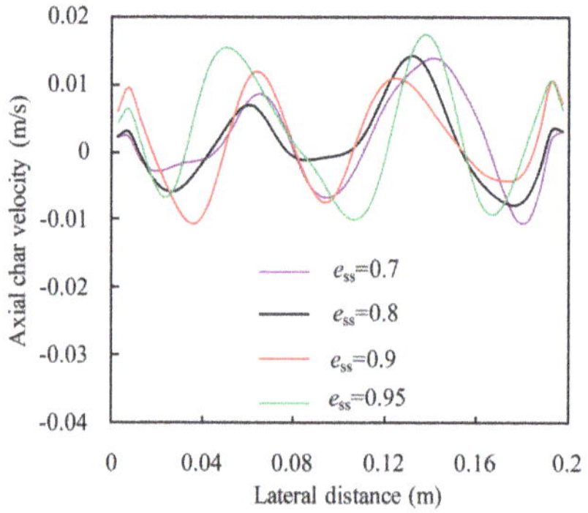

Figure 4. Lateral profiles of the time-averaged axial velocity of the flotsam at Z = 0.153 m.

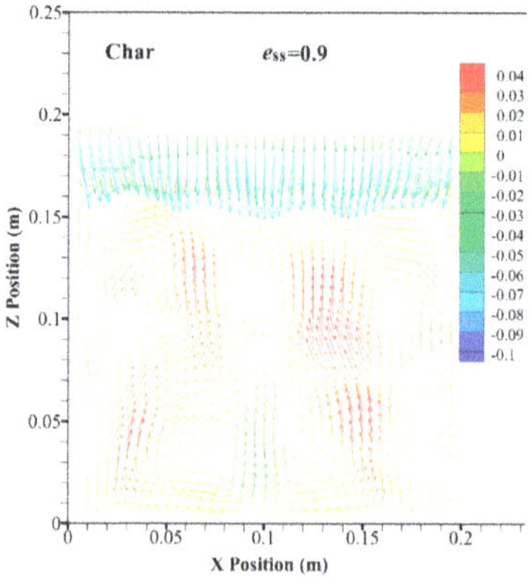

Figure 5. Time-averaged particle velocity distribution of the flotsam for e_{ss} = 0.9 (the color legend represents the axial velocity of the particles).

The small bubbles are generally generated near the wall and the larger bubbles form near the centerline region of the bed as small bubbles move upwards [25]. Based on the analysis of positive axial velocity of particles, from Figures 5 and 6, it clearly shows that the motion of particles obtained in the simulation has the same characteristics.

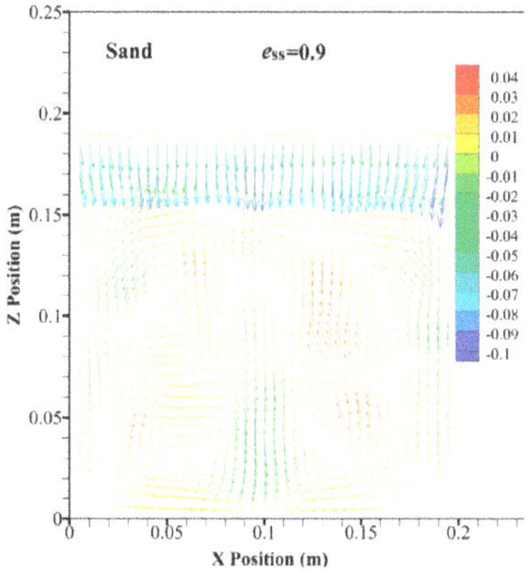

Figure 6. Time-averaged particle velocity distribution of the jetsam for e_{ss} = 0.9 (the color legend represents the axial velocity of the particles).

Due to the similar flow patterns for the flotsam and jetsam, as shown in Figures 5 and 6, in the cases with same restitution coefficient, the lateral profile of the time-averaged axial velocity of the jetsam at Z = 0.153 m is similar to the profile for the flotsam. Some studies [26] have shown that the central region of the profile for jetsam changed more gently and the minimum value appeared near the wall when the layer height was at the top of the bed, which was above the jetsam-rich layer. The shape of the profile mentioned in these studies is similar with the shape of the curve obtained in the case of e_{ss} = 0.95 in this study, as shown in Figure 4.

In the cases of e_{ss} = 0.7, 0.8, and 0.95, the flow patterns of the flotsam are shown in Figures 7–9, respectively. The vortex structures for different restitution coefficients are similar. However, due to the lesser dissipation of granular energy, with the increase in restitution coefficient, the vortex intensity increases. At the bottom of the bed, due to the effect of the inlet gas, the restitution coefficient does not much affect the vortex intensity. However, for the two vortices at the top of the bed, with increase in the height of the bed, the dissipation of granular energy is accumulated. Hence, the effect of restitution coefficient on the flow pattern of particles at the top of the bed is more evident. It can be seen that the vortex characteristic of the two vortices at the top of the bed, as shown in Figure 9, is more obvious when the restitution coefficient increases from 0.9 to 0.95.

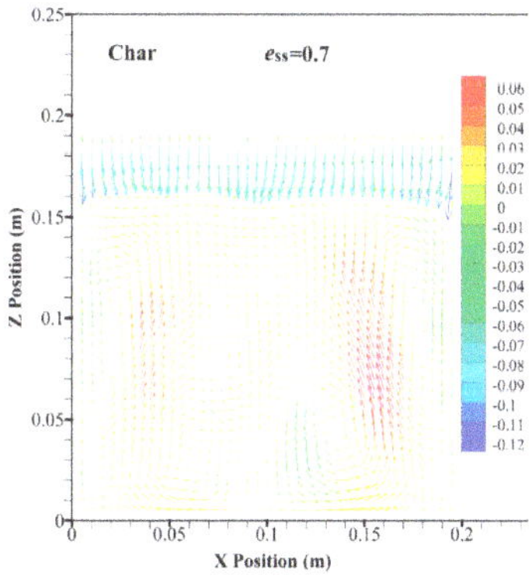

Figure 7. Time-averaged particle velocity distribution of the flotsam for e_{ss} = 0.7 (the color legend represents the axial velocity of the particles).

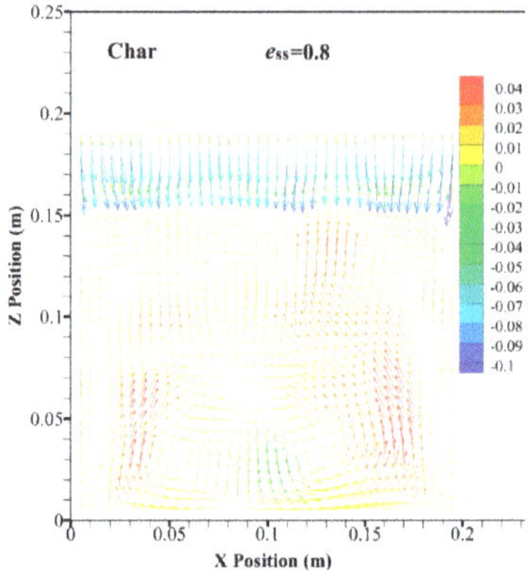

Figure 8. Time-averaged particle velocity distribution of the flotsam for e_{ss} = 0.8 (the color legend represents the axial velocity of the particles).

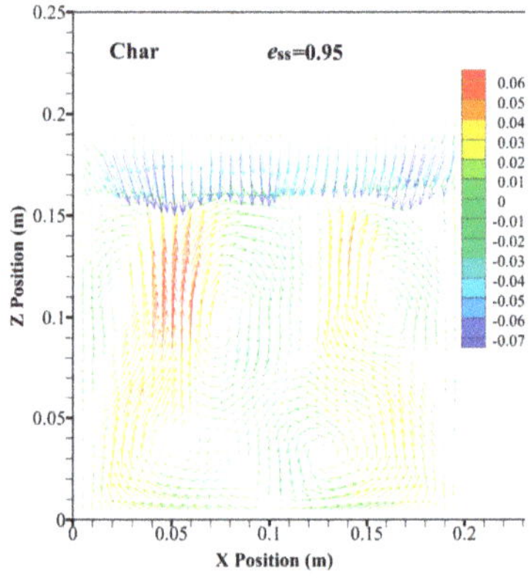

Figure 9. Time-averaged particle velocity distribution of the flotsam for e_{ss} = 0.95 (the color legend represents the axial velocity of the particles).

4.3. Granular Temperature Distribution

Granular temperature is a concept in the field of the kinetic theory of granular flow, and it is used for assessing fluctuating energy of the particles suspended in the gas flow. The lateral profiles of the time-averaged granular temperature of the flotsam at a height of 0.153 m are shown in Figure 10. When the restitution coefficient is no more than 0.9, the granular temperature of the flotsam closer to the wall is high and decreases towards the central region of the bed. Meanwhile, the curve becomes almost flat at the center of the bed. However, for the case of e_{ss} = 0.95, at the center of the bed, the curve is no longer flat. Two peaks of granular temperature can be found, which is related to the big values of the axial velocity, as shown in Figures 4 and 9. With restitution coefficient increases, the values of the granular temperature increase, since there is less dissipation of the randomly fluctuating kinetic energy of particles. For a restitution coefficient below than 0.9, the granular temperature increases slightly. However, the granular temperature increases dramatically when the restitution coefficient increases from 0.9 to 0.95, especially for the center region of the bed.

Figure 11 presents the axial profiles of the time-averaged granular temperature of the flotsam. In the region of the gas-solid interface, the granular temperature is very high, due to the collapse of bubbles and granular splash [27]. The value of granular temperature increases as the restitution coefficient increases. When the restitution coefficient increases from 0.7 to 0.8, the granular temperature along the axial direction increases slightly. However, with increase in restitution coefficient from 0.8 to 0.9, the granular temperature at the top of the bed increases obviously. Due to the similar flow patterns for the flotsam and jetsam, it can be seen than the effects of the restitution coefficient with different values on the granular temperature are also different.

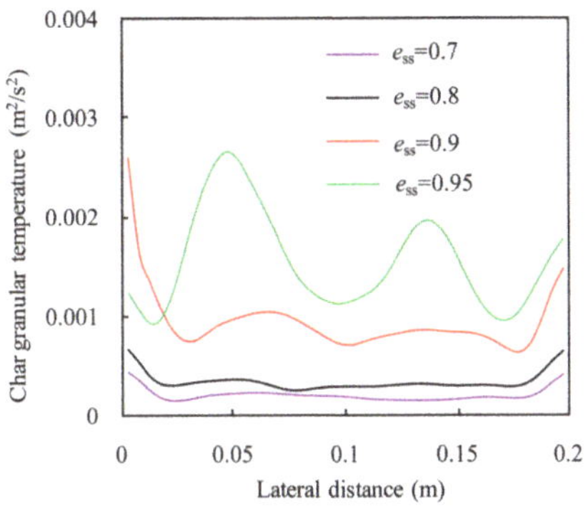

Figure 10. Lateral profiles of the time-averaged granular temperature of the flotsam at Z = 0.153 m.

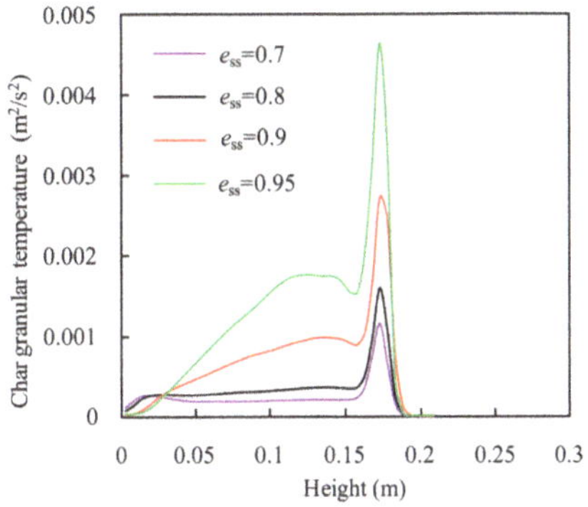

Figure 11. Axial profiles of the time-averaged granular temperature of the flotsam.

4.4. Distribution of Particle Volume Fraction

The effects of restitution coefficient on the instantaneous volume fraction of char and sand particles are shown in Figures 12 and 13, respectively. Closer to the gas-solid interface, the volume fraction of sand particles is relatively low, which is just the opposite of the value of the char particles. This shows that a strong segregation phenomenon exists in the fluidization process of a binary particle system. With increase in the restitution coefficient, more char particles accumulate at the top of the bed, which shows that the tendency of segregation is more remarkable.

More studies [26,28] are concerned with the distribution characteristics of jetsam, however, in the gasification process, the flotsam will react with gas. Hence, the study on distribution characteristics of flotsam is much more significant. Lateral distributions of the time-averaged volume fraction of the flotsam at a height of 0.153 m are plotted in Figure 14. For e_{ss} = 0.7, 0.8, and 0.9, the effect of the restitution coefficient on the volume fraction of the flotsam is not obvious, hence, the profiles are very similar. However, when the restitution coefficient increases to 0.95, the values of volume fraction increase dramatically.

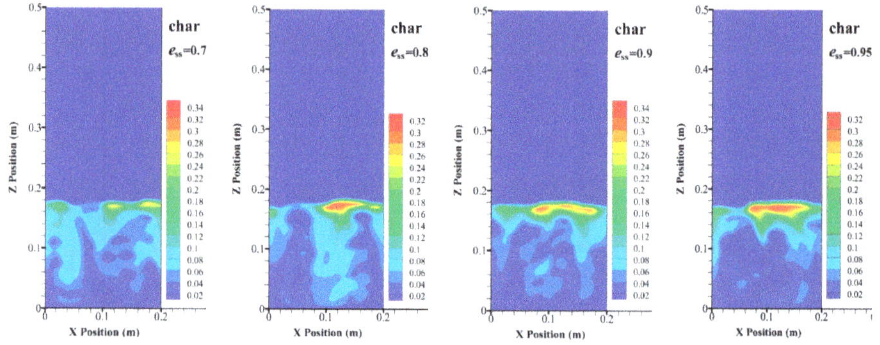

Figure 12. Distributions of instantaneous volume fraction of char particles.

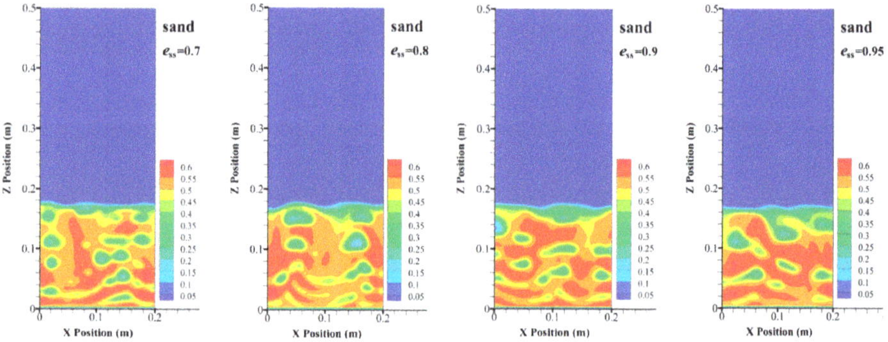

Figure 13. Distributions of instantaneous volume fraction of sand particles.

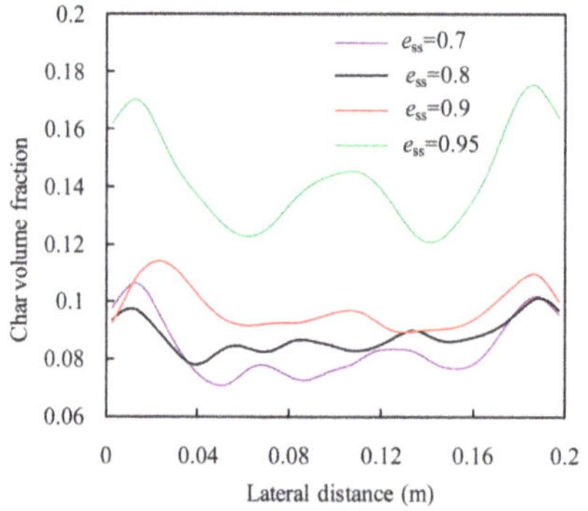

Figure 14. Lateral profiles of the time-averaged particle volume fraction of the flotsam at Z = 0.153 m.

The effect of the restitution coefficient on the distribution of char mass fraction in the axial direction is shown in Figure 15. The char mass fraction is defined as the ratio of the char mass corresponding to the specific height range to total char mass in the bed. With increase in restitution coefficient, the char mass fraction in the upper part of the bed gradually increases, whereas the char mass fraction in the lower part of the bed decreases. In the case of e_{ss} = 0.95, the predicted profile well fits the experimental result. In the binary particles system with less difference between particles in density or volume fraction [12], they found that there was a small difference between the simulation results and the experimental data only for low restitution coefficient such as 0.8. However, for the binary particles system with huge difference in density and volume fraction between particles, the predicted degree of segregation compared with experimental result may be too low, when a relatively small value of restitution coefficient is adopted. From Figure 15, it can be seen that the mass fraction profile of char particles along the height direction for e_{ss} = 0.95 is quite close to the experimental result.

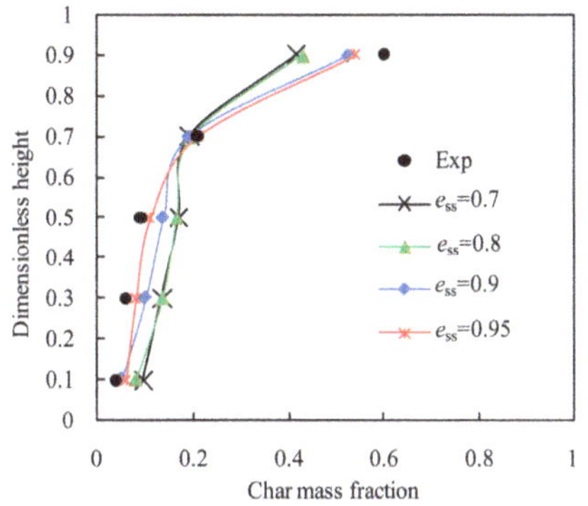

Figure 15. Effect of restitution coefficient on the axial mass fraction profile of the flotsam.

Figure 16 presents the axial profiles of the time-averaged particle volume fraction of the jetsam. Initially, the particle volume fraction along the bed height shows a slight increase, due to the effect of the inlet gas, and then gradually decreases. Closer to the gas-solid interface, the particle volume fraction decreases more sharply. This shows that the splash zone formed by the collapse of bubbles is a small area, which is consistent with previous studies [29,30]. The result from Figure 16 shows that the restitution coefficient had little effect on the axial profile of the jetsam volume fraction.

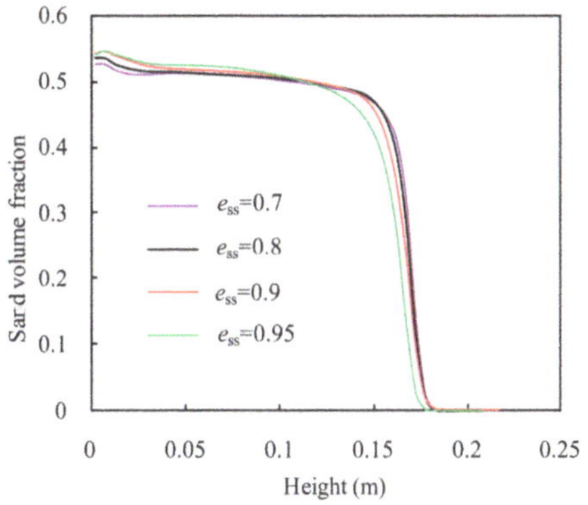

Figure 16. Axial profiles of the time-averaged particle volume fraction of the jetsam.

There are some studies about the comparison of results between 2D simulation and 3D simulation in the rectangular fluidized bed for monoparticle flow. For a pseudo-2D fluidized bed [31], the result has shown that no major differences are observed between 2D and 3D simulations in predicting the mean pressure drop and bed expansion. Just for bubble diameter and rise velocity, the 3D simulations are better agreement with experiments than the corresponding 2D simulations, whereas, for bubble aspect ratio, the 2D simulation has a better agreement with the experimental data. Xie et al. [32] also used an Eulerian-Eulerian model to investigate the differences between 2D and 3D simulations of a rectangular fluidized bed. They found that for a bubbling fluidized bed a satisfactory qualitative agreement between 2D and 3D simulations is observed. Hence, for monoparticle flow, though 2D simulations have certain limitations, they can provide reasonable results compared to experimental observations. Meanwhile, due to the requirement for lower computational resources, 2D simulation is widely used. However, for a binary mixture system in the rectangular fluidized bed, there are few related studies. Geng et al. [33] investigated the difference between the results of 2D and 3D simulations for a binary mixture system in a pseudo-2D rectangular bubbling fluidized bed. The results showed that the flotsam (coal particles) is nearly constant along the height direction, which totally deviates from the experimental observation. This means that 2D simulation is not suitable for modeling a pseudo-2D fluidized bed, which is entirely different from the conclusion for monoparticle flow. They also found that when the thickness of the rectangular bed was larger than a critical value (20 mm in this study), 2D simulation can provide a reasonable result. To sum up, the difference between 2D simulation and 3D simulation might change with the composition of the particle system. For a binary particle system, the relationship between the critical value in the thickness direction of the rectangular bed and the dimension parameters of the bed or the flow parameters may be important, and how to determine the relationship may be a key task in a follow-up study.

5. Conclusions

In the present work, the effect of the restitution coefficient on the numerical results for a binary particles system in a bubbling fluidized bed with the huge difference between the particles in terms of density and volume fraction has been studied based on two-fluid model along with the kinetic theory of granular flow. The effect of the restitution coefficient on the particle velocity, particle flow patterns, and mass fraction distribution varies in the different regions of the bed. At the bottom of the bed, the restitution coefficient does not affect the flow characteristic of particles significantly. However, in the top region of the bed, due to the cumulative effect of the dissipation of granular energy, the restitution coefficient has an obvious influence on the flow characteristic of the particles.

With an increase in the restitution coefficient, the degree of segregation increases. However, it does not change linearly with the restitution coefficient. Considering the effect of the restitution coefficient on the degree of segregation and flow pattern of particles in the top region of the bed, the restitution coefficient can be categorized into two classes: restitution coefficients of 0.7 and 0.8 can be included in one class, whereas the restitution coefficients of 0.9 and 0.95 can be included in another class.

For a binary particles system with the huge density and volume fraction difference between the particles, two vortices at the bottom of the bed and two vortices at the top of the bed are observed in the flow pattern distribution. The time-averaged flow pattern of particles in this study is very different from the result obtained for the system with similar values between particles in density and volume fraction, in which only two vortices can be found in the entire region of the dense bed layer.

Author Contributions: Xinjun Zhao did the simulation and wrote the manuscript. Qitai Eri and Qiang Wang supervised the investigation and checked the model.

Conflicts of Interest: The authors declare no conflict of interest.

References

1. Fan, X.; Yang, Z.; Parker, D.J. Impact of solid sizes on flow structure and particle motions in bubbling fluidization. *Powder Technol.* **2011**, *206*, 132–138. [CrossRef]
2. Mohd Salleh, M.A.; Kisiki, N.H.; Yusuf, H.M.; Ab Karim Ghani, W.A.W. Gasification of Biochar from Empty Fruit Bunch in a Fluidized Bed Reactor. *Energies* **2010**, *3*, 1344–1352. [CrossRef]
3. Liu, H.; Zhao, Y.; Ding, J.; Gidaspow, D.; Wei, L. Investigation of mixing/segregation of mixture particles in gas-solid fluidized beds. *Chem. Eng. Sci.* **2007**, *62*, 301–317.

4. Coroneo, M.; Mazzei, L.; Lettieri, P.; Paglianti, A.; Montante, G. CFD prediction of segregating fluidized bidisperse mixtures of particles differing in size and density in gas-solid fluidized beds. *Chem. Eng. Sci.* **2011**, *66*, 2317–2327. [CrossRef]
5. Sun, J.; Wang, J.; Yang, Y. CFD investigation of particle fluctuation characteristics of bidisperse mixture in a gas-solid fluidized bed. *Chem. Eng. Sci.* **2012**, *82*, 285–298. [CrossRef]
6. Loha, C.; Chattopadhyay, H.; Chatterjee, P.K. Euler-Euler CFD modeling of fluidized bed: Influence of specularity coefficient on hydrodynamic behavior. *Particuology* **2013**, *11*, 673–680. [CrossRef]
7. Bakshi, A.; Altantzis, C.; Bates, R.B.; Ghoniem, A.F. Eulerian–Eulerian simulation of dense solid–gas cylindrical fluidized beds: Impact of wall boundary condition and drag model on fluidization. *Powder Technol.* **2015**, *277*, 47–62. [CrossRef]
8. Wang, Y.; Zou, Z.; Li, H.; Zhu, Q. A new drag model for TFM simulation of gas-solid bubbling fluidized beds with Geldart-B particles. *Particuology* **2014**, *15*, 151–159. [CrossRef]
9. Johnson, P.C.; Jackson, R. Frictional-collisional constitutive relations for granular materials with application to plane shearing. *J. Fluid Mech.* **1987**, *176*, 67–93. [CrossRef]
10. Chao, Z.; Wang, Y.; Jakobsen, J.P.; Fernandino, M.; Jakobsen, H.A. Derivation and validation of a binary multi-fluid Eulerian model for fluidized beds. *Chem. Eng. Sci.* **2011**, *66*, 3605–3616. [CrossRef]
11. Bai, W.; Keller, N.K.G.; Heindel, T.J.; Fox, R.O. Numerical study of mixing and segregation in a biomass fluidized bed. *Powder Technol.* **2013**, *237*, 355–366. [CrossRef]
12. Tagliaferri, C.; Mazzei, L.; Lettieri, P.; Marzocchella, A.; Olivieri, G.; Salatino, P. CFD simulation of bubbling fluidized bidisperse mixtures: Effect of integration methods and restitution coefficient. *Chem. Eng. Sci.* **2013**, *102*, 324–334. [CrossRef]
13. Mostafazadeh, M.; Rahimzadeh, H.; Hamzei, M. Numerical analysis of the mixing process in a gas-solid fluidized bed reactor. *Powder Technol.* **2013**, *239*, 422–433. [CrossRef]
14. Zhong, H.; Gao, J.; Xu, C.; Lan, X. CFD modeling the hydrodynamics of binary particle mixtures in bubbling fluidized beds: Effect of wall boundary condition. *Powder Technol.* **2012**, *230*, 232–240. [CrossRef]
15. Xue, Q.; Heindel, T.J.; Fox, R.O. A CFD model for biomass fast pyrolysis in fluidized-bed reactors. *Chem. Eng. Sci.* **2011**, *66*, 2440–2452. [CrossRef]
16. Askarishahi, M.; Salehi, M.-S.; Molaei Dehkordi, A. Numerical investigation on the solid flow pattern in bubbling gas–solid fluidized beds: Effects of particle size and time averaging. *Powder Technol.* **2014**, *264*, 466–476. [CrossRef]
17. Sande, P.C.; Ray, S. Mesh size effect on CFD simulation of gas-fluidized Geldart A particles. *Powder Technol.* **2014**, *264*, 43–53. [CrossRef]
18. Wang, Y.; Chao, Z.; Jakobsen, H.A. A sensitivity study of the two-fluid model closure parameters (β, e) Determing the main gas-solid flow pattern characteristics. *Ind. Eng. Chem. Res.* **2010**, *49*, 3433–3441. [CrossRef]
19. Zhong, H.; Lan, X.; Gao, J.; Zheng, Y.; Zhang, Z. The difference between specularity coefficient of 1 and no-slip solid phase wall boundary conditions in CFD simulation of gas-solid fluidized beds. *Powder Technol.* **2015**, *286*, 740–743. [CrossRef]
20. Lun, C.C.K.; Savage, S.B.; Jeffrey, D.J.; Chepurniy, N. Kinetic theories for granular flow: Inelastic particles in Couette flow and slightly inelastic particles in a general flow field. *J. Fluid Mech.* **1984**, *140*, 223–256. [CrossRef]
21. Park, H.C.; Choi, H.S. The segregation characteristics of char in a fluidized bed with varying column shapes. *Powder Technol.* **2013**, *246*, 561–571. [CrossRef]
22. Gelderbloom, S.J.; Gidaspow, D.; Lyczkowski, R.W. CFD Simulations of Bubbling/Collapsing Fluidized Beds for Three Geldart Groups. *AIChE J.* **2003**, *49*, 844–858. [CrossRef]
23. Chen, J.; Yu, G.; Dai, B.; Liu, D.; Zhao, L. CFD simulation of a bubbling fluidized bed gasifier using a bubble-based drag model. *Energy Fuels* **2014**, *28*, 6351–6360. [CrossRef]
24. Chao, Z.; Wang, Y.; Jakobsen, J.P.; Fernandino, M.; Jakobsen, H.A. Multi-fluid modeling of density segregation in a dense binary fluidized bed. *Particuology* **2012**, *10*, 62–71. [CrossRef]
25. Verma, V.; Deen, N.G.; Padding, J.T.; Kuipers, J.A.M. Two-fluid modeling of three-dimensional cylindrical gas–solid fluidized beds using the kinetic theory of granular flow. *Chem. Eng. Sci.* **2013**, *102*, 227–245. [CrossRef]

26. Zhong, H.; Lan, X.; Gao, J.; Xu, C. Effect of particle frictional sliding during collisions on modeling the hydrodynamics of binary particle mixtures in bubbling fluidized beds. *Powder Technol.* **2014**, *254*, 36–43. [CrossRef]
27. Yang, S.; Luo, K.; Fang, M.; Fan, J. LES–DEM investigation of the solid transportation mechanism in a 3-D bubbling fluidized bed. Part II: Solid dispersion and circulation properties. *Powder Technol.* **2014**, *256*, 395–403. [CrossRef]
28. Di Renzo, A.; Di Maio, F.P.; Girimonte, R.; Vivacqua, V. Segregation direction reversal of gas-fluidized biomass/inert mixtures—Experiments based on particle segregation model predictions. *Chem. Eng. J.* **2015**, *262*, 727–736. [CrossRef]
29. Li, T.; Grace, J.; Bi, X. Study of wall boundary condition in numerical simulations of bubbling fluidized beds. *Powder Technol.* **2010**, *203*, 447–457. [CrossRef]
30. Zhao, Y.; Lu, B.; Zhong, Y. Influence of collisional parameters for rough particles on simulation of a gas-fluidized bed using a two-fluid model. *Int. J. Multiph. Flow* **2015**, *71*, 1–13. [CrossRef]
31. Asegehegn, T.W.; Schreiber, M.; Krautz, H.J. Influence of two- and three-dimensional simulations on bubble behavior in gas-solid fluidized beds with and without immersed horizontal tubes. *Powder Technol.* **2012**, *219*, 9–19. [CrossRef]
32. Xie, N.; Battaglia, F.; Pannala, S. Effects of using two-versus three-dimensional computational modeling of fluidized beds. *Powder Technol.* **2008**, *182*, 1–13. [CrossRef]
33. Geng, S.; Jia, Z.; Zhan, J.; Liu, X.; Xu, G. CFD modeling the hydrodynamics of binary particle mixture in pseudo-2D bubbling fluidized bed: Effect of model parameters. *Powder Technol.* **2016**, *302*, 384–395. [CrossRef]

energies

Article

Study on the Effects of Evaporation and Condensation on the Underfloor Space of Japanese Detached Houses Using CFD Analysis

Wonseok Oh [1,*] and Shinsuke Kato [2]

[1] Graduate School of Engineering, Department of Architecture, The University of Tokyo, 4-6-1 Komaba, Meguro-ku, Tokyo 153-8505, Japan
[2] Institute of Industrial Science, The University of Tokyo, 4-6-1 Komaba, Meguro-ku, Tokyo 153-8505, Japan; kato@iis.u-tokyo.ac.jp
* Correspondence: oh-ws@iis.u-tokyo.ac.jp; Tel.: +81-3-5452-6430

Academic Editor: Bjørn Hjertager
Received: 27 April 2017; Accepted: 8 June 2017; Published: 13 June 2017

Abstract: The purpose of this study is to determine the effects of evaporation and condensation on the underfloor space of Japanese detached houses. In this underfloor space, natural ventilation is applied. A typical Japanese wooden detached house is raised 0.3–0.5 m over an underfloor space made of concrete. The bottom of the underfloor space is usually paved with concrete, and the ceiling which is directly underneath the indoor occupant zone is made of wood. Computational fluid dynamics (CFD) analysis is applied to calculate the rates of the evaporation and condensation generated inside the underfloor under two conditions, namely, a constant (fixed) outdoor environmental condition and a fluctuating environmental condition. In the constant condition, we verified the effects of the outdoor humidity, ventilation rate, and ratio of wetted surface (RWS, ω) on the evaporation and condensation inside the underfloor space. In this condition, the rate of evaporation and condensation was quantified considering the varying outdoor humidity between 0 to 100%, and the RWS (ω = 1 or 0). In addition, the influence of the different ventilation rates at 1.0 m/s for normal and 0.05 m/s for stagnant wind velocities were confirmed. Under fluctuating environmental conditions, the outdoor conditions change for 24 h, so the RWS varies. Therefore, the rate of evaporation and condensation, the amount of the condensed water, and the area of condensation were confirmed. The results were as follows: with a high airflow rate on the underfloor space, the evaporation and condensation phenomenon occurs continuously and is easily affected by outdoor humidity, while under low airflow rate conditions, only the condensation appeared steadily. If the wind velocity is strong, the convective mass transfer on a surface becomes large. In a condition of the outdoor humidity and the airflow rate on underfloor are high, condensation mainly occurs in a corner of the underfloor space due to high evaporation by convection in the mainstream of the airflow. By contrast, when the airflow rate is low, condensation occurs along the air stream. Accordingly, this information could be employed as design considerations for the underfloor space at the architectural design stage.

Keywords: underfloor ventilation; evaporation and condensation; computational fluid dynamics (CFD); ratio of wetted surface (RWS)

1. Introduction

The aim of ventilation of the underfloor space is to prevent moisture generated in the ground from entering the living area and to improve the heating and cooling efficiency of the building, given the large heat capacity of the ground. Moisture is the main problem in wooden buildings, as it causes rotting of the wood and an unpleasant smell inside the building. Even so, the underfloor space is generally ignored by the residents, as it is not only inconspicuous, but also difficult to clean. Generally, moisture originates from the ground (soil), which absorbs water after rain, or from groundwater, and enters the building through the ventilation openings. It is important to block the source of the moisture in order to control the humidity level and to prevent the generation of condensation inside the underfloor space. Therefore, it is crucial not only to prevent ground moisture from entering, but also to eliminate any moisture forming inside from rainwater or high humidity. Since these phenomena influence the health of the occupants and the durability of the buildings, methods, such as paving of the ground or mechanical ventilation are applied occasionally. However, it is useful to predict the climate of the underfloor space in the attempt to control the humidity and to apply an appropriate method during construction, as installing such alternative methods could be costly.

In previous research, the relative humidity, temperature, and pressure inside the underfloor space were measured, the ground moisture evaporation rate was calculated, and subsequently the effects of natural and mechanical ventilation were confirmed [1]. The results showed that the average value of moisture evaporation with unpaved ground was 3.6 g h^{-1} m^{-2} in a naturally ventilated condition and 5.7 g h^{-1} m^{-2} in a mechanically ventilated condition. Research also indicated that the optimum air change rate on the underfloor space was 1–3 air change rate per hour (ACH) throughout the year [2]. Experiments have also been conducted with benzene in the underfloor space to identify the potential health risks associated with toxic soil vapors penetrating into the living spaces [3,4]. These studies employed point measurements. However, the internal humidity was excluded as a factor in the research. Several studies have attempted to confirm the presence of internal humidity by employing computational fluid dynamics (CFD) to determine the distribution of such internal humidity. However, no research has been conducted on the internal condensation of the underfloor space. In the current study, we attempt to determine the effects of internal condensation and evaporation and the probable area where condensation would occur relevant to the natural ventilation. Research on such aspects has not been conducted before.

A Japanese detached house can be simply illustrated as in Figure 1. The height of the underfloor space is 0.3–0.5 m from ground level. As openings are dispersed around the underfloor space for natural ventilation, outdoor moist air flows into the underfloor space, which gives rise to condensation and evaporation. The bottom of the underfloor space is usually paved with concrete and the ceiling, which is directly underneath the indoor occupant zone, is made of wood.

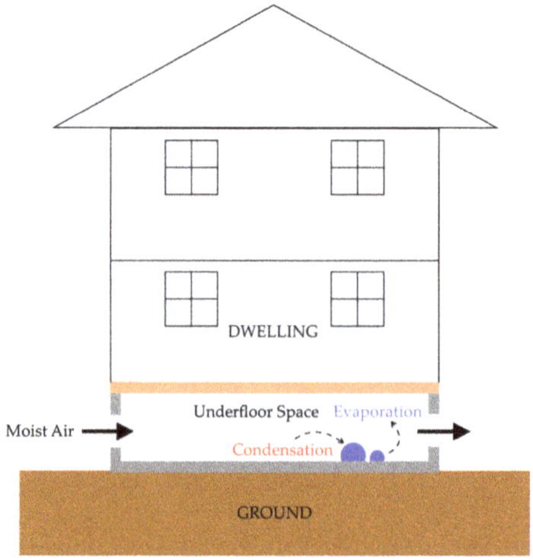

Figure 1. Underfloor ventilation in a Japanese detached house.

There are two main causes of condensation on the underfloor environment, namely, moisture deriving from the outside environment that condenses on the cold surface of the underfloor, and re-condensation that occurs when the condensed water inside the underfloor space evaporates. Once the condensation has been evaporated by ventilation, it becomes another moisture generation source inside the underfloor, raising the indoor humidity and significantly reducing the evaporation effect. The airflow and temperature distribution of the internal environment cause condensation and evaporation locally. Consequently, simple point measurements of the humidity cannot adequately explain the general condensation phenomena on the underfloor space or predict the probable occurrence of condensation. Furthermore, as the inside of the underfloor has a complicated shape, it is important to understand which parts would be vulnerable to condensation and to give due consideration to these vulnerable regions in the design of the house. Generally, in Japan, a detached house is a wooden building and it has an underfloor space, where its bottom touches the ground surface, which is usually paved with concrete to block water seepage from the ground. Natural ventilation is usually employed through underfloor air vents, shown in Figure 2, and mechanical ventilation is generally not introduced.

Figure 2. Underfloor openings for the natural ventilation in Tokyo, Japan.

As many previous studies have conducted research on the conditions of unpaved ground and natural ventilation, we attempted to confirm the internal condensation and evaporation effects on the underfloor space under conditions of paved ground and natural ventilation. It is difficult to determine the distribution of internal humidity by employing experimental methods; therefore, we chose CFD analysis to determine the climate inside the underfloor space. Additionally, the aim of the study is to understand the degree of evaporation and condensation arising because of underfloor ventilation and the distribution of the internal humidity, as well as to identify the area at risk of condensation.

Generally, In CFD, turbulent flows can be predicted through three approaches: direct numerical simulation (DNS), large-eddy simulation (LES), and Reynolds-averaged Navier-Stokes (RANS) equation simulated with turbulence models [5]. DNS gives highly reliable results directly solving Navier-Stokes equations but the required calculation time will be very long. The LES approach is an intermediate modeling technique between DNS and RANS. Turbulent motion can be divided into large eddies and small eddies, where LES is a method that solves the Navier-Stokes equations with filtered large-scale eddies of the turbulent flow except for small-scale eddies. LES can give us more detailed information on turbulence than RANS. However, it still requires a considerable calculation cost. Using RANS, turbulence motion can be quickly predicted and it is the most practical approach.

Computational fluid dynamics (CFD) is also the most effective way to predict the airflow in a building. Predicting airflow patterns in an indoor room by using a CFD analysis was first applied by Nielsen [6]. For the indoor environment prediction, the RANS k–ε turbulence model was used for many years and shows the acceptable results according to many researchers. [7–10]. The standard k–ε turbulence model showed reasonable results to predict the airflows in indoor environment [11]. Moreover, realizable k–ε models usually provide much improved results for swirling flows and flows involving separation compared to the standard k–ε model [5,12,13].

1.1. Weather Data in Tokyo, Japan

The extended Automated Meteorological Data Acquisition System (AMeDAS, 1981–2000) was used as standard-year weather data to estimate the underfloor climate. Tokyo city in Japan was selected as the representative location for our simulation. The detailed weather conditions are described in Figures 3 and 4. Figure 3 shows the correlation of the cumulative relative humidity per month in Tokyo. The wettest months of the year are June to September and a high-humidity climate prevails for more than half of this period, with the relative humidity being more than 75%.

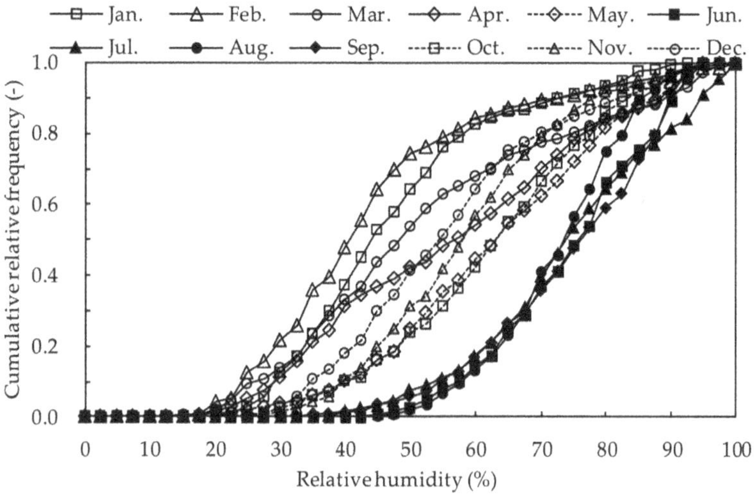

Figure 3. The correlation of the cumulative relative humidity and relative humidity per a month during a year in Tokyo, Japan.

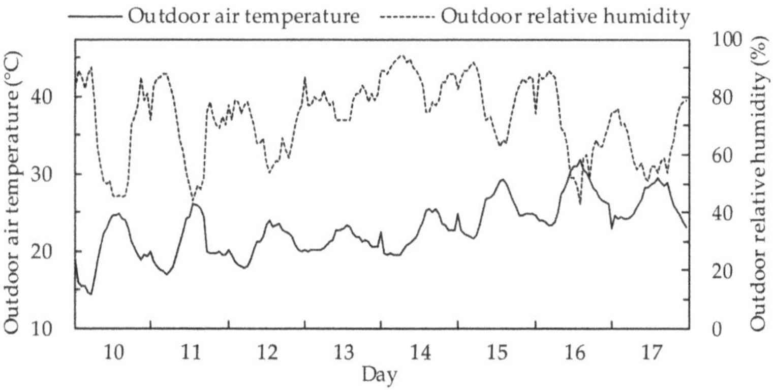

Figure 4. Outdoor air temperature and relative humidity of Tokyo according to the AMeDAS standard weather data (10–17 June).

To consider the worst probable weather situation, the data for 10–17 June were employed in our study, as these days represent the high humidity season in Tokyo. During the rainy season, although the external humidity increases significantly, the temperature of the internal environment remains lower than that of the external environment. This is because the influence of solar radiation is insignificant as the sky is always overcast. Therefore, the risk of internal condensation during this period is expected to increase significantly.

1.2. Research Procedure

In the design of the underfloor space, it is important to consider the interior environmental characteristics and the likely location and time of condensation occurring. However, since the internal airflow is complicated, it is difficult to predict the areas where condensation could occur. Therefore, we employed CFD analyses to determine and confirm the effects of evaporation and condensation relevant to natural ventilation. The underfloor space has dispersed ventilation openings for natural ventilation, with the pattern of the airflow being complex, as this area is divided into several zones by internal partition walls. Evaporation and condensation phenomena arise because of the difference of the partial pressure of water vapor caused by the difference between the surface temperature and the adjacent air temperature right near the surface. It is difficult to confirm this phenomenon inside the whole underfloor space through experiments. This study was conducted relevant to two different circumstances, namely, a constant outdoor environmental condition for 1 h and a fluctuating outdoor environmental condition for 24 h. The rate of evaporation and condensation generated inside the underfloor space was calculated and confirmed relevant to the first condition to determine the influence of outdoor environmental factors. The influence of natural ventilation was determined by quantifying the rate of evaporation and condensation with the internal surface being completely wet (ω = 1) and completely dry (ω = 0), and with the outdoor humidity increasing at a range of 0 to 100%, with a 10–20% interval. Under this condition, the influence of different ventilation rates was also confirmed by verification at a general wind speed condition of 1.0 m/s and a stagnant wind condition at 0.05 m/s. In a fluctuating condition, the ambient air temperature near the internal surface temperature also fluctuated continuously, and evaporation and condensation easily occurred because of the temperature difference. Accordingly, based on these fluctuating conditions, the rate of evaporation and condensation over time and the accumulated volume of condensed water on the surface were determined. The information obtained from our research could be utilized at the architectural design stage to pinpoint the spots vulnerable to and the conditions conducive to the generation of condensation inside the underfloor space.

2. Natural Ventilation on Underfloor Space

Since the geometry of the underfloor space is complex and it has many vents all around, it is difficult to predict the internal airflow patterns. Generally, as the wind blows from the south during a summer season in Japan, it is considered only two conditions of wind speed. During a calculation, the variation of wind direction and wind speed were disregarded in this study. The ventilation rate from inside and outside pressure difference at each opening was calculated.

2.1. Outline of CFD Analysis

The underfloor ventilation rate was calculated by realizable k–ε model with the semi-implicit method for pressure-linked equation (SIMPLE) algorithm and the second-order upwind-convection differencing scheme. The performance of realizable k–ε turbulence model is superior to that of other k–ε turbulence models [13]. Transport equations of airflow are solved for turbulence kinetic energy k (m^2/s^2) and its dissipation ε (m^2/s^3). The turbulent eddy viscosity ν_t is calculated as follows:

$$\nu_t = C_\mu \frac{k^2}{\varepsilon} \quad (1)$$

Transport equation of k and ε is expressed as:

$$\frac{\partial(\rho k)}{\partial t} + \frac{\partial(\rho k u_i)}{\partial x_i} = \frac{\partial}{\partial x_j}\left[\left(\mu + \frac{\mu_t}{\sigma_k}\right)\frac{\partial k}{\partial x_j}\right] + G_k + G_b - \rho\varepsilon - Y_M + S_k \quad (2)$$

and:

$$\frac{\partial(\rho\varepsilon)}{\partial t} + \frac{\partial(\rho\varepsilon u_i)}{\partial x_i} = \frac{\partial}{\partial x_j}\left[\left(\mu + \frac{\mu_t}{\sigma_k}\right)\frac{\partial \varepsilon}{\partial x_j}\right] + \rho C_1 S_\varepsilon - \rho C_2 \frac{\varepsilon^2}{k + \sqrt{\nu\varepsilon}} + C_{1\varepsilon}\frac{\varepsilon}{k} C_{3\varepsilon} G_b + S_\varepsilon \quad (3)$$

where G_k is the generation of turbulence kinetic energy due to the mean velocity gradients, G_b is the generation of turbulence kinetic energy due to buoyancy effect, Y_M is the contribution of the fluctuating dilatation in compressible turbulence to the overall dissipation rate, σ_k, σ_ε are the turbulent Prandtl numbers for k and ε and S_k, S_ε are source terms. The model constants are $C_{1\varepsilon} = 1.44$, $C_2 = 1.9$, $\sigma_k = 1.0$, and $\sigma_\varepsilon = 1.2$. The standard wall functions were applied to calculate the airflow near the wall.

2.2. Simulated Model

The simulation model was selected based on the Institute for Building Environment and Energy Conservation (IBEC) standard model which guides design and evaluation of a typical Japanese house [14]. At the bottom of the building is the underfloor space, which has air vents all around to supply natural ventilation.

The control volume for the calculation was selected properly, i.e., larger than the building model, as shown in Figure 5. Because the wind direction and speed were considered only for the constant southern wind, the south side was set as the inlet and the north side as the outflow (zero gradient condition).

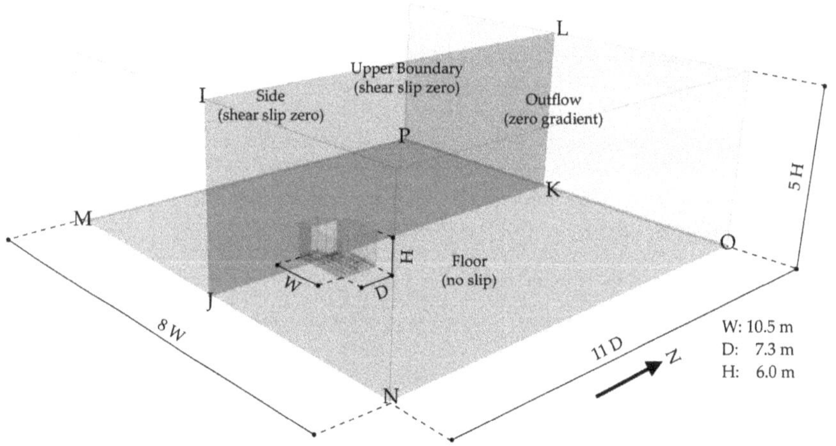

Figure 5. Residential building and control volume for calculating the ventilation rate in sub floor ventilation area.

The inlet wind from the south was created from the power law model to the mean velocity profile in Equation (4):

$$U_z = U_0(z/h)^{0.25} \quad (4)$$

where U_0 represents wind speed, and U_z is the wind speed at a height of z. The specified velocity U_0 takes into account two conditions, namely, a typical 1 m/s condition and a stagnant wind speed of 0.05 m/s, and the reference height h was assumed 1 m. The power law exponent was determined as 0.25 [15]. The logarithm-raw was applied on a surface of the building and the floor as a boundary condition. The side boundary (east, west) and upper boundary conditions were chosen as symmetry conditions (shear slip zero). The calculation was conducted by realizable k–ε turbulence model using a fine grid with 38 million hexahedral cells. Predicting airflow and dispersion around a building, numerical calculation by realizable k–ε turbulence models showed good agreement with the wind tunnel experiment than standard k–ε turbulence model [16]. Plan IJKL is a cross section of the left part of the opening on the southern wall of the building. Plan MNOP refers to the area 0.35 m above ground level.

2.3. Anslysis Results

The results for the two wind speed conditions, namely, 1.0 m/s and 0.05 m/s, are as follows: the wind direction is from the south, with the wind flowing into the underfloor space through two openings in the wall on the southern side, from where it is dispersed into each internal zone and exhausted through the other five openings.

The normalized velocity distribution around the building and inside the underfloor is shown in Figure 6. The airflow rate was calculated by the wind speed of the outdoor and the internal pressure difference of each opening, with the result shown in Table 1. In addition, this result was used as a boundary condition to calculate where the moist air from outside would be drawn into the internal underfloor space.

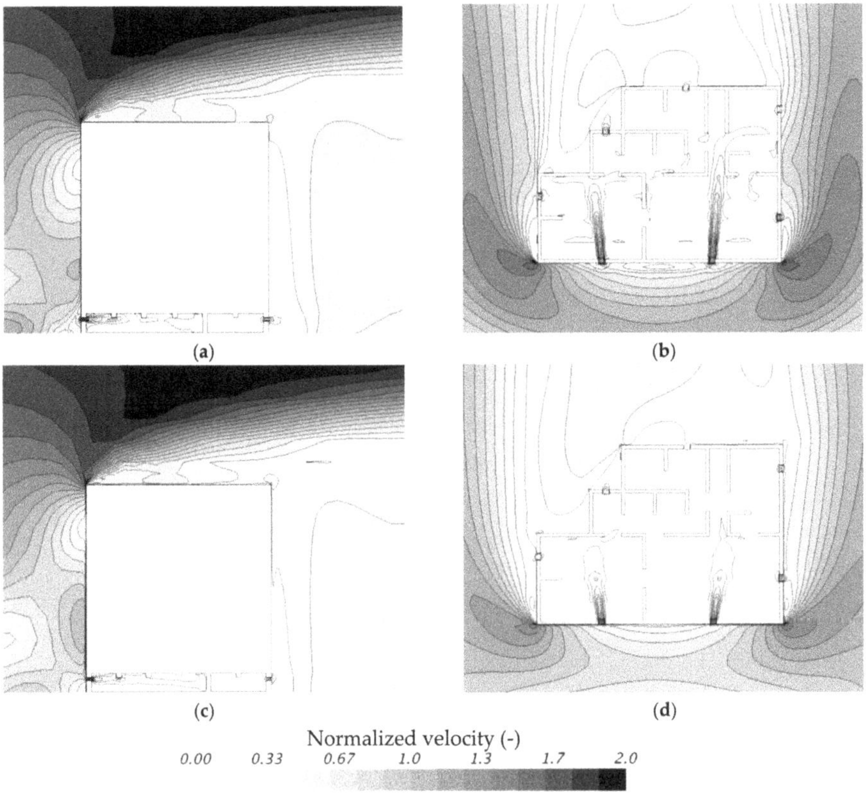

Figure 6. The normalized velocity $\frac{U}{U_0}$ distribution of around the building and inside of the underfloor space in condition of $\frac{U_z}{U_0} = \left(\frac{z}{h}\right)^{1/4}$ velocity profile: (**a**) Scalar velocity distribution in section IJKL, where the velocity condition $U_0(h)$ is 1.0 m/s; (**b**) Scalar velocity distribution in section MNOP, where the velocity condition $U_0(h)$ is 1.0 m/s; (**c**) Section velocity distribution on section IJKL, where the velocity condition $U_0(h)$ is 0.05 m/s; (**d**) Section velocity distribution on MNOP, where the velocity condition $U_0(h)$ is 0.05 m/s.

Table 1. The result of the Airflow rate through the underfloor openings.

Opening	Air Flow Rate ΔQ, m^3/h	
	1 m/s, South	0.05 m/s, South
E1	−75.3 *	−0.3 *
E2	−33.9 *	−1.9 *
W1	−54.5 *	−2.2 *
S1	112.4	4.9
S2	112.9	5.0
N1	−46.9 *	−1.7 *
N2	−33.5 *	−0.8 *

* A negative value means that the airflow is from inside to outside.

3. Evaporation and Condensation Effects on the Underfloor Space

3.1. Outline of Evaporation and Condensation Estimation Method

By modeling the water thickness calculation, the evaporation and condensation phenomena on the surface of the wall could be estimated. This involves the estimation of an additional scalar transport equation for the mass fraction of the water vapor. A source–sink term is considered for the scalar of the condensation and evaporation of the water layer, as well as the latent heat required for the transition. In the instance of a difference between the water vapor content of the water layer on the surface and the cell next to this surface, the model would calculate a rate of evaporation or condensation, depending on the conditions. This model assumes the following [17], namely, the vapor content in the air does not affect the thermal properties of the vapor–air mixture, and the water vapor mass is ignored with respect to the total mass in a cell.

The rate of mass transfer per unit surface, in kg s^{-1} m^{-2}, is given by:

$$\dot{m} = \rho_g \cdot \beta_g \cdot CG \cdot \ln \frac{CG(1 - C_s)}{CS(1 - C_g)} \quad (5)$$

where ρ_g is the density (kg m^{-3})

$$\beta_g = C_{emp} \cdot D_v / L \cdot 0.045 \cdot Re^{0.8} \cdot Sc^{0.43} \ (m/s) \quad (6)$$

C_{emp} is the empirical constant 0.9 (-)
L is a characteristic length (cubic root of cell volume next to fog layer boundary) (m)
D_v is diffusion of vapor in air (m^2/s)
Re is the Reynolds number (-)
Sc is the Schmidt number (-)

$$\begin{aligned} CG &= 0.622 + 0.378C_g \\ CS &= 0.622 + 0.378C_s \end{aligned} \quad (7)$$

C_g is the actual concentration of vapor in air
C_s is the saturation concentration of vapor in air

For a given temperature, the saturation pressure p_s is calculated from the following expression:

$$p_s(T) = 611.85e^{\frac{17.502(T+273.15)}{T-32.25}} \quad (8)$$

The change rate of the water thickness over time can be confirmed from the simulation. In addition, the mass change rate can be obtained from the calculated area of each cell, the water-thickness change rate, and water density. A decrease of the mass change rate indicates that the evaporation phenomenon has occurred and an increase means the condensation phenomenon has occurred. Finally, by determining the mass change rate, the rate of evaporation and condensation of the underfloor space can be calculated.

3.2. Simulated Model

The simulation model was selected based on the IBEC standard, which includes design standards for a Japanese detached house [14]. A two-story residential building is considered in this research, with seven openings on the underfloor space, as shown in Figure 7, which depicts the inside of the underfloor space. The walls and floor are made of concrete, with a basic thickness of 150 mm. The floor of the room corresponding to the ceiling of the underfloor space is made of a wood, supported by a wooden beam section of 105 mm × 105 mm, which is expressed as a dotted line. Figure 8 indicates the air movement passage inside the underfloor space and Figure 9 indicates the ventilation passage exposed to the outside, which is described in Figure 2.

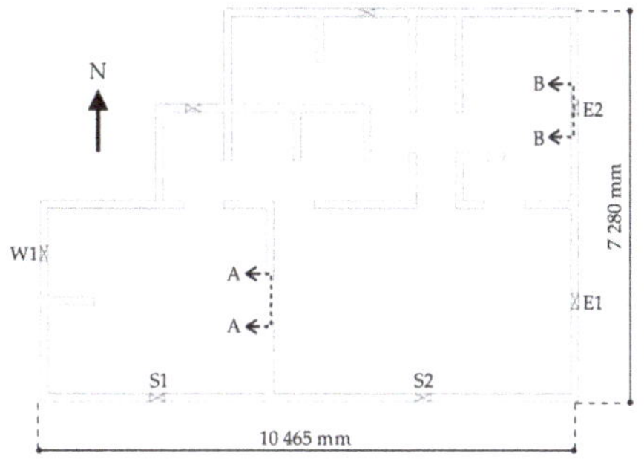

Figure 7. Inside of the underfloor ventilation area.

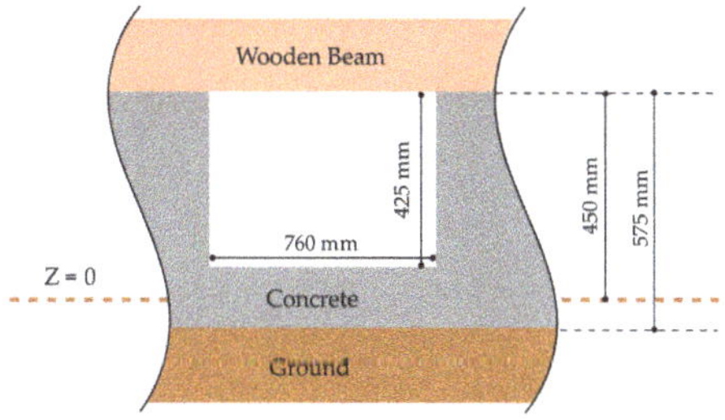

Figure 8. The section view of A-A.

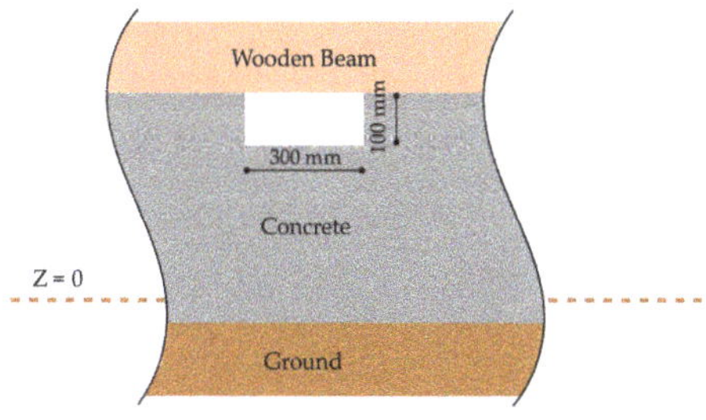

Figure 9. The section view of B-B.

The temperature of underfloor space is low even in the summer season, as it is mostly influenced by the ground temperature, which is not affected by solar radiation and does not have significant temperature changes. Our understanding is that the concrete surface on underfloor space of typical house is maintained low temperature, 20–22 °C during rainy season in Japan. The characteristics of the material should be considered because the surface temperature of the material is time dependent. The material properties for the calculations refer to the American Society of Heating, Refrigerating, and Air-Conditioning Engineers (ASHRAE) data in Table 2.

Table 2. The material properties of underfloor ventilation area and physical conditions.

Material	Density, (kg/m^3)	Specific Heat, (J kg^{-1} K^{-1})	Thermal Conductivity (W m^{-1} K^{-1})	Thickness (m)
Concrete	2240	750	0.53	0.15
Ground (soil)	1500	800	0.33	0.85
Floor (wood)	370	1900	0.11	0.012

The opposite side of the ceiling on underfloor space which represents the floor of the dwelling zone is assumed to remain constant at a temperature of 25 °C, as is the indoor environment. According to the weather data, the temperature of 1 m underground had almost no change in daytime. Therefore, it is assumed that the temperature of the concrete surface contacted with the ground is constant at 18.7 °C during the rainy season. As shown in Figure 10, the room temperature and the ground temperature is constant and the thermal conduction acts purely one-dimensional by the material characteristics of Table 2. The moist air from the outside was set to cause the evaporation and condensation phenomenon on the inner surface of underfloor space. Although some studies have considered the risk of annual condensation considering the absorption-desorption properties of the building materials [18], in this study, only the short-term evaporative condensation due to internal and external environmental conditions was focused on, so that the moisture transfer inside the materials was not considered.

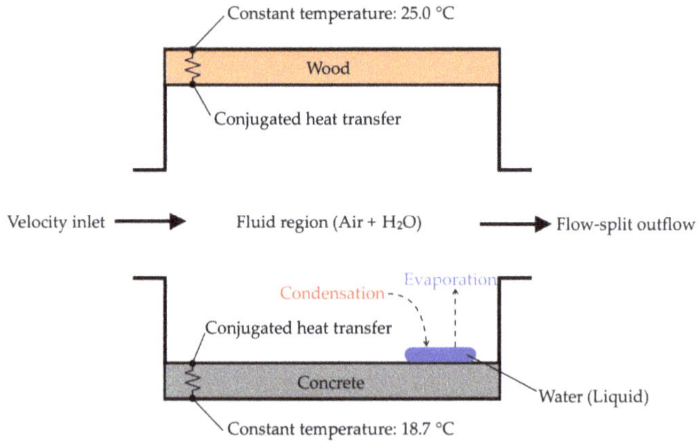

Figure 10. Sketch of the computational domain showing the relevant boundary conditions.

The pressure-velocity coupling for the airflow solution was conducted with the SIMPLE algorithm [19]. The convection and diffusion terms were integrated, using the QUICK difference scheme [20]. The boundary and calculation conditions are shown in Table 3. The parameters of the outdoor environment, which include wind speed (V_S), air temperature (T_S), and humidity (W_S) were applied in openings, the inlet boundary condition. The initial condition of the internal space was set equal to the temperature of the input initial condition. Wind speed conditions of 1.0 m/s and 0.05 m/s were considered, based on the ventilation rates calculated in the previous section. The condition of the inner wall was set to adiabatic and the emissivity was assumed as 0.95.

Table 3. Boundary and calculation conditions.

Turbulence model	Realizable k–ε turbulence model
Algorithm	SIMPLE
Convection	Upwind second-order difference
Numerical schemes	QUICK
Time dependence	Transient calculation
Room size	Described in Figure 9
Inlet	Velocity (V_S), Temperature (T_S), Humidity (W_S)
Outflow	Calculated with mass balance
Concrete-side wall	Standard log-law, emissivity = 0.95
Floor-side wall	Standard log-law, emissivity = 0.95
Initial conditions	Concrete: 20 °C, Floor: 25 °C

The emissivity of inside wall is supposed to be 0.95 and the simulation modeling of underfloor space is illustrated in Figure 11, which is composed of 48,000 hexahedral cells and covered with five prism layers. The value of y+ is lower than 1 on all over the surface of wall. All meshes were calculated by the realized k–ε turbulence model which performs better [13], and enhanced wall treatment, that is to say, two layer all y+ wall treatment which is a hybrid approach that seeks to recover the behaviors of the other two wall treatments in the limit of fine or coarse meshes. When the mesh resolution increases, the result showed better accuracy, while the calculation time has become longer. On the other hand, the results were shown quite similar in case of fine and coarse meshed model [21,22].

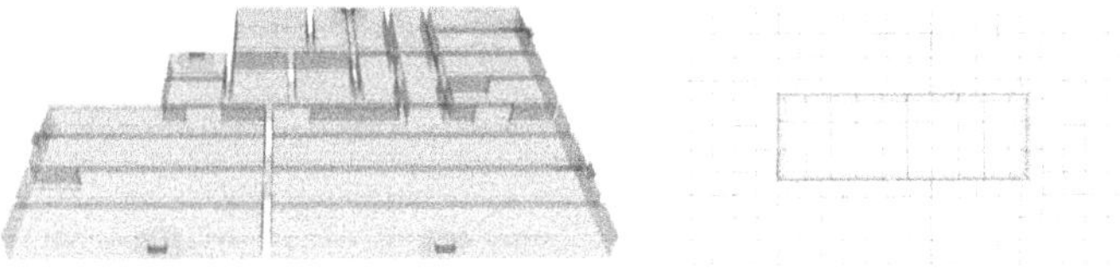

Figure 11. Outline of prism layers and hexahedral meshes prism layers on underfloor space and its opening.

3.3. Wetted Area Ratio

When there is a source of moisture generation inside, the partial pressure of the water vapor in the air is shown to differ, which affects the evaporation and condensation phenomena. Therefore, it is necessary to consider the wetted area ratio on the inner surface:

$$\text{Ratio of wetted surface (RWS, } \omega) = \frac{\text{Wetted surface area } (\text{m}^2)}{\text{Concrete surface area } (\text{m}^2)} \tag{9}$$

When water is present partially on the surface, the distribution of the internal partial pressure of water vapor differs, depending on its position due to the evaporation. Therefore, even if the ratio of wetted surface is the same, different aspects of evaporation and condensation phenomena are expected to be seen depending on the location of the wet area. However, as it is difficult to generalize the effect of condensation according to the position of the wet area, therefore, only the conditions where the RWS is 1 (all surfaces are completely wet) and the RWS is 0 (all surfaces are completely dry) were considered. Furthermore, a sufficient amount of water was taken into account, so that the ratio of the RWS would not change during the simulation (water thickness = 0.01 m).

If the evaporation phenomenon occurred on all the surfaces under the condition ω = 1, the maximum evaporation effect from underfloor ventilation could be assumed confirmed.

If condensation occurred on all the surfaces in the condition ω = 0, the maximum condensation effect from underfloor ventilation could be assumed confirmed. The wind speed was calculated based on the rate of ventilation at 1.0 m/s and 0.05 m/s. The temperature was set at 25 °C, considering the average temperature during the rainy season in June. In addition, the influence of relative humidity was confirmed by varying it within the range 0–100%, with 20% increasing intervals. A summary is presented in Table 4 of the detail conditions of each study case to confirm the evaporation and condensation effect on the internal underfloor space relevant to the outdoor environmental conditions. The effects of underfloor ventilation relevant to the outdoor environment were investigated by determining the overall average rate of evaporation and condensation inside the underfloor space after a time elapse of 1 h in each study condition. The mass change from evaporation and condensation was calculated for each cell of the surface through the above process.

Table 4. Initial conditions of the film thickness of water on surface of the concrete for the CFD analysis.

Outdoor Condition (25 °C)			Initial Water Thickness on Surface of the Concrete (m)	
Humidity (%)	**Wind Speed (m/s)**	**Wind Direction**	ω = 1	ω = 0
0	1.0	South	0.01	0
10			0.01	0
30			0.01	0
50			0.01	0
70			0.01	0
90			0.01	0
100			0.01	0
0	0.05		0.01	0
10			0.01	0
30			0.01	0
50			0.01	0
70			0.01	0
90			0.01	0
100			0.01	0

3.4. Analysis Results

The results of evaporation and condensation on underfloor space when outdoor environment conditions remained constant for an hour can be identified from the thickness of the water. Accordingly, the results, depending on the differences in the outdoor environmental conditions (wind speed, humidity), were compared by calculating the change per unit time (mg m^{-2} h^{-1}) through the analytical method, as shown in Figure 12.

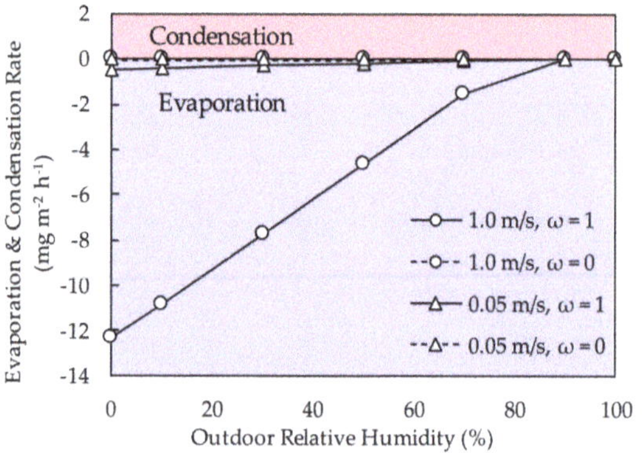

Figure 12. Evaporation and condensation effects relevant to the outdoor environmental conditions and RWS on the surface of the concrete on underfloor space.

Evaporation and Condensation Effect

The drier the air that flowed into the underfloor space and the larger the rate of ventilation, the faster the moisture evaporated from the surface to be exhausted to the outside, leading to the rate of evaporation increasing.

Under conditions of 1.0 m/s wind speed and 0% humidity, and when the entire surface was completely wet (ω = 1), evaporation occurred on all the surfaces, resulting in the largest evaporation rate of 12.3 mg m^{-2} h^{-1}. Considering that the floor area covers 59 m^2, the total evaporation rate of the internal underfloor space was 726 mg/h. When the outdoor humidity was 90%, the evaporation phenomenon reversed and condensation occurred, and, with humidity of 90–100%, the result was 0.4×10^{-3}–0.5×10^{-3} mg m^{-2} h^{-1}. The total internal condensation rate was indicated as 0.24–0.03 mg/h.

Under the condition of ω = 1, the outdoor wind speed decreased 1.0 m/s to 0.05 m/s, and the underfloor space shows a lower evaporation rate. This is because the exhausted volume of water that evaporated from the internal surface was smaller than the 1.0 m/s resulting from ventilation was. The higher the humidity the less the difference between the partial pressure of the water vapor on the water surface and the adjacent air would be, so that the occurrence of evaporation was reduced. Accordingly, with a 0–70% outdoor humidity condition, the evaporation effect was 0.5–0.05 mg m^{-2} h^{-1} per underfloor space on average. On the other hand, with a condition of 90% outdoor humidity, a state of equilibrium was reached, with no evaporation or condensation occurring. When outdoor humidity of 100% was introduced, the evaporation phenomenon reversed and 0.7×10^{-4} mg m^{-2} h^{-1} condensation occurred.

As there was no source of internal water generation when ω = 0, the evaporation phenomenon would not occur under humidity conditions of 0–70%. However, with 90–100% humidity, the moisture introduced from the outside air condensed on the cold surface of the concrete at a condensation rate of 0.03 g/h. The condensation phenomenon was slight, as indicated by the result of 0.4×10^{-2} mg/h. Furthermore, under conditions of 90–100% humidity and ω = 0, the condensation was 0.7×10^{-4} mg m^{-2} h^{-1}, which means all the internal surfaces were dry. As the temperature of the concrete was lower than was the air temperature of the internal underfloor space, the occurrence of condensation had higher significance when all the surfaces were dry (ω = 0) than when they were all wet (ω = 1). This is because the surface temperature of the dry concrete was lower than was the temperature of the water that formed on this surface; therefore, less condensation occurred. As regards the result shown in Table 5, (+) and (−) represent condensation and evaporation, respectively.

Table 5. Water thickness on the surface of the concrete relevant to the outdoor conditions.

Outdoor Condition (25 °C)			Evaporation and Condensation Rate (mg m^{-2} h^{-1})	
Humidity (%)	**Wind Speed (m/s)**	**Wind Direction**	ω = 1	ω = 0
0	1.0	South	−12.3	0
10			−10.8	0
30			−7.8	0
50			−4.7	0
70			−1.6	0
90			0.4×10^{-3}	0.5×10^{-3}
100			0.5×10^{-3}	0.5×10^{-3}
0	0.05		−0.5	0
10			−0.4	0
30			−0.3	0
50			−0.2	0
70			−0.05	0
90			0	0.7×10^{-4}
100			0.7×10^{-4}	0.7×10^{-4}

(+): Positive value corresponds to the condensation. (−): Negative value corresponds to the evaporation.

4. Daily Changes in Fluctuating Environmental Condition

In the second instance, we studied a fluctuation of outdoor environmental conditions over time, as the RWS status varied with the introduction of varying parameter values over time relevant to the outdoor environment. Accordingly, the effect of the ventilation rate of the underfloor space was confirmed by verifying the rate of evaporation and condensation, the amount of the condensed water, and the region where the condensation was likely to occur. Such information could be utilized in the architectural design of residences with underfloor spaces.

When the air conditions changed with time, the indoor conditions of evaporation and condensation fluctuated correspondingly. Furthermore, depending on the airflow, the surface temperature distribution, and moisture distribution, and the position where the condensation can occur would change, so as the RWS. When the condensed water evaporated, it acted as a generation source of water vapor, which affected the occurrence of the evaporation and condensation phenomena. Accordingly, our study confirmed variations over time in the rate of evaporation and condensation, the weight of condensed water, and the area where condensation would likely occur relevant to a rainy day during the rainy season, when the average humidity was high.

4.1. Weather Data

In this part of the study, we considered the outdoor environment during the rainy season, with the highest humidity levels and a high risk of condensation. Figure 13 shows the changes in daily temperature and relative humidity in Tokyo during the rainy season, and absolute humidity is expressed in Figure 14.

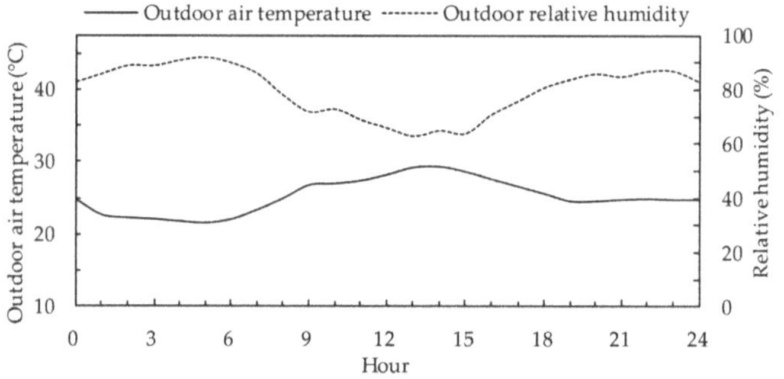

Figure 13. Variations in the daily outdoor air temperature and relative humidity in Tokyo, Japan, according to the AMeDAS standard weather data (15 June).

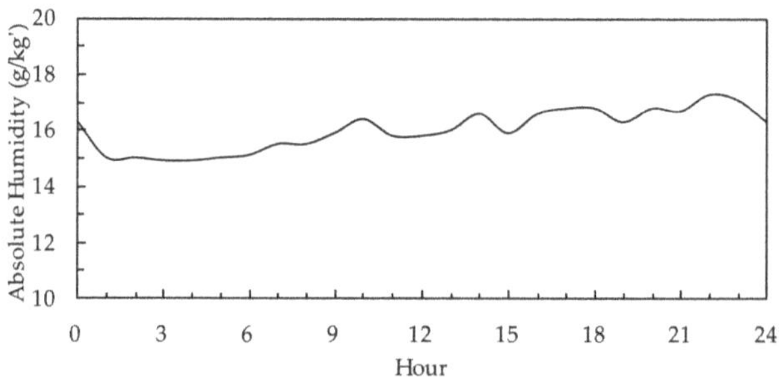

Figure 14. Daily outdoor absolute humidity variations of Tokyo, Japan by AMeDAS standard weather data (June 15).

The average temperature measured was 25 °C, and the highest temperature was measured between 12:00 and 15:00. The relative humidity with the increasing temperature was approximately 65%. The average daily humidity was 74%.

Wind speed conditions of the outdoor air were assigned uniformly by using the calculated ventilation rate of the underfloor, as described in Table 1, assuming a general wind speed and a stagnant wind speed. However, this time, changes in wind direction and wind speed as time elapses were not considered.

4.2. *Analysis Method*

As shown in Figure 15, the average rate of evaporation and condensation, the amount of condensed water on the surface, and the change in the condensation area over 24 h were confirmed in our study. As changes in wind speed and direction were not taken into consideration, we utilized the values in Table 1, representing the results of ventilation rates at 1.0 m/s and 0.05 m/s of the southern wind. In this instance, unlike the uniform condition of the outdoor environment, RWS changed continuously according to the elapse of time. The simulation calculations were conducted for 24 h, with the method reflected in Table 3. The environmental conditions of the outdoor air complied with the temperature and humidity data of the rainy season in Tokyo, as represented in Figure 13. The initial RWS condition of the internal surface was set at 0, which means that all the surfaces were dry.

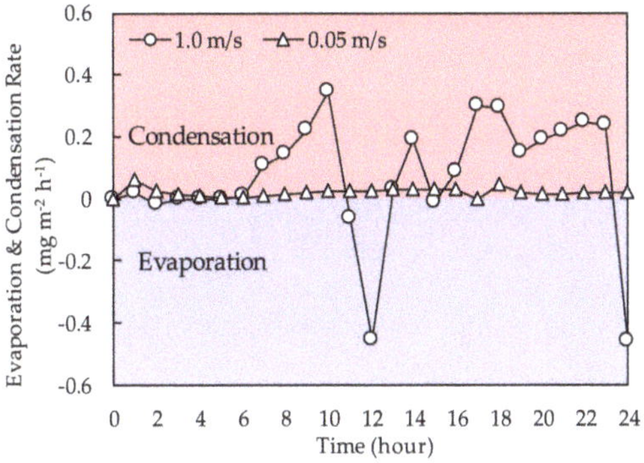

Figure 15. Variation in the average evaporation and condensation rate on the surface of the underfloor space during a day.

4.3. *Analysis Results*

The evaporation and condensation rates, the amount of the condensed water, and the area of condensation of the underfloor space were observed at hourly intervals over time to identify the effects of the ventilation rate of the underfloor space.

4.3.1. Evaporation and Condensation Effect

The distribution of temperature and pressure inside the underfloor was not uniform because of the influx of outdoor air introduced through the ventilation openings. The value of RWS changed continuously because of the condensation phenomenon. Furthermore, as evaporation and condensation occur differently, depending on the location, the rates of evaporation and condensation was confirmed through the average rate of the entire underfloor space.

When the amount of evaporation was greater than that of condensation, it is expressed as the total average condensation, but when the amount of condensation was greater than that of evaporation, it is expressed as the total average evaporation. The internal humidity of the underfloor would be influenced by the outdoor humidity when the volume of introduced outdoor air increased with an increase in the ventilation rate. In the instance of the outdoor air velocity being 1.0 m/s, the absolute humidity increased with the rising temperature after 6:00 in the morning. The rate of internal condensation continued to increase until 11:00, resulting in a maximum condensation rate of 0.34 mg m^{-2} h^{-1} at 10:00. From 11:00, the condensation rate gradually decreased and the condensed water on the surface evaporated between 11:00 and 12:00, whereas the evaporation rate increased significantly at 12:00 to 0.46 mg m^{-2} h^{-1}. During the rainy season, the humidity remains high throughout the day. Therefore, during the calculated period, the absolute humidity increased with the rise in temperature and the condensation phenomenon appeared again after 13:00 and continued until 23:00. Furthermore, at 24:00, the humidity decreased and evaporation occurred at a largest value of 0.46 mg m^{-2} h^{-1} for the day measured. In a condition of stagnant wind velocity (0.05 m/s), it was not influenced significantly by the outdoor conditions. This means that not only was the outdoor moisture not introduced well but also that the internal moisture was not exhausted well. As time elapsed, the condensation phenomenon continued as the humidity from outside entered the underfloor space; however, the evaporation did not occur for 24 h. Furthermore, the evaporation and condensation phenomena were shown to be in equilibrium at 15:00.

4.3.2. Evaporation and Condensation Effect

We confirmed the overall characteristics of the internal underfloor space over time according to the average value of the amount of condensed water occurring in the entire underfloor space. The occurrence of condensation initially commenced when all the internal surfaces were in the dry state and it subsequently accumulated water inside the underfloor space.

As shown in Figure 16, in a condition of wind speed of 1.0 m/s, the condensation continued until 10:00, with evaporation starting to occur between 11:00 and 12:00, and the amount of condensed water starting to decrease. Between 13:00 and 23:00, condensation continued, with the amount of condensed water being the highest at an average value of 2.22 mg/m^2. At 24:00, the rate of evaporation was high, resulting in an average cumulative condensation of 1.76 mg/m^2 per day. Condensation occurred continuously with a wind velocity of 0.05 m/s, assuming a stagnant environment, and an amount of 0.48 mg/m^2 of cumulative condensed water was observed at 24:00.

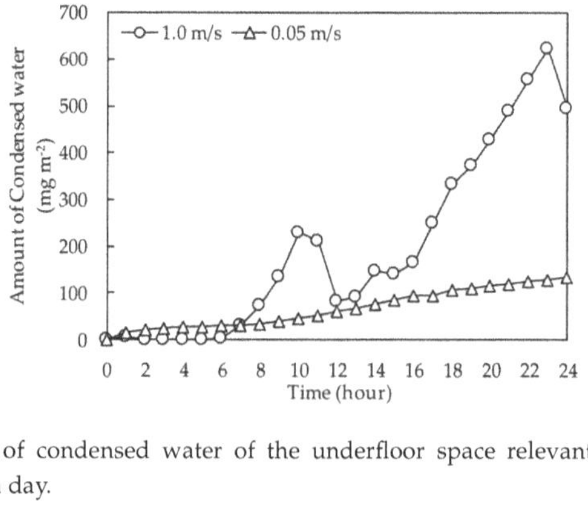

Figure 16. Amount of condensed water of the underfloor space relevant to evaporation and condensation during a day.

Convective mass transfer on surface increases as the air velocity increased. The air temperature increase causes the absolute humidity rise in the moist air due to the change of saturated vapor pressure. These results affect the amount of moist introduced from the outdoor air. As the higher ventilation rate, inflow rate of water vapor into underfloor space increase. This means that the convective mass transfer in a condition of 0.05 m/s is not a dominant comparing to 1.0 m/s. Before 6 o'clock, the evaporation at 1.0 m/s was stronger than the condensation due to water vapor inflow. On the other hand, at 0.5 m/s, the condensation by convection was stronger than the evaporation. In the case of 1.0 m/s, condensation due to water vapor flow is stronger than evaporation due to forced convection as the absolute humidity flowing into the inside increases after 6 o'clock. For this reason, the amount of condensed water in the underfloor space is reversed by 1.0 m/s and 0.05 m/s after 7 o'clock.

4.3.3. Condensed Water Distribution on Surface of Concrete

The area where condensation occurred differed according to the ventilation rates. As shown in Figure 17, at a wind speed of 1.0 m/s, condensation did not occur where the airflow is strong, but did occur nearby the mainstream of the strong airflow, and was likely to occur in the corners of the underfloor space. However, in the northern zone, where the airflow does not reach, condensation hardly occurred at all. The evaporation phenomenon was likely to occur in an area of strong airflow, as mass transfer easily occurred and condensation rarely occurred. In addition, the convective heat transfer from the surface was more pronounced because of the strong airflow when the outdoor air flowed in; therefore, condensation was less likely to occur. In contrast, in a stagnant condition of 0.05 m/s wind speed, condensation occurred mostly near the zone where the airflow was generated. This is because the velocity of the incoming airflow was extremely low and the convective heat transfer on the surface was not large; therefore, the moisture of the incoming outdoor air condensed at the area where the airflow was formed.

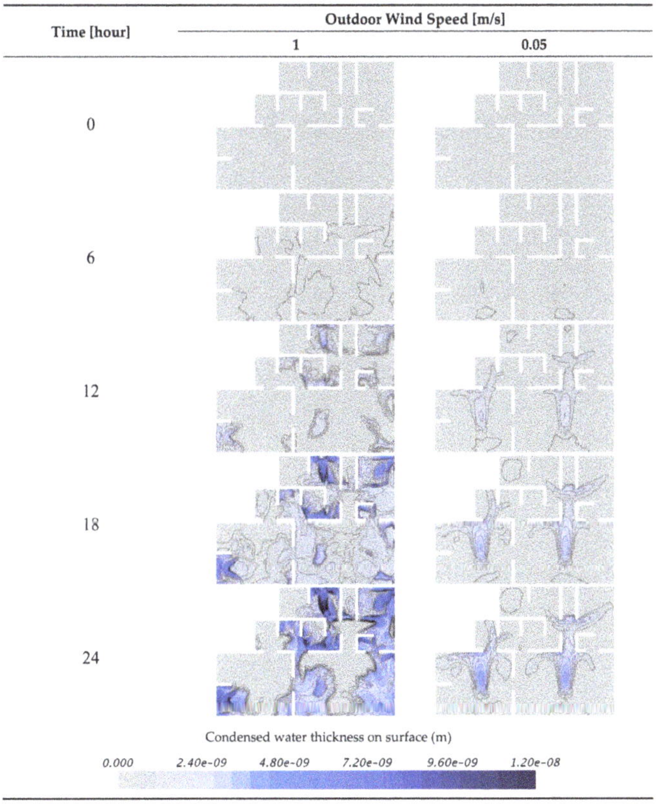

Figure 17. Variation in the condensed water thickness on the surface of the concrete during a day.

5. Discussion

The RWS of the initial internal environment was assumed zero in fluctuating conditions. Under such conditions, dry air was introduced into the underfloor space and the evaporation phenomenon did not occur, even if ventilation were performed. Therefore, the result could not be confirmed and could be different from the actual phenomenon. However, as the relative humidity of the airflow in the initial condition was 80%, the evaporation phenomenon was considered unlikely to occur originally. It is necessary to check the temperature distribution by calculations over a long time in order to confirm an environment similar to the actual environment. However, both the initial surface temperature and the internal temperature of the material were assumed uniform because of the limitation of the calculation load. Therefore, it could differ from the actual environmental conditions.

6. Conclusions

In the case where condensed water exists inside the subfloor, the rate of evaporation was shown to be greater when the airflow rate was high. However, under extreme conditions where the outdoor humidity is 90% or more, condensation occurs instead of evaporation. When the outdoor humidity was less than 90%, the condensation did not occur, regardless of the airflow rate. This means that keeping the inside of the underfloor space free from condensed water and dry is important. In the situation when the outdoor humidity is high, such as the rainy season, internal condensation in the underfloor space easily occurs due to natural ventilation no matter what the airflow rate is. If the airflow rate is high, it is easily affected by outdoor humidity, and the evaporation and condensation phenomena are continuously changing. On the other hand, under low airflow rate conditions, only the condensation appeared steadily. If the wind velocity is strong, the convective mass transfer on a surface becomes large. When the outdoor humidity is high and the airflow rate is high, condensation mainly occurs in a corner of the underfloor space due to high evaporation by convection in the mainstream of the airflow. By contrast, when the airflow rate is low, condensation occurs along the air stream. If condensation occurs once in the corner where the airflow has not reached well, it is difficult to evaporate again. Therefore, it is necessary to control the ventilation rate according to the characteristics of the external environment and internal RWS.

1. We confirmed the influence of the difference in the humidity when the outdoor wind speed was assumed 1.0 m/s and the outdoor air flowing into the room remained constant according to the calculated ventilation rate. In the instance of a RWS value of 1, evaporation occurred actively and the evaporation rate was 12.3 mg m^{-2} h^{-1}. The total volume of the internal evaporation rate was 726 mg/h. The evaporation phenomenon did not occur in a condition of 0–70% humidity, where the value of RWS was 0 and the entire surface was dry. When humidity was high at 90–100%, the total condensation rate of the underfloor was 0.03 g/h.
2. We confirmed the influence of the difference in humidity when the outdoor wind speed was assumed stagnant at 0.05 m/s and the outdoor air flowing into the room remained constant according to the calculated ventilation rate. When the value of RWS was 1, the average evaporation effect per area was determined as 0.5–0.05 mg m^{-2} h^{-1}, and the total volume of the evaporation rate was 29.5–3.0 mg/h. In a condition of 90% humidity of the introduced outdoor air, a state of equilibrium was indicated, with neither evaporation nor condensation occurring. With 100% humidity, the evaporation phenomenon reversed and a condensation phenomenon of 0.7×10^{-4} mg m^{-2} h^{-1} occurred. However, the condensation phenomenon was slight, as indicated by the result of 0.4×10^{-2} mg/h. When the RWS value was zero, with all the internal surfaces dry and humidity of 90–100%, the total condensation rate of the underfloor space was 0.03 g/h.
3. When the outdoor wind speed was 1.0 m/s and the outdoor environmental conditions fluctuated for 24 h, the outdoor humidity remained high and condensation continued. However,

with the temperature rising or humidity decreasing, evaporation occurred between 11:00 and 12:00, and at 24:00. The maximum condensation rate was 0.34 mg m^{-2} h^{-1} at 10:00 and the maximum evaporation rate was 0.46 mg m^{-2} h^{-1} at 12:00 and at 24:00.

4. In a stagnant environment, with an outdoor wind speed of 0.05 m/s, the rate of natural ventilation into the underfloor was low and the effect of the outdoor environment was therefore insignificant. Consequently, condensation continued for 24 h and evaporation did not occur at all.
5. Under a condition of 1.0 m/s wind speed, evaporation and condensation occurred continuously throughout the day and the cumulative rate of condensation of the underfloor space was 1.76 mg/m^2 per day. Under a condition of 0.05 m/s wind speed, the cumulative rate of condensation of the underfloor space from natural ventilation was 0.48 mg/m^2 per day.
6. When the wind speed was 1.0 m/s, condensation did not occur near the strong airflow, but did occur in the vicinity of the airflow, and easily occurred in the corners. Furthermore, as there was less moisture transfer through the airflow, condensation hardly occurred where the airflow could not reach. At a wind speed of 0.05 m/s, assuming a stagnant condition, moisture was transferred to the inside by the airflow and condensation occurred in accordance with the airflow.

Author Contributions: Wonseok Oh performed the simulations, analyzed the simulation results, designed the simulations and wrote the paper. Shinsuke Kato reviewed the paper. All authors have read and approved the final manuscript.

Conflicts of Interest: The authors declare no conflict of interest.

References

1. Kurnitski, J. Ground moisture evaporation in crawl spaces. *Build. Environ.* **2001**, *36*, 359–373. [CrossRef]
2. Kurnitski, J.; Kurnitski, J. Crawl space air change, heat and moisture behaviour. *Energy Build.* **2000**, *32*, 19–39. [CrossRef]
3. Ferguson, C.C.; Krylov, V.V.; McGrath, P.T. Contamination of indoor air by toxic soil vapours: A screening risk assessment model. *Build. Environ.* **1995**, *30*, 375–383. [CrossRef]
4. Krylov, V.V.; Ferguson, C.C. Contamination of indoor air by toxic soil vapours: The effects of subfloor ventilation and other protective measures. *Build. Environ.* **1998**, *33*, 331–347. [CrossRef]
5. Zhai, Z.J.; Zhang, Z.; Zhang, W.; Zhang, Z.; Zhang, W.; Chen, Q.Y. Evaluation of various turbulence models in predicting airflow and turbulence in enclosed environments by CFD: Part 1—Summary of prevalent turbulence models. *HVAC R Res.* **2007**, *13*, 853–870. [CrossRef]
6. Nielsen, P.V. Flow in Air Conditioned Rooms: Model Experiments and Numerical Solution of the Flow Equations. Ph.D. Thesis, Technical University of Denmark, Nordborg, Denmark, 1974.
7. Murakami, S.; Kato, S. Numerical and experimental study on room airflow—3-D predictions using the k-ε turbulence model. *Build. Environ.* **1989**, *24*, 85–97. [CrossRef]
8. Nielsen, P.V. The description of supply openings in numerical models for room air distribution. *Indoor Environ. Tech.* **1991**, *R9251*, 963–971.
9. Nielsen, P.V. Computational fluid dynamics and room air movement. *Indoor Air* **2004**, *7*, 134–143. [CrossRef] [PubMed]
10. Gan, G. Evaluation of room air distribution systems using computational fluid dynamics. *Energy Build.* **1995**, *94*, 665–675. [CrossRef]
11. Zhai, Z.J. Application of computational fluid dynamics in building design: Aspects and trends. *Indoor Built Environ.* **2006**, *15*, 305–313. [CrossRef]
12. Van Maele, K.; Merci, B. Application of two buoyancy-modified k–ε turbulence models to different types of buoyant plumes. *Fire Saf. J.* **2006**, *41*, 122–138. [CrossRef]
13. Shin, T.H.; Liou, W.W.; Shbbir, A.; Yang, Z.; Zhu, J. A new k-ε eddy viscosity model for high Reynolds number turbulent flows. *Comput. Fluids* **1995**, *24*, 227–238.
14. *CASBEE for Home (Detached House) Technical Manual 2007 Edition;* Japan Sustainable Building Consortium & Japan Green Build Council: Tokyo, Japan, 2007; pp. 195–215.

15. Counihan, J. Adiabatic atmospheric boundary layers: A review and analysis of data from the period 1880–1972. *Atmos. Environ.* **1975**, *9*, 871–905. [CrossRef]
16. Tominaga, Y.; Stathopoulos, T. Numerical simulation of dispersion around an isolated cubic building: Comparison of various types of k–ε models. *Atmos. Environ.* **2009**, *43*, 3200–3210. [CrossRef]
17. Sandhu, K.S. *Predicting the Windscreen Demisting Performance Using CAE*; Elsevier: Amsterdam, The Netherlands, 2011; pp. 401–410.
18. Cho, W.; Iwamoto, S.; Kato, S. Condensation risk due to variations in airtightness and thermal insulation of an office building in warm and wet climate. *Energies* **2016**, *9*, 875. [CrossRef]
19. Patankar, S.V.; Spalding, D.B. A calculation procedure for heat, mass and momentum transfer in three-dimensional parabolic flows. *Int. J. Heat Mass Transf.* **1972**, *15*, 1787–1806. [CrossRef]
20. Leonard, B.P. A stable and accurate convective modelling procedure based on quadratic upstream interpolation. *Comput. Methods Appl. Mech. Eng.* **1979**, *19*, 59–98. [CrossRef]
21. Mulvany, N.; Tu, J.Y.; Chen, L. Assessment of two-equation turbulence modelling for high Reynolds number hydrofoil flows. *Int. J. Numer. Methods Fluids* **2004**, *45*. [CrossRef]
22. Salim, S.M.; Cheah, S. Wall Y strategy for dealing with wall-bounded turbulent flows. In Proceedings of the International Multi Conference of Engineers and Computer Scientists, Hong Kong, China, 18–20 March 2009; Volume II.

energies

Article

Parametric Investigation Using Computational Fluid Dynamics of the HVAC Air Distribution in a Railway Vehicle for Representative Weather and Operating Conditions

Christian Suárez [1], Alfredo Iranzo [2,*], José Antonio Salva [1], Elvira Tapia [2], Gonzalo Barea [3] and José Guerra [2]

[1] AICIA, Andalusian Association for Research & Industrial Cooperation, Camino de los Descubrimientos s/n, Edf. Escuela Superior de Ingenieros de Sevilla, 41092 Seville, Spain; chss@us.es (C.S.); jsalva@us.es (J.A.S.)
[2] Thermal Engineering Group, Energy Engineering Department, School of Engineering, University of Seville, Camino de los Descubrimientos s/n, 41092 Seville, Spain; etapia@us.es (E.T.); jjguerra@us.es (J.G.)
[3] Hispacold, C/Pino Alepo 1, Polígono Industrial El Pino, 41016 Seville, Spain; gbarea@hispacold.es
* Correspondence: airanzo@us.es; Tel./Fax: +34-954487471

Academic Editor: Bjørn H. Hjertager
Received: 26 June 2017; Accepted: 19 July 2017; Published: 25 July 2017

Abstract: A computational fluid dynamics (CFD) analysis of air distribution in a representative railway vehicle equipped with a heating, ventilation, air conditioning (HVAC) system is presented in this paper. Air distribution in the passenger's compartment is a very important factor to regulate temperature and air velocity in order to achieve thermal comfort. A complete CFD model, including the car's geometry in detail, the passengers, the luminaires, and other the important features related to the HVAC system (air supply inlets, exhaust outlets, convectors, etc.) are developed to investigate eight different typical scenarios for Northern Europe climate conditions. The results, analyzed and discussed in terms of temperature and velocity fields in different sections of the tram, and also in terms of volumetric parameters representative of the whole tram volume, show an adequate behavior from the passengers' comfort point of view, especially for summer climate conditions.

Keywords: heating, ventilation, air conditioning; computational fluid dynamics; railway vehicle; heat transfer; thermal comfort; tram

1. Introduction

Air distribution in passenger's compartment is a very important factor to regulate temperature and air velocity in order to achieve thermal comfort. The study of air distribution plays an important role in the design of new HVAC (heating, ventilation, and air conditioning) equipment and also in the evaluation of existing solutions. Technological solutions such as innovations in air-conditioning and other forms of cooling or ventilation can improve environmental conditions, which is beneficial for human health, comfort, and productivity.

Saving energy and providing thermal comfort are two important goals of HVAC systems. Regarding the first goal, the European Project ECORailS (Energy Efficiency and Environmental Criteria in the Awarding of Regional Rail Transport Vehicles and Services, 2011) [1] for enhanced energy-efficiency of regional rail passenger services is a good example, as its main objective is to reduce the specific energy consumption of European regional passenger rail transport by 15% by 2020. While it is relatively easy to define and estimate the energy consumption of HVAC systems, the evaluation of thermal comfort is much more challenging. The American Society of Heating, Refrigerating and Air-Conditioning Engineers (ASHRAE) defines thermal comfort as "condition of mind that expresses satisfaction with the thermal environment" [2]. As it depends on personal factors (such as metabolic rate or clothing) and environmental factors (such as air velocity, air temperature, air temperature stratification, radiant temperature, radiant temperature asymmetry, relative humidity, or turbulence intensity in the occupied zone), the definition and evaluation of thermal comfort is subjective and complex.

European standard EN-14750-1:2006 [3], relative to air conditioning for urban and suburban rolling stock, establishes a classification of the railway vehicles depending on the number of standing passengers, the average travel time, and the average time between two vehicle stops. According to that definition a railway vehicle is category A for suburban/regional transport and B for the rest. It also defines different comfort parameters and requirements that depend on the vehicle category and are related to the temperature and velocity within the railway vehicle.

Keeping fixed the parameters related to personal factors and the indoor geometry, thermal comfort is directly related to air conditions: velocity, temperature, and temperature stratification. It is evident that different diffusers and different locations of supply inlets and exhaust outlets will affect air distribution and, consequently, also the distribution of the cited thermal comfort parameters. Moreover, for a given HVAC system operating with different conditions (air supply temperatures, air supply flow, climate conditions etc.) these parameters will also change. Hence, it is necessary to understand quantitatively how these different operating conditions will affect local thermal comfort.

In public transport, such as railway vehicles, thermal comfort is even more important than in building applications. In fact, standing passengers inside a cabin for several minutes/hours do not have opportunities of moving. Passenger comfort has to be one of the most important elements in the design of public transport in order to persuade people to choose it instead of other means of transport. This is the reason why, for several years, many studies have focused on the air distribution system in the passenger's cabin.

As discussed by Liu et al., in its literature review paper [4] (focused on the particular case of commercial airliner cabins), two main approaches are available for analyzing air distribution: experimental measurements inside the cabin (with different equipment such as hotwire anemometers and hot-sphere anemometers, particle tracking velocimetry, particle streak velocimetry, particle image velocimetry, and ultrasonic anemometry) and numerical simulations (mainly CFD simulations). R. Lieto [5] studied, both numerically and experimentally, the indoor climate in city busses. He contrasted his numerical results with measurements of the air speed in some selected points in two different situations (a moving vehicle with the door closed and the vehicle at a bus stop with open doors). Zhu et al. [6] also studied, numerically and experimentally, the micro-environmental conditions in public transportation buses, focusing on thermal comfort and also on air quality. In other work, Zhu et al. [7] studied numerically the risk of airborne influenza infection in a bus microenvironment with different ventilation system configurations.

There are more publications regarding air distribution in aircraft cabins that are also focused on thermal comfort and air quality. Bianco et al. [8] studied numerically-transient simulations of the thermal and fluid dynamic fields in the cabin of an executive aircraft, considering 2D and 3D models. In other papers [9–14] the studies have been devoted to air quality and the distribution of pollutants, mainly CO and CO_2.

Although comfort conditions in public transport, such as buses or aircraft cabin, planes have been widely discussed in recent years, to the best of the knowledge of the authors, there are only a few papers specifically related to railway vehicles. Thompson et al. [15] presented a review of cooling systems used in railway vehicles and the advantages of using heat storage in the complex environment of an underground railway. Show [16] discussed a ventilation criterion for ensuring a clean environment, with low carbon-dioxide concentration (lower than 0.1% in air). To obtain such acceptable indoor air-quality inside the train compartment, he mentions two possibilities: to increase the fresh-air supply rate for dilution or to design a better air-distribution system. However, none of these studies in railway vehicles include an analysis of air distribution and its relationship with the passengers' thermal comfort.

To the best of the knowledge of the authors, perhaps due to the complex and time-consuming process of the geometry definition and the high required computational time in the simulations, there are not any previous studies related specifically to air distribution railway vehicles with passengers using CFD techniques. Even though all the modes of transportation share some similarities, such as the air conditioning and the comfort conditions, there are also some differences in the functionality and the passengers' distribution that make these transports different from a thermal analysis point of view. Moreover, as concluded in Yan et al. [17] the computational manikin model approach representing the passengers in the vehicle can affect the temperature fields, and an excessive degree of simplification can be incapable of predicting accurate results.

Experimental measurements, which are often considered reliable to analyze air distribution and thermal comfort, can be technically very difficult and expensive. It is very difficult to obtain results under realistic thermo-fluid conditions or detailed geometry. On the other hand, CFD simulations offer a good alternative. Due to their flexibility and efficiency a lot of scenarios can be studied with a relatively low cost. Furthermore, the results in terms of temperature and velocity fields can be obtained for the whole volume of interest and not only for a specific measured point.

In the present work, a detailed CFD analysis of air distribution and the temperature and velocity profiles in different sections of a representative railway vehicle is performed (including the passengers), in order to discuss the results in terms of thermal comfort.

2. Problem Definition and Methodology

A detailed three-dimensional CFD model of a railway vehicle formed by four cars is developed using the software ANSYS-CFX (ANSYS, Inc., Southpointe, Canonsburg, PA, USA). The tram (39.475 m length and 2.650 m width) is divided in two parts (part one: cars C1 and N, and part two: cars R and C2). In each car there are two doors and eight windows. Although cars C1 and C2 are identical, cars N and R are geometrically different.

In the model, a representative HVAC system and a random distribution of passengers are integrated. In all the studied cases, the tram remains stopped and the driver's cabin is not included. A 3D view of the railway vehicle is shown in Figure 1.

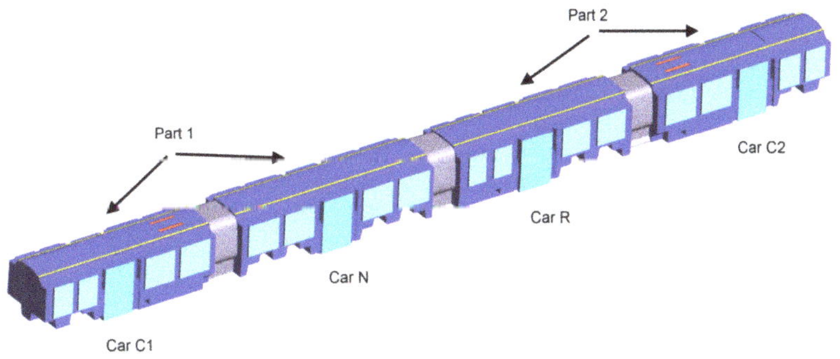

Figure 1. 3D railway vehicle model view.

The boundary conditions related to the HVAC system are shown in Figure 2, for part one of the tram. The conditioned air (3000 m^3/h in each part of the tram) is supplied by two longitudinal forced ventilation inlets located at the ceiling of the tram. In the extreme cars (C1 and C2) there are two forced ventilation exhausts. Air is homogenously supplied along the tram, except near the forced exhaust vents located at the ceiling, where vents are closed in order to avoid the short-circuiting of the supply air.

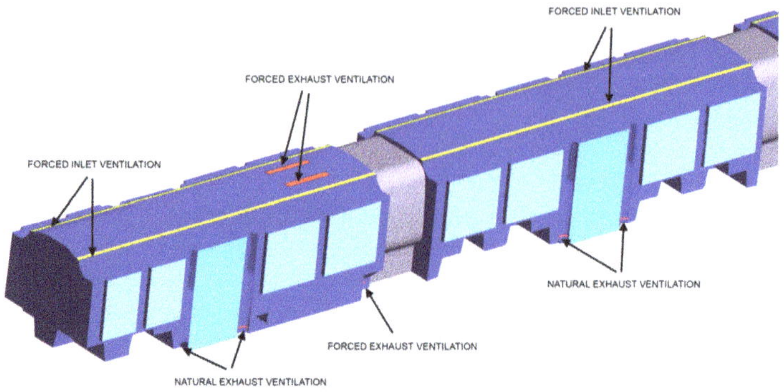

Figure 2. Inlets and exhaust openings, part 1.

In each door there are two natural exhaust openings. In cars C1 and C2 there are, in total, two forced exhaust ventilation systems located at the walls of 150 m^3/h each.

In addition, the HVAC system includes 16 convectors (four in each car) that only work during winter climate conditions. These convectors are of two different models: type 1 (supplied flow 230 m^3/h) and type 2 (supplied flow 200 m^3/h). The number of the different convectors in each car is shown in Table 1.

Table 1. Convector types in each car.

Model	Tram (C1-N-R-C2)	Car C1	Car N	Car R	Car C2
Type 1	10	2	4	2	2
Type 2	6	2	0	2	2
Total	16	4	4	4	4

The cases studied are based on two different passengers' occupation scenarios, one including the passengers (220 in total, 120 standing and 100 seated) and the other without passengers. In the cases including occupation, it is considered a sensible heat load of 78 W/person (17.16 kW in total). There are two lighting rows in the longitudinal direction on each side of the ceiling. The total lighting sensible heat load in the tram is 2.67 kW (homogeneously distributed in the four cars). The overall heat loss coefficient through the tram envelope (including walls, windows, floor and ceiling) is assumed to be 2 $W/(m^2K)$.

Eight different parameters of working conditions have been studied for typical weather conditions in Northern Europe, in an attempt to investigate air distribution, temperature, and velocity fields inside the tram and its effects on thermal comfort. The temperature conditions are summarized in Table 2 for different scenarios.

Table 2. Temperature working conditions.

Case	Season	Passengers (Yes/No)	Supply Temperature (°C)	Exterior Temperature (°C)	Convector Type 1 Supply Temperature (°C)	Convector Type 2 Supply Temperature (°C)
1	SUMMER	No	20.3	40	-	-
2	SUMMER	No	12.4	28	-	-
3	SUMMER	Yes	12.4	28	-	-
4	SUMMER	Yes	12.1	27	-	-
5	WINTER	No	32.6	0	41	32
6	WINTER	No	22.4	−16	43	35
7	WINTER	Yes	22.4	−16	43	35
8	WINTER	Yes	21.6	−6	43	35

Cases 1–4 and 5–8 correspond to summer and winter weather conditions, respectively. Scenarios without passengers (cases 1, 2, 5, and 6) and with passengers (cases 3, 4, 7, and 8) are considered. In all cases, supplied air flow temperature, exterior temperature, and the convectors' supplied temperature are indicated (the last only for winter cases). Solar radiation is not included in the model. The air flow conditions for each part of the tram are described in Table 3.

Table 3. Air flow conditions for each part of the tram.

Case	Supplied Flow (m^3/h)	Fresh Flow (m^3/h)	Forced Exhaust Ceiling v(m^3/h)	Forced Exhaust Walls (m^3/h)	Convector Supply Flow Part 1 (m^3/h)	Convector Supply Flow Part 2 (m^3/h)
1	3000	-	3000	-	-	-
2	3000	1100	1900	150	-	-
3	3000	1100	1900	150	-	-
4	3000	1650	1350	150	-	-
5	3000	-	3000	-	1780	1720
6	3000	1100	1900	150	1780	1720
7	3000	1100	1900	150	1780	1720
8	3000	1650	1350	150	1780	1720

In all the studied scenarios the fresh air flow for each part of the tram is indicated. The difference between the 3000 m^3/h supplied flow in each part of the tram and the fresh flow is the forced flow exhausted by the ceiling extractions (recirculated air flow).

The rest of the air is exhausted by the forced wall extractions located at the walls in cars C1 and C2 and by the natural openings located at the doors. Particularly, in cases 1 and 5 the HVAC system does not supply fresh air, which means that in each part of the tram, the 3000 m^3/h supplied air flow is exhausted by the ceiling extractions. The forced wall extractions located at the walls of cars C1 and C2 are active in all cases except for cases 1 and 5. Finally, the convectors' supplied air flow in the two different parts of the tram (only for winter cases) is also indicated.

For the defined operating conditions, a 3D CFD model is used to predict the temperature and velocity fields of air inside the tram. With this methodology, the conservation equations of mass, momentum and energy (Navier-Stokes equations) are solved using a computer-based tool over the region of interest, with specified conditions on the boundaries of the tram.

The mesh is formed by tetrahedral elements in the fluid volume and prismatic elements near the walls in order to obtain the correct refinement of the viscous kinematic and thermal boundary layer. For the inlet and outlet regions of air ventilation, the mesh has been generated with smaller elements in order to obtain more precision and detail in the characteristics of the air movement in this critical zone for the velocity field.

In the analysis it is considered a steady-state regime, and the sheer stress transport (SST) turbulence model is used. Air is considered to behave like an ideal gas.

The passengers are modeled in two different positions: standing or sitting. For the purpose of this study, the design capacity of the tram is 220 passengers, 120 standing and 100 sitting. In order to reduce the number of nodes and the time consumption during the simulations, a simplified geometry model of the passengers has been defined. A sketch of the passengers' distribution in car C1 is shown in Figure 3.

A similar passenger distribution is defined in the rest of the cars.

Figure 3. Passenger's distribution in car C1.

3. Results

Results are presented in three different sections. Firstly, the air distribution within the tram volume is described briefly, with a focus on the main inlets and outlets of air and pointing out the singularities of the velocity fields in its proximities. Secondly, results are discussed in terms of the temperature fields in different representative planes in the tram (shown in Figure 4).

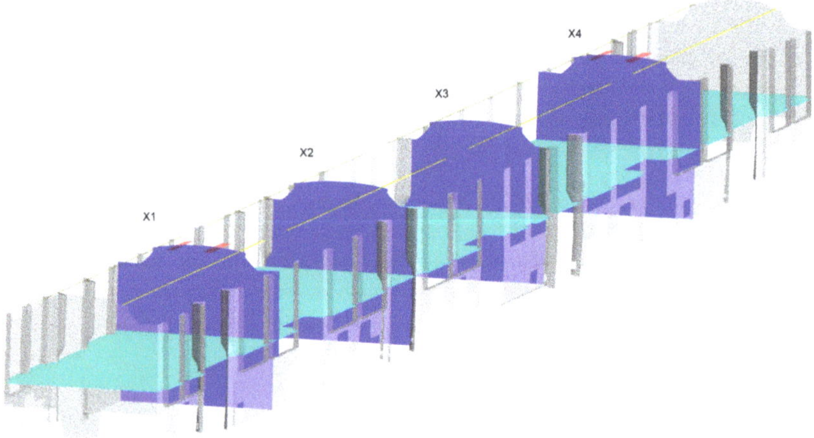

Figure 4. Selected horizontal and vertical planes.

The selected horizontal plane, representative for the occupancy plane, is at a height of +1.1 m. Four different vertical planes (X1, X2, X3, and X4) are selected, one in each car of the tram. While planes X1 and X4 are representative of the forced extraction ventilation located at the ceiling of cars C1 and C2, respectively, planes X2 and X3 are representative of the forced inlet ventilation and the natural exhaust ventilation located at doors of cars N and R, respectively. Finally, the results are also quantitatively analyzed in the tram volume. In order to avoid areas of local discomfort such as the proximities of the tram envelope, the luminaires, or the convectors, it is defined an interior comfort volume that does not consider these zones to evaluate the comfort parameters. In the present work, where the design of the HVAC system is defined in advance, these comfort parameters are useful to compare multiple scenarios with different working conditions, but these parameters could be also used to analyze the impact in air distribution and thermal comfort of different HVAC systems.

3.1. Description of Air Distribution in the Tram

The conditioned air is supplied by the longitudinal inlets located at the cars' ceiling (and in the case of winter season, also by the convectors). Air is homogenously supplied by the longitudinal inlets except near the forced exhaust vents located at the ceiling, where the vents are closed in order to avoid a short circuit in the air distribution. The vents are also closed in the joints between cars. While most of the air is exhausted by the forced ventilation system located at the ceiling in cars C1 and C2, the rest is driven out of the tram by the natural openings located at each door and the forced exhaust ventilation systems located at the walls of cars C1 and C2.

In Figure 5a, the velocity fields for case 1 (without passengers) in the vertical section containing the ceiling extractions in car C1 (plane X1) is shown. It is observed an ascendant air flow and the maximum velocities are reached in the proximities of the ceiling extractions. In Figure 5b, the velocity fields for the same case in the vertical section containing the longitudinal supply inlets and the natural openings located at the doors in car N (plane X2) is shown. It is observed that air is driven from the supply inlets located at the ceiling to the doors' natural openings located near the floor. A circular air flow pattern is reached and a relatively small amount of air is driven out of the tram by the natural exhaust ventilation located at the doors.

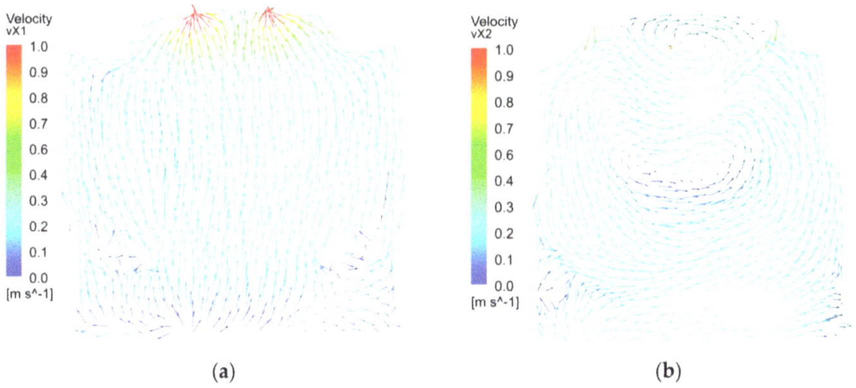

Figure 5. Velocity fields, case 1, plane X1 (**a**) and plane X2 (**b**).

Similar air flow distribution is observed for cases with occupation, even though the air distribution is more complex. Figure 6 shows the velocity fields for Case 3 (with passengers) in the same sections described in the previous figure.

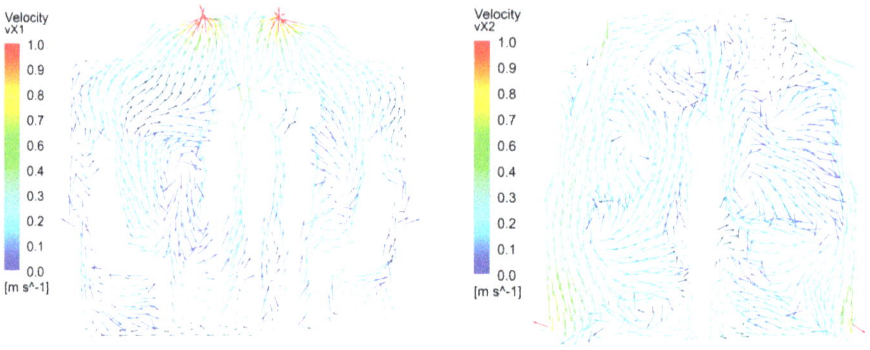

(**a**) (**b**)

Figure 6. Velocity fields, Case 3, plane X1 (**a**) and plane X2 (**b**).

In Figure 6a, it is observed again that air is driven to the ceiling forced extractions, reaching maximum velocities values near them. In Figure 6b, air is driven from the supply inlets located at the ceiling to the doors' natural openings located near the floor. Again, the maximum velocities are reached in the proximities of inlets and outlets. Even though the conditioned air is not supplied near the ceiling extractions, as observed in Figures 5a and 6a, air distribution and circulation inside the tram is adequate, with no death zones without air circulation.

3.2. Temperature Distribution

According to the European standard EN-14750-1:2006 [3], the difference of temperature at a horizontal plane of a height of +1.1 m must be lower than 8 °C for B category railway vehicles. The same condition must be satisfied at any vertical plane (X direction). In this work, the requirement is quantified in terms of the percentage area that is in the range $T_{avg} \pm 4$ °C, where T_{avg} is the average temperature at the particular plane. With this definition, a percentage of 100% means that the whole plane satisfies the requirement. The temperature field in the cited horizontal plane is presented in Figures 7 and 8 for the summer cases (Cases 1–4) and the winter cases (Cases 5–8) respectively.

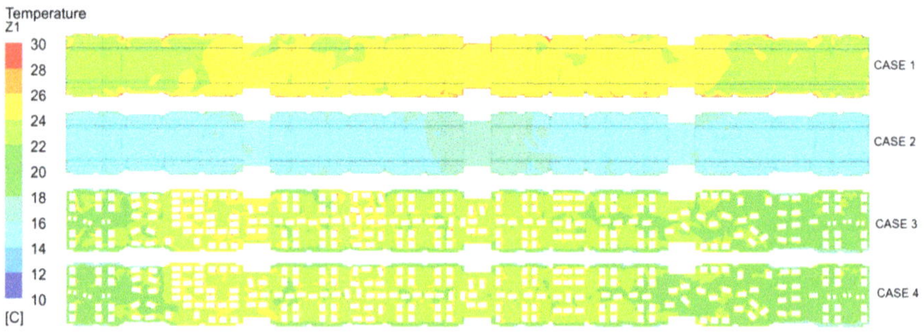

Figure 7. Temperature fields in the horizontal plane at +1.1 m, summer cases (1–4).

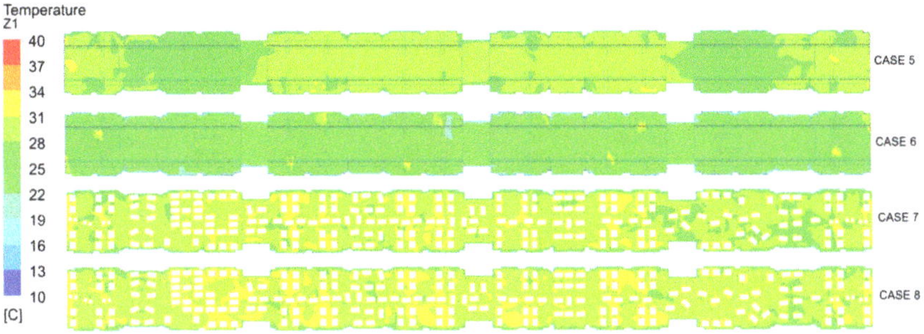

Figure 8. Temperature fields in the horizontal plane at +1.1 m, winter cases (5–8).

The quantitative results at the horizontal plane, summarized in Table 4, show a homogeneously-distributed temperature around the average temperature in all cases, especially for summer cases 1–4 (percentages above 98%) in comparison with winter cases 5–8 (percentages above 96%).

Table 4. Temperature parameters at surfaces.

Comfort Parameters (Plane Z +1.1 m)		
Case	**T_{avg} (°C)**	**T_{avg} ± 4 °C (%)**
1	24.4	100
2	15.7	100
3	22.7	100
4	22.4	98
5	28.6	100
6	24.3	99
7	29.4	97
8	29.9	96

As an example, Figures 9 and 10 show the temperature fields in the vertical planes X1, X2, X3, and X4 for Cases 3 and 7 (representative cases with passengers for summer and winter seasons, respectively).

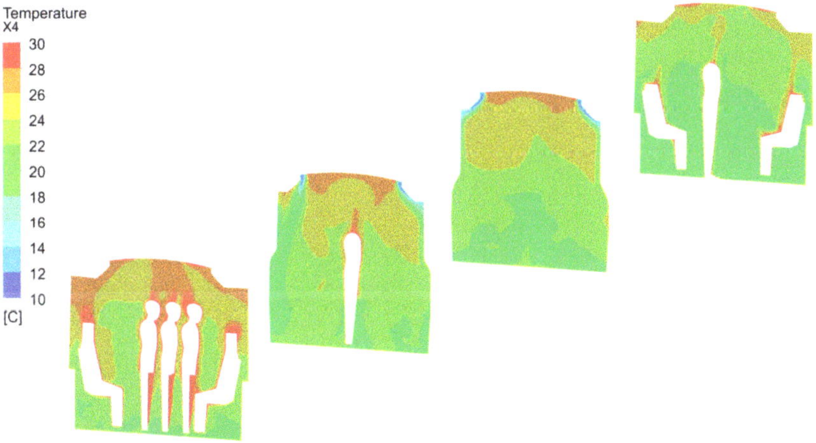

Figure 9. Temperature fields at vertical planes X1, X2, X3, and X4, summer Case 3.

Figure 10. Temperature fields at vertical planes X1, X2, X3, and X4, winter case 7.

In the representative summer case (Figure 9), it is observed that maximum temperatures are reached near the passengers' proximities and minimum temperatures are reached near the supply inlets. In the representative winter case (Figure 10) the maximum temperatures are also reached near the passengers' proximities and also in the convectors' zone of influence, while the minimum temperatures are reached near the tram envelope (due to the high gradient of temperature with the exterior).

The quantitative results at the vertical planes for the eight cases are summarized in Table 5.

Table 5. Temperature parameters at vertical planes.

Case	Comfort Parameters (Planes X1, X2, X3, and X4)							
	Plane X1 (Car C1)		Plane X2 (Car N)		Plane X3 (Car R)		Plane X4 (Car C2)	
	T_{avg} (°C)	$T_{avg} \pm 4$ °C (%)	T_{avg} (°C)	$T_{avg} \pm 4$ °C (%)	T_{avg} (°C)	$T_{avg} \pm 4$ °C (%)	T_{avg} (°C)	$T_{avg} \pm 4$ °C (%)
1	24.7	100	24.4	100	24.9	100	24.5	100
2	15.8	100	15.7	100	15.6	100	15.6	100
3	24.7	100	24.4	100	24.9	100	24.5	100
4	24.6	99	23.6	98	23.2	99	22.6	100
5	27.3	100	27.4	86	28.2	87	27.4	100
6	23.9	99	23.0	91	24.5	95	23.8	98
7	30.7	95	28.8	83	28.7	90	28.7	96
8	31.7	96	29.2	84	29.7	92	29.8	97

Results show a homogeneously distributed temperature around the average temperature in all cases, especially for summer cases (1–4), with percentages above 98%. In winter cases (5–8), in which the temperature gradients are higher, the percentages in the range are lower (percentages above 83%). In these winter cases, the results are better for sections X1 and X4 (percentages above 95%) in comparison with planes X2 and X3. This is explained by the fact that sections X1 and X4 are located at the extreme cars in which the forced ceiling extractions cause a more homogeneous air distribution.

3.3. Temperature and Velocity within the Tram Volume

The previously-analyzed sections were selected in a first approach as representative of the whole tram thermal behavior. These sections included hot areas (such as the proximities of the lighting or the passengers) and other cold areas (such as the proximities of the ceiling air supply inlets or the tram envelope in winter cases) that are not strictly situated in the occupancy zones and can distort the results. Taking into account the whole volume of the tram, the volume of these zones is not negligible. In order to obtain a more realistic approach of the tram thermal behavior in the zone where the passengers are located, the comfort volume is defined as the space between heights +0.5 m and +1.7 m, excluding the convectors' influence zone and also the proximities of the walls of the tram.

Temperature and velocity requirements are quantified in terms of the volume percentage that satisfies the given requirement in the defined comfort volume.

In the case of the temperature requirement, it is calculated the percentage volume in the range $T_{avg} \pm 4$ °C, where T_{avg} is the average temperature at the volume.

In the case of the velocity requirement, it is quantified the percentage volume in which the velocity is lower than the maximum allowed (v_{max}), where v_{max} is defined according to EN-14750-1:2006 [3], and depends on the railway vehicle category and also on the average temperature in the tram.

This maximum allowed velocity v_{max} is shown in Table 6 for different average temperatures for B category trams:

Table 6. EN-14750-1 velocity requirements depending on the average temperature (category B).

T_{avg} (°C)	Maximum Allowed Velocity (m/s)
18	0.30
22	0.35
25	0.70
28	1.4
30	2.0
35	4.0

Taking into account this definition, the results in the defined volume are summarized in Table 7.

Table 7. Velocity requirement depending on the average temperature (category B).

Case	Comfort Parameters (Volume)				
	Temperature		Velocity		
	T_{avg} (°C)	$T_{avg} \pm 4$°C (%)	v_{max} (m/s)	v_{av} (m/s)	$v < v_{max}$ (%)
1	24.5	100	0.60	0.19	100
2	15.8	100	0.30	0.15	99
3	23.1	99	0.42	0.17	97
4	22.9	94	0.42	0.16	96
5	29.0	97	1.60	0.20	100
6	23.7	95	0.54	0.20	98
7	29.0	89	1.60	0.19	100
8	29.6	89	1.80	0.19	100

The quantitative results in the defined comfort volume show a good behavior in terms of the defined comfort parameters for temperature and velocity. In the case of air temperature, results show a homogeneously-distributed temperature around the average temperature in all cases, especially for summer cases 1–4 (percentages above 94%) in comparison with winter cases 5–8 (percentages above 89%). In the case of air velocity, results in the defined comfort volume are excellent. In winter cases 5–8, where the maximum allowed velocities are higher (due to the higher average temperatures in the tram) the percentages are 100% except for case 6 (98%). In summer cases 1–4, the percentages are above 96%.

4. Conclusions

A CFD analysis of air distribution in a railway vehicle equipped with a specific HVAC system has been presented for eight different scenarios representative of typical summer and winter Northern European climate conditions.

Firstly, a description of the CFD model focusing on the HVAC system characteristics is presented and the results are discussed in terms of air distribution. Results show that air distribution and circulation inside the tram is adequate, with no death zones without air circulation.

Temperature fields in different representative horizontal and vertical sections of the tram are also analyzed in terms of the area percentage that is in the range $T_{avg} \pm 4$ °C, where T_{avg} is the average temperature at the particular plane. The quantitative results at the selected sections show a homogeneously distributed temperature around the average temperature in all cases, especially for summer cases 1–4 (percentages above 98%) in comparison with winter cases 5–8 (percentages above 83%). These worse results for winter conditions are explained by the fact that in these cases the difference of temperature between the hot areas (such as the convectors' influence zone or the passengers' proximities) and the cold areas (the proximities of the tram envelope, with very low exterior temperatures) is higher in comparison with the summer cases.

To obtain a more realistic approach representative of the whole tram volume (not depending on the selected sections) the temperature and velocity results are quantified in a comfort volume, defined as the space between heights +0.5 m and +1.7 m, excluding the convectors' influence zone and also the proximities of the walls of the tram. This comfort volume represents the volume occupied by the passengers. The quantitative results in the defined comfort volume corroborate the previous temperature surface results. Again, better results are observed especially for summer cases 1–4 (percentages above 94%) in comparison with winter cases 5–8 (percentages above 89%). In the case of air velocity, the results for the defined comfort volume are excellent for both winter and summer climate conditions.

The main contribution of this work to the state of the art is the detailed CFD analysis of a representative railway vehicle for summer and winter conditions, considering a realistic occupancy level. As indicated in the introduction section, to the best of the knowledge of the authors there are no previous studies specifically related to air distribution in railway vehicles with passengers. The present work introduces this analysis and establishes a methodology and starting point for future studies. It is expected that HVAC systems designers and integrators will increasingly require CFD analysis of the final systems (railway vehicles and other means of public transport) in order to finely adjust the system configuration and provide more satisfactory comfort levels to final users. It is, therefore, necessary to establish an appropriate methodology for such CFD analysis. The design of the air distribution system in the railway coach will also require CFD analysis of the air ducts for ensuring better performance (lower pressure drop and better flow distribution).

Finally, solar radiation is not included in the model in the present work. Even though for winter weather conditions this assumption is conservative in terms of thermal comfort, further analysis must be carried out in future works that analyze the influence of solar radiation in the results, especially for summer weather conditions.

Author Contributions: Christian Suárez ran and post-processed the complete set of simulation cases, analyzed the results, and wrote the manuscript. Alfredo Iranzo critically analyzed the simulation results and revised the manuscript. José Antonio Salva and Elvira Tapia conducted the pre-processing of the simulations cases. Gonzalo Barea established the operating conditions according to the applicable normative and José Guerra critically revised the whole work and coordinated the tasks. All authors read and approved the final manuscript.

Conflicts of Interest: The authors declare no conflict of interest.

References

1. Energy Efficiency and Environmental Criteria in the Awarding of Regional Rail Transport Vehicles and Services. Guidelines for Public Transport Administrations in Europe. ECORailS Project, 2011. Available online: http://www.ecorails.eu/index.php (accessed on 1 September 2016).
2. *ASHRAE Standard 55–1981: Thermal Environmental Conditions for Human Occupancy*; American Society of Heating, Refrigerating, and Air-Conditioning Engineers, Inc.: Atlanta, GA, USA, 1981.
3. *EN 14750-1:2006: Railway Applications—Air Conditioning for Urban and Suburban Rolling Stock—Part 1: Comfort Parameters*; BSI: London, UK, 2006.
4. Liu, W.; Mazumdar, S.; Zhang, Z.; Poussou, S.B.; Liu, J.; Lin, C.-H.; Chen, Q. State-of-the-art methods for studying air distributions in commercial airliner cabins. *Build. Environ.* **2012**, *47*, 5–12.
5. De Lieto Vollaro, R. Indoor Climate Analysis for Urban Mobility Buses: A CFD Model for the Evaluation of thermal Comfort. *Int. J. Environ. Prot. Policy* **2013**, *1*, 1–8. [CrossRef]
6. Zhu, S.; Demokritou, P.; Spengler, J. Experimental and numerical investigation of micro-environmental conditions in public transportation buses. *Build. Environ.* **2010**, *45*, 2077–2088. [CrossRef]
7. Zhu, S.; Srebric, J.; Spengler, J.; Demokritou, P. An advanced numerical model for the assessment of airbone transmission of influenza in bus microenvironments. *Build. Environ.* **2012**, *47*, 67–75. [CrossRef]
8. Bianco, V.; Manca, O.; Nardini, S.; Roma, M. Numerical investigation of transient thermal and fluidynamic fields in a executive aircraft cabin. *Appl. Therm. Eng.* **2009**, *29*, 3418–3425. [CrossRef]
9. Dygert, R.; Dang, T. Mitigation of cross-contamination in an aircraft cabin via localized exhaust. *Build. Environ.* **2010**, *45*, 2015–2016. [CrossRef]
10. Mazumbar, S.; Poussou, S.; Lin, C.; Isukapalli, S.; Plesniak, M.; Chen, Q. Impact of scaling and body movement on contaminat transport in airliner cabins. *Atmos. Environ.* **2011**, *45*, 6019–6028. [CrossRef]
11. Poussou, S.; Mazumdar, S.; Plesniak, M.; Sojka, P.; Chen, Q. Flow and contaminant transport in an airliner cabin induced by a moving body: Model experiments and CFD predictions. *Atmos. Environ.* **2010**, *44*, 2830–2839. [CrossRef]
12. Yan, W.; Zhang, Y.; Sun, Y.; Li, D. Experimental and CFD study of unsteady airbone pollutant transport within an aircraft cabin mock-up. *Build. Environ.* **2009**, *44*, 34–43. [CrossRef]
13. Zhang, T.; Chen, Q. Novel air distribution systems for commercial aircraft cabins. *Build. Environ.* **2007**, *42*, 1675–1684. [CrossRef]
14. Zhang, Z.; Chen, X.; Mazumdar, S.; Zhang, T.; Chen, Q. Experimental and numerical investigation of airflow and contaminant transport in an airliner cabin mockup. *Build. Environ.* **2009**, *44*, 85–94. [CrossRef]
15. Thompson, J.A.; Maidment, G.G.; Missenden, J.F. Modelling low-energy cooling strategies for underground railways. *Appl. Energy* **2006**, *83*, 1152–1162. [CrossRef]
16. Chow, W.K. Ventilation of enclosed train compartments in Hong Kong. *Appl. Energy* **2002**, *71*, 161–170. [CrossRef]
17. Yan, Y.; Li, X.; Yang, L.; Tu, J. Evaluation of manikin simplification methods for CFD simulations in occupied indoor environments. *Energy Build.* **2016**, *127*, 611–626. [CrossRef]

energies

Article

Study of a High-Pressure External Gear Pump with a Computational Fluid Dynamic Modeling Approach

Emma Frosina [1,*], Adolfo Senatore [1] and Manuel Rigosi [2]

1 Department of Industrial Engineering, University of Naples Federico II, Via Claudio, 21-80125 Naples, Italy; senatore@unina.it
2 Casappa S.p.A., Via Balestrieri 1, Lemignano di Collecchio, 43044 Parma, Italy; rigosim@casappa.com
* Correspondence: emma.frosina@unina.it; Tel.: +39-081-768-3511

Received: 30 May 2017; Accepted: 27 July 2017; Published: 31 July 2017

Abstract: A study on the internal fluid dynamic of a high-pressure external gear pump is described in this paper. The pump has been analyzed with both numerical and experimental techniques. Starting from a geometry of the pump, a three-dimensional computational fluid dynamics (CFD) model has been built up using the commercial code PumpLinx®. All leakages have been taken into account in order to estimate the volumetric efficiency of the pump. Then the pump has been tested on a test bench of Casappa S.p.A. Model results like the volumetric efficiency, absorbed torque, and outlet pressure ripple have been compared with the experimental data. The model has demonstrated the ability to predict with good accuracy the performance of the real pump. The CFD model has been also used to evaluate the effect on the pump performance of clearances in the meshing area. With the validated model the pressure inside the chambers of both driving and driven gears have been studied underlining cavitation in meshing fluid volume of the pump. For this reason, the model has been implemented in order to predict the cavitation phenomena. The analysis has allowed the detection of cavitating areas, especially at high rotation speeds and delivery pressure. Isosurfaces of the fluid volume have been colored as a function of the total gas fraction to underline where the cavitation occurs.

Keywords: high-pressure external gear pump; numerical modeling; cavitation

1. Introduction

As is well known, external gear pumps are widely used in the fluid power field. These pumps are used in both mobile and fixed applications such as in agriculture, construction machines and hydraulic presses. These machines have many advantages like their compactness and low cost, combined with relatively high efficiency and remarkable reliability, a wide range of operating conditions and structural simplicity. Usually, these pumps are designed to work over a wide range of speeds and high delivery pressures as positive displacement units in hydraulic systems.

The working principle of external gear pumps is very simple. The pump consists of two gears. One is connected to the shaft and is called driving gear; the other one is free and is called the driven gear. These gears are meshed with each other. Chambers are created by coupling the gears with sliding bushing blocks, one for each gear. Two plates connect the gears with the pump housing where, usually, ports are located.

Even if the geometry of these pumps is relatively simple, they are object of many studies addressing the improvement of their performance. This research is mainly focused on the study of all the causes of losses inside the components, through internal fluid-dynamics investigations. Several numerical models are available in literature. These studies (analytical, experimental and modeling) are focused on the prediction of the performance of this pump typology.

Thiagarajan et al. [1] demonstrated improvements of the lubrication ability of external gear machines by designing a micro-surface linear wedge added to the lateral surfaces of the gear teeth. The author showed the reduction of torque loss by adopting a computational fluid dynamics (CFD) numerical approach. The results were confirmed by experimentation.

Vacca et al. [2] studied the operation of spur external gear units with a numerical approach. Then, Pellegri et al. [3] applied the new numerical approach described in [2] with a CFD model, whereby CFD solved the film flow and the lumped parameter model evaluated the overall operation of an external gear pump.

Borghi et al. [4,5], predicted the volumetric efficiency of gear pumps using a mathematical model. The model has been validated with experimental data. Mancò et al. [6] have studied an external gear pump using a lumped parameter approach. Also for this research, model results have been compared with experimental data validating the model.

The improvement of the pump performance can be achieved only by correctly designing the internal fluid dynamic of both lateral plates. These grooves designed in the plates can have several functions such as reducing the noise inside the pump, however grooves connecting volumes at different pressures, can significantly affect the volumetric efficiency of gear pumps if not well designed [5]. The effects of grooves inside plates have also been widely studied by Koc et al. [7]. Borghi et al. [8] also did an interesting analysis on the transient pressure in the meshing volume of external gear pumps.

Many literature studies propose examining gear pumps using 2D modeling approaches with deforming mesh and volume remeshing. However, for the study of pumps like external gear ones, even if the 2D modeling approaches give interesting results, they cannot predict the internal flow behavior like 3D models [9,10]. This depends on the fact that during the operation of an external gear pump, the flow is really complex because of the rotation speeds (typically in the range 500–3000 rpm) and high pressure.

These conditions can cause turbulent flows and sometimes phenomena like cavitation. Yoon et al. [10] did an interesting study on external gear pumps using a three-dimensional approach (with an immersed solid method). The model included the decompression slots that cannot be simulated with a 2D approximation. The model has been used to make interesting considerations on the internal pressure peak, local cavitation, and delivery pressure ripple [11]. They have verified the influence of the gear tip and lateral clearances on the pump performance. Those gaps become crucial in the study of these pumps that, as said, work at a very low speed (500 rpm). Castilla et al. [12], have also built up a complete 3D model of an external gear pump developed with an OPENFOAM Toolbox. No simulation models have been found for the study of cavitation.

In this paper, the authors have studied principally the flow behavior in the chambers and in the meshing zone. The implementation in the numerical model of all the groove and leakages is important to achieve the best accuracy in the prediction of working conditions. The present study is focused on the analysis of the internal fluid-dynamic of an external gear pump using a tridimensional numerical approach. As said, in the literature there are few studies done with three dimensional simulation tools, and as far as we know, no study has been carried out to fully describe an external gear pump under all operating conditions, even cavitation. The purpose of this research is finding an extremely efficient and fast computational methodology able to predict the overall performance of the components and to allow, by means of monitoring certain points, detecting parameters otherwise difficult to experimentally measure. In addition, this analysis describes a methodology that could be widely used to predict when cavitation occurs inside an external gear pump. This point is of extreme interest for engineers involved in the pump design and optimization phases.

The pump under investigation is the Casappa KP30 (Casappa S.p.A., Parma, Italy), shown in Figure 1a. The pump geometry is shown in the exploded vision in Figure 1b. The pump main features are listed in Table 1. An accurate CFD model has been built up using the commercial code PumpLinx® (Simerics Inc., Bellevue, WA, USA). The entire model of the pump has been created and then validated with experimental data collected by the pump manufacturer.

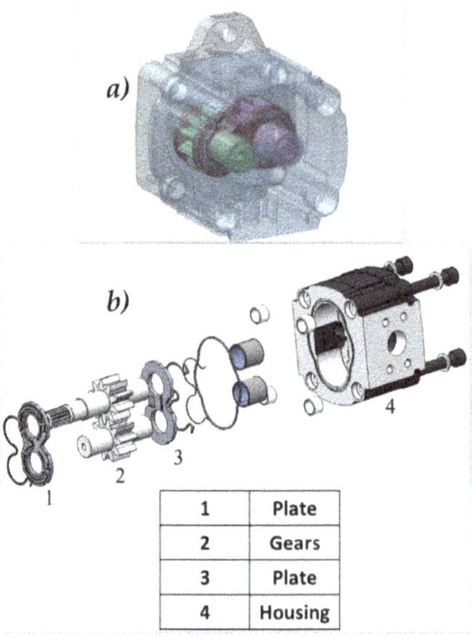

Figure 1. (**a**) Casappa KP30; (**b**) exploded vision of the pump drawing.

Table 1. Pump main features—Casappa KP30.

Pump Main Features	Value
Nominal displacement	44 cm^3
Inlet pressure range	0.7–3 bar
Max. continuous pressure	P1 = 250 bar
Max. intermitted pressure	P2 = 270 bar
Max. peak pressure	P3 = 290 bar
Rotational speed	Min = 350 min^{-1} Max = 3000 min^{-1}

Flow-rate comparisons have been done at three different delivery pressures and oil temperatures while varying the pump speeds. The model demonstrates an accuracy below 2% for all the analyzed working conditions. The validation of the model has been also done on the pressure ripples. Monitoring points have been located in one chamber of each gear. The post-processing of the pressure inside the chambers underlined the areas under low pressure, especially at high pump speeds.

An analysis on the total volume gas fraction has been performed in order to verify if in the chambers of both gears the pressure goes below the saturation value causing cavitation. For this reason, the model has been implemented with an accurate submodel for the prediction of cavitating conditions.

Particular attention has been paid to the connection plates interposed between the ports and gears. In Figure 2, the geometry of one of them is shown. As said, in the grooves drawn inside plates, during the gear rotation, the pressure reduction can cause cavitation. For this reason, the design phase of plates is important and a CFD modeling approach can help optimize those geometries achieving the best solution. The groove created at the delivery side (underlined with the green rectangle in Figure 2, allows the trapping of the fluid in the high-pressure volume, reducing, as a consequence, the pressure spike in the meshing area. On the other hand, the groove instead allows the filling of the chambers from the suction port. In this way, the pressure drop is reduced, limiting the occurrence of cavitation. Those grooves have also the function to reduce the noisiness of the pump during the operation.

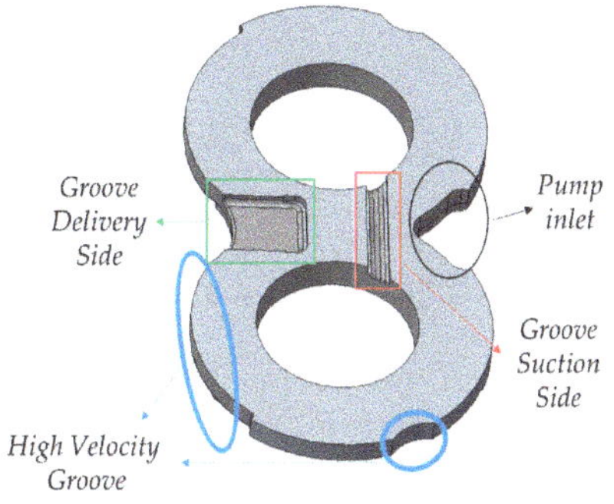

Figure 2. Grooves inside the lateral plates.

The design of grooves is important because they also cause a loss of flow-rate and, as a consequence, of the pump efficiency. Other grooves have been designed on the plates. These volumes are called high velocity grooves (underlined with the ellipse in blue in Figure 2) and connect a defined number of chambers with the high pressure side. This solution affects the forces acting on the gears. In Figure 2 there is also a rectangular vacancy (underlined with the ellipse in black) that allows the connection to the section port of the pump. This fluid volume will be shown better in Figure 3b. In the second section of the paper, the numerical model of the pump is described.

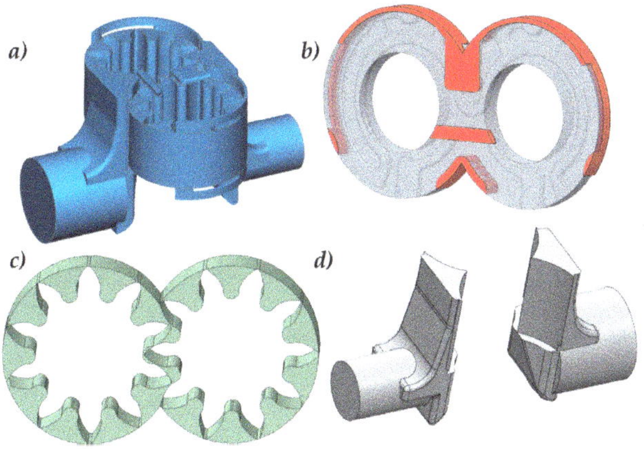

Figure 3. Extracted fluid volume of the external gear pump, (**a**) entire fluid volume; (**b**) fluid volume of the plate; (**c**) fluid volume in rotation; (**d**) fluid volume of the ports.

2. Numerical Model Description

In this section, a numerical model of the studied external gear pump is described. The model has been built up using the commercial code PumpLinx®. Starting from the real CAD geometry of the pump already shown in Figure 1, the fluid volume has been extracted and then meshed. The extracted fluid volume of the entire pump is presented in Figure 3.

The fluid volume has been meshed using a body-fitted binary tree approach. The grid generated with a body-fitted binary tree approach is accurate and efficient because the parent-child tree architecture allows for an expandable data structure with reduced memory storage.

In this architecture of the grid, the binary refinement is optimal for transitioning between different length scales and resolutions within the model. The majority of cells are cubes, which is the optimum cell type in terms of orthogonally, aspect ratio, and skewness, thereby reducing the influence of numerical errors and improving the speed and accuracy. The grid can also tolerate inaccurate CAD surfaces with small gaps and overlaps. Therefore, a body-fitted binary tree approach has demonstrate to be the best methodology to mesh the complex geometry of the pump under analysis. Thanks to the approach adopted, it is possible to create a grid able to take into account the complexity of the geometry, especially in the boundary layer. In fact, also on the surfaces the grid density has been increased without excessively increasing the total cell count. This has been realized using cubic cells, which, as said, are the optimum cell type in terms of orthogonally and improve speed and accuracy. For this reason, in the regions of high curvature and small details the mesh been subdivided and cut to conform it to the surface. From the volume in Figure 3 the mesh has been generated and is shown in Figure 4.

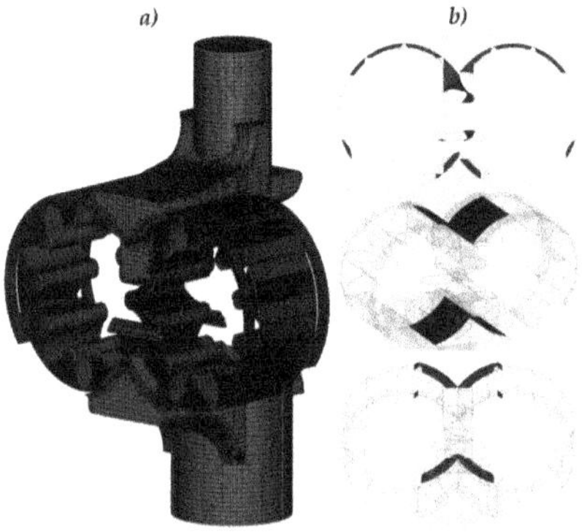

Figure 4. (**a**) Binary tree mesh; (**b**) mismatched grid interface (MGI) between volumes.

A maximum cell size has been chosen for the grid. This parameter defines the maximum cell size in all the fluid volume. In the same way, a minimum cell size has been fixed to limit the minimum dimensions of all cells. Fixing this parameter means that no cell side can be smaller than the minimum cell size. Another important parameter must to be set, which defines the size of cells on surfaces.

Different techniques are available for the treatment of a moving mesh. For positive displacement pumps, it is necessary to use a moving/sliding methodology whereby the stationary and moving volumes are meshed separately. The code, therefore, allows for the simultaneous treatment of moving (gear chambers) and stationary (like suction and delivery ports and grooves in the plate geometry) fluid volumes. Each volume has been connected to the others via an implicit interface called mismatched grid interface (MGI). Each MGI, due to deformations and motion, is updated at each time-step.

2.1. Study of the Gap between Teeth

Before describing the model results obtained it is important to proceed with an analysis on the sizing of the gaps between teeth. This study becomes crucial to correctly predict the performance of an external gear pump. Therefore, in the following a preliminary study done choosing the clearances between teeth of gears is presented.

As shown in Figure 5, the minimum gap between teeth is 5.3 μm. This gap value has been defined after a analyzing the pump geometry and the angles θ and ϕ in Figure 6). The angle θ defines the "driving gear" (Gear 1 in Figure 6) position while the angle ϕ is referred to the "driven gear" (Gear 2 in Figure 6). The angle ϕ is defined as function of the angle θ and the number of teeth z of each gears. This angle ideally is calculated as follows:

$$\phi = \theta + \frac{360}{2z} \tag{1}$$

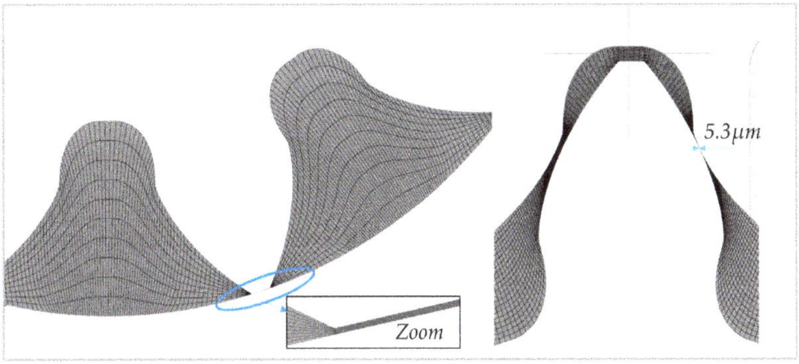

Figure 5. Mesh of the rotating fluid volume and size of the minimum gaps.

Looking at Figure 6, in this configuration gears are perfectly meshing but not in contact. However this meshing is not real, because, as well know, there is no gap between gears in a real working pump. Therefore, in order to correctly reproduce real working conditions, where contact occurs, Equation (1) must be implemented including the shift angle ϕ. Equation (1) becomes:

$$\phi = \theta + \frac{360}{2z} \pm \phi \tag{2}$$

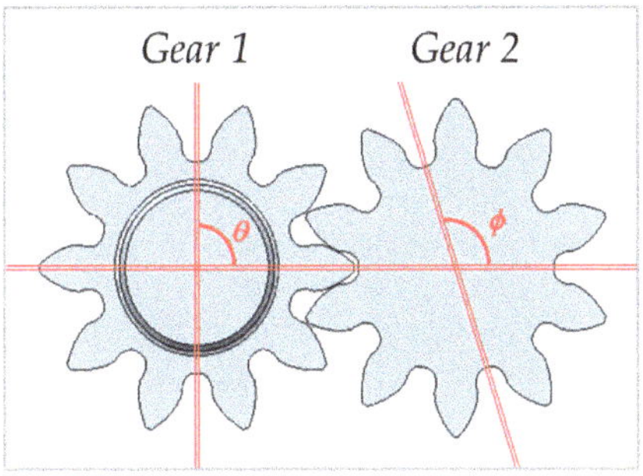

Figure 6. Front view of the gears; angle θ and ϕ.

Therefore, from an ideal gap shown in Figure 6 simulation have been run with three values of ϕ and gap in order to achieve the best accuracy of the model (Figure 7c). The final value of the gap between gears is 5.3 μm.

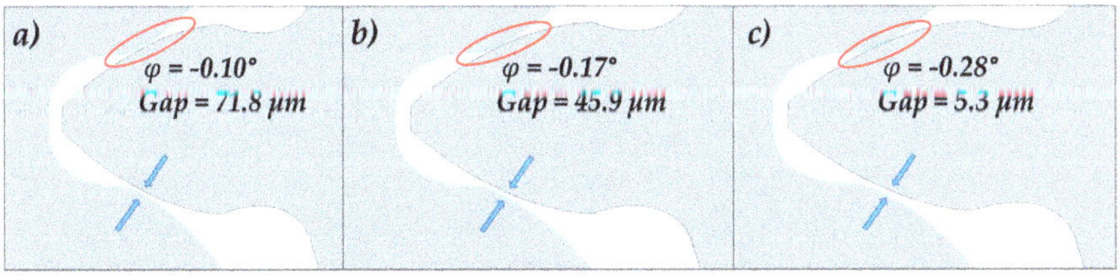

Figure 7. Three different pump configurations: (**a**) ϕ of $-0.10°$ and gap size of 71.8 μm; (**b**) ϕ of $-0.17°$ and gap size of 45.9 μm and (**c**) ϕ of $-0.28°$ and gap size of 5.3 μm.

2.2. Mesh Sensitivity Analysis

A mesh sensitivity analysis has been done on the model by changing the size of the grid. This analysis should be always done in a CFD study to define the best grid for each application and to correctly predict the real working conditions of the pump. Therefore, in this section a mesh sensitivity showing the effects on the model results of variation for example of the clearances between teeth and sliding bushing blocks is presented.

Fluid volumes have been slit in six parts: inlet port, outlet port, outlet high speed grooves, inlet low noise grooves, outlet low noise grooves e rotor. Each volume has been connected to others, as said, via the "MGI" interface (see Figure 4b). Table 2 clarifies the importance of the mesh sensitivity analysis in a CFD study.

A first step has been done analyzing the influence of the grid on the boundary layers. Two different meshes have been created by changing the grid parameters in the boundary layers. These models are called "Mesh 1" and "Mesh 2" where the second one has a finer grid size.

In Table 2, the percentages of error between the suction and delivery flow evaluated by the model are reported. The simulations in Table 2 have been run at 1500 rpm and for two delivery pressures: 50 and 200 bar. As shown in the Table 2, the ΔQ% of both models "Mesh 1" and "Mesh 2", as expected, increases with the delivery pressure.

Table 2. Mesh sensitivity varying the mesh size in the boundary layers, simulations at 1500 rpm.

Delivery Pressure	ΔQ% Mesh 1	ΔQ% Mesh 2
$p_{delivery} = 50$ bar	0.72%	0.24%
$p_{delivery} = 200$ bar	1.78%	1.36%

As shown in the table, improving the quality of the mesh in the boundary layers, the ΔQ% between suction and delivery ports becomes closer to each other. In fact, the reductions from simulations "1" and "2", for both pressure levels, are significant.

After the analysis on the boundary layers, the mesh has been studied improving the quality. Nine models have been created increasing the number of cells. In Figure 8, a comparison of the suction and delivery mass flows is shown. Looking at the graph, it is clear that only with a grid above 200,000 cells does the solution become stable. Each point in Figure 8 corresponds to a model built up with a different grid size.

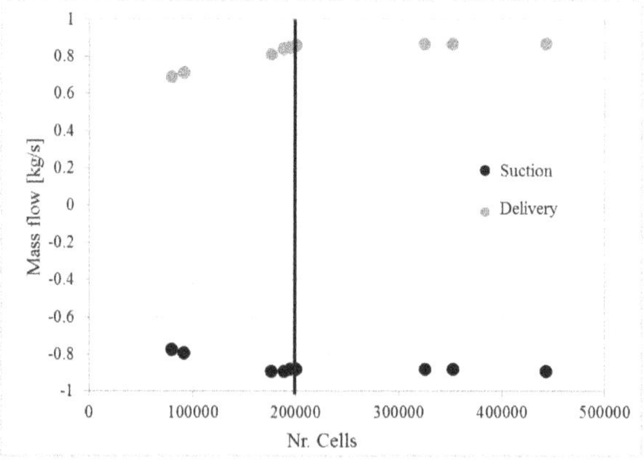

Figure 8. Mesh sensitivity—flow-rate vs cells number at 1500 rpm and a delivery pressure of 50 bar.

Simulations have been run, as already said, has been run on an Intel® Xeon® X5472 CPU @ 3.00 GHz (two processors) with 24 GB RAM. As we know, upon increasing the number of cells the simulation time increases as well. From the analyzed case, the time goes from half an hour for the first point in Figure 8 to 10 h of the last model of almost 500,000 cells. Therefore, in order to achieve the best compromise between accuracy and computational time, the final model consisted of 200,000 cells which is the point underlined in Figure 8.

2.3. Model Setup

The standard *k-ε* turbulence model has been adopted because for this specific problem it has proven to be accurate. The *k-ε* model is a robust method demonstrated to provide good engineering results. As well known, in the literature there are other resolution methods more accurate than the *k-ε* model and RNG *k*-model (RNG, LES, or DES, etc.). However since the losses due to the viscous stresses are negligible compared to pressure forces the *k-ε* model can be chosen because it is numerically robust, computationally efficient and it provides good accuracy. In fact, in this application, where the computational time is close to 10 hours per revolution, the adoption of a higher order turbulence model would have increased the computational time with no relevant improvement of the results. The authors of this paper have already used this strategy for many other similar analyses confirming the accuracy of the solutions obtained [13–21].

An analysis on the convergence criteria have been done to achieve the best accuracy results. "R_i" is the residual drop referred to the "i-th volume". It is defined as the difference between the obtained results, for the selected volume, during two subsequent iterations. The correct solution convergence is realized when all the residuals go below the defined R_i.

For all the simulations, the convergence criteria for the pressure, velocity and the vapor mass fraction is 0.01 as shown in Figure 9a. Figure 9a, in fact, presents the trend of the residual drop for each variable in a single time-step. As shown, residuals go below 0.01 (settled $R_{i.}$).

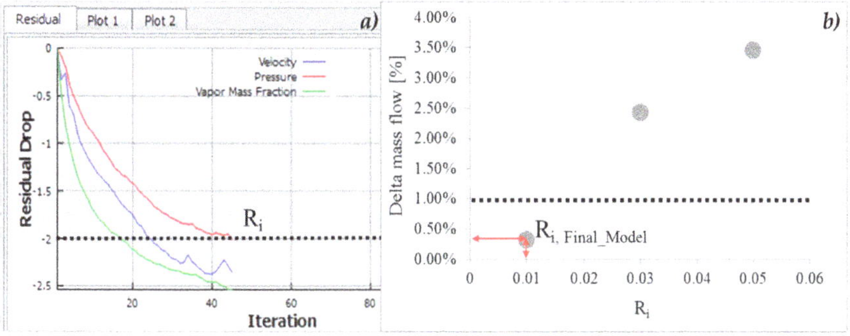

Figure 9. (**a**) Pressure, velocity and vapor mass fraction—residual drop analysis R_i; (**b**) error on the mass flow vs residual drop R_i.

These values have been demonstrated to be suitable for this application. However, a study on the convergence criteria has been also performed varying the R_I. As example, in Figure 9b the R_i is shown as function of the delta mass fraction (%) between the pump inlet and outlet mass flows. As shown, the R_i has been varied from 0.01 to 0.05 with steps of 0.02.

The maximum error on the delta mass fraction between the pump inlet and outlet mass flows has been fixed at 1%. It has been demonstrated that the best solution can been achieved with a R_i of 0.01 (Figure 9b). Other analyses have been run reducing the R_i for the to lower value without improving the delta mass flow percentage.

Before running simulations, it is important to underline that in this study the model has been built up including an accurate submodel able to predict cavitation. In the following subsection the study of cavitation inside an external gear pump has been done. In fact, one of the targets of this research was to detect cavitation, in order to predict the worst working conditions for the pump.

Since one of the aims of the research is the investigation on the ability of the pump to avoid cavitation, the model has been equipped with a robust submodel, available in the code, able to predict with high accuracy cavitating conditions. The chosen cavitation module accounts for all three important real liquid properties: cavitation, aeration, and liquid compressibility. This model builds on the work of Singhal et al. [22]. A module is used to predict the aeration and cavitation in a liquid system, therefore allowing for the calculation of cavitation effects, when the pressure in a specified zone of the fluid domain falls below the saturation pressure and vapor bubbles form and then collapse as the pressure rises again [22]. Many physical models for the formation and transport of vapor bubbles in liquids are available in literature, but only few computational codes offer robust cavitation models. This is due to the difficulty on handling gas/liquid mixtures with very different densities. Even small pressure variations may cause numerical instability if they are not optimally treated [22].The original cavitation model proposed by Singhal et al. describes the vapor distribution using the following formulation [13–16,22]:

$$\frac{\partial}{\partial t}\int_{\Omega(t)}^{0} \rho f \mathrm{d}\Omega + \int_{\sigma}^{0} \rho((v-v_{\sigma})n) f \mathrm{d}\sigma = \int_{\sigma}^{0}\left(D_f + \frac{\mu_t}{\sigma_f}\right)(\nabla f n)\mathrm{d}\sigma + \int_{\Omega}^{0}(R_e - R_c)\mathrm{d}\Omega \tag{3}$$

where D_f is the diffusivity of the vapor mass fraction and σ_f is the turbulent Schmidt number. In the present study, these two numbers are set equal to the mixture viscosity and unity, respectively. The vapor generation term, R_e, and the condensation rate, R_c, are modeled as [13,21]:

$$R_e = C_e \frac{\sqrt{k}}{\sigma_l}\rho_l \rho_v \left[\frac{2}{3}\frac{(p-p_v)}{\rho_l}\right]^{\frac{1}{2}}(1 - f_v - f_g) \tag{4}$$

$$R_c = C_c \frac{\sqrt{k}}{\sigma_l}\rho_l \rho_v \left[\frac{2}{3}\frac{(p-p_v)}{\rho_l}\right]^{\frac{1}{2}} f_v \tag{5}$$

in which the model constants are C_e = 0.02 and C_c = 0.01.

The final density calculation for the mixture is done by [13–16,22]:

$$\frac{1}{\rho} = \frac{f_v}{\rho_v} + \frac{f_g}{\rho_g} + \frac{(1 - f_v - f_g)}{\rho_l} \tag{6}$$

The model by Singhal et al. has been extended to include non-condensable gases, finite rate and equilibrium dissolved gas. The fluid model, in addiction, accounts for liquid compressibility. This is critical to accurately model pressure wave propagation in liquids. The liquid compressibility is found to be very important for a high-pressure system and the systems in which water hammer effects are relevant.

3. Model Validation

The pump has been tested on a dedicated test bench at Casappa S.p.A. The test bench is shown in Figure 10. Simulations have been run under the same conditions tested in the lab with pump speeds of 1000, 1500 and 2500 rpm, oil temperatures of 50 and 80 °C and delivery pressures of 50, 150 and 250 bar.

The test bench layout is shown in Figure 10b. There are two strain gauge sensors P1 and P2 (both from WIKA®, Lawrenceville, GA, USA, scale: 0–40 bar and 0.25% FS accuracy) and P2 at the delivery side (by WIKA®, with a scale of 0–400 bar and an accuracy of 0.25% FS). The pressure sensor P4 is piezoelectric (made by KISTLER®, Winterthur, Switzerland, scale: 0–1000 bar,

140 kHz natural frequency, 0.8% FS accuracy) while the traducer P5 is piezoresistive (ENTRAN®, Strainsense Ltd, Milton Keynes, UK, scale: 0–350 bar, 450 kHz natural frequency, compensated temperature, 0.5% FSO accuracy).

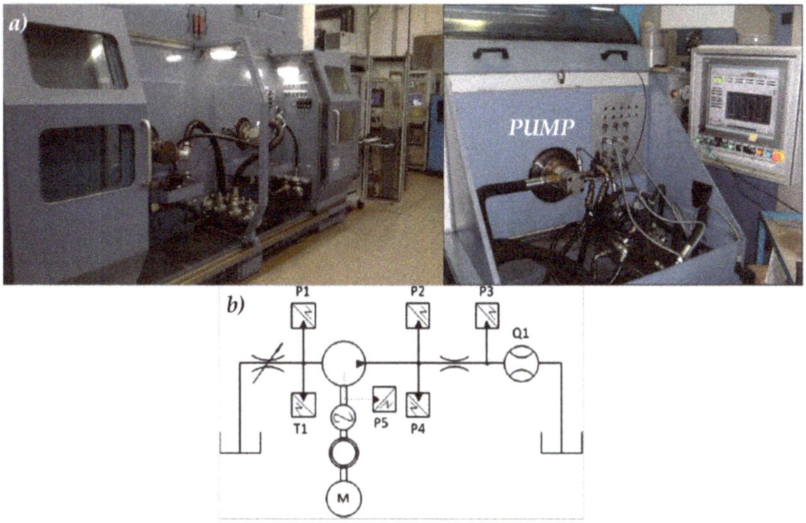

Figure 10. (**a**) Test bench of Casappa S.p.A.; (**b**) test bench layout.

The flow-rate mater Q1 is a VSE® VS1 (VSE.flow, Neuenrade, Germany), scale 0.05–80 l/min, 0.3% measured value accuracy. There is an HEIDENHAIN®, ERN120 θ encoder (HEIDENHAIN, Traunreut, Germany), 3600 r/min, a limit velocity 4000 r/min and a period accuracy of 1/20.

In Figure 11a–c, a first comparison between model and experimental data is shown for three different pump speeds: 1000, 1500 and 2500 rpm. In the graph, the continuous line are the experimental data while the dashed ones are the simulation results.

Simulations and test, as said, have been done for three delivery pressure conditions: 50, 150 and 250 bar. The comparisons have been done on the torque requested on the driving gear (gear 1 in Figure 6).

By analyzing Figure 11, it is possible to appreciate that the model results are really close to the experimental data. However, there is a gap between model and experimental data that depends also on the fact that the model does not consider torque due to mechanical friction.

The experimental data in thick lines represents the absorbed torque by the pump measured by a torque meter. This means that experimental data takes into account, of course, the "hydraulic" component of torque (volumetric efficiency) even of all the friction (mechanical efficiency). This is the reason why the simulation results are lower than the experimental data. In Figure 12 the delta torque between experimental and model results is presented. This ΔC represent the toque lost due to friction.

Model validation has been done also comparing at the delivery flow-rate varying the delivery pressure. This comparison is shown in Figure 13 at the oil temperatures of 50 and 80 °C and at pump speeds of 1500 and 2500 rpm.

It can be noticed that delivery flow rate decreases by increasing the delivery pressure for both the model and experimental data; the error percentage between model and experimental data is always below 2%.

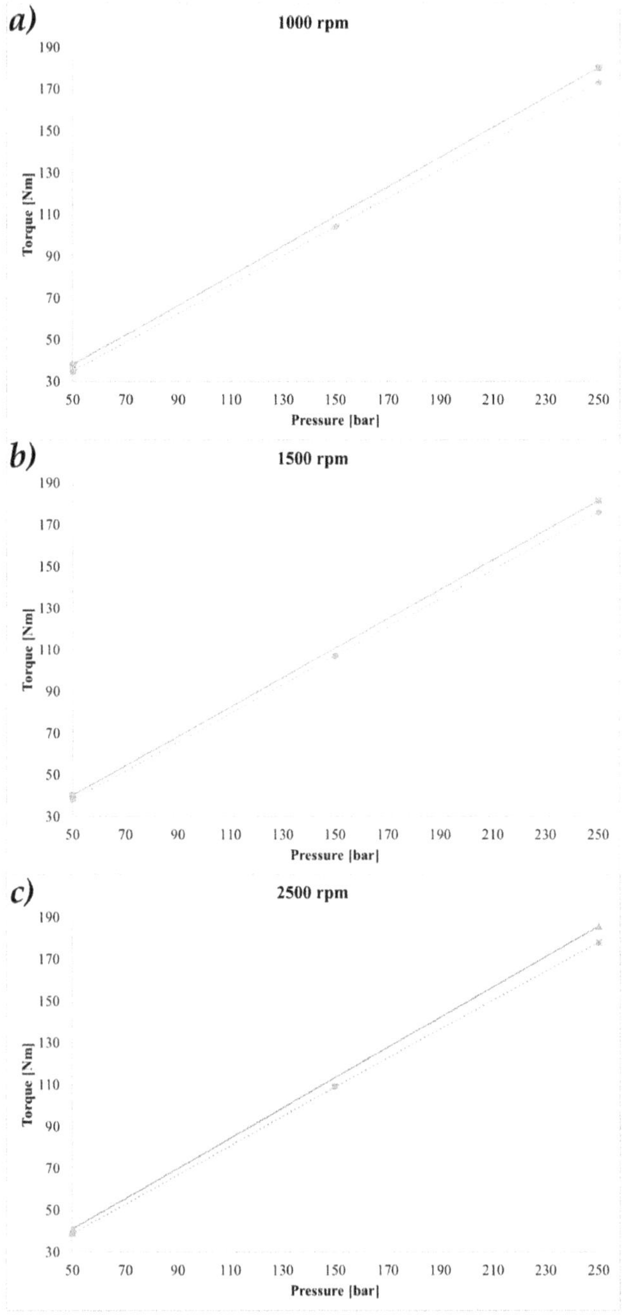

Figure 11. Absorbed torque—comparison between model and experimental data. (**a**) 1000 rpm; (**b**) 1500 rpm and (**c**) 2500 rpm.

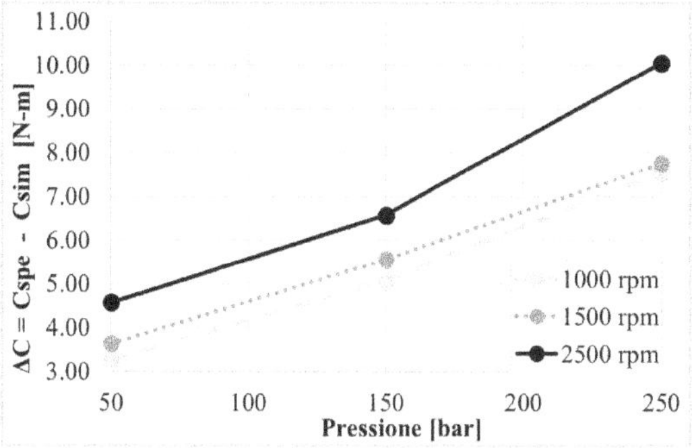

Figure 12. Delta torque between experimental data and simulation data at 1000, 1500 and 2500 rpm.

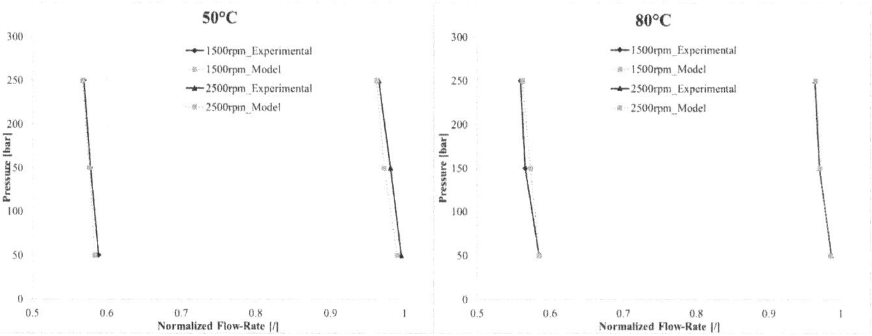

Figure 13. Pressure vs. delivery flow-rate comparison at 50 and 80 °C.

4. Model Results

The validated model has been analyzed in order to study the pump in depth. A three-dimensional CFD modeling approach, as said, has many advantages. In fact, considering the real geometry of the pump, it is useful to visualize parameters that are impossible or onerous to obtain experimentally such as the velocity magnitude of the fluid in leachates.

Figure 14 shows a first result of the numerical model with the velocity magnitude countering through the leakages between gears and the static body. The area under investigation in the pictures is shown in Figure 14c. In fact, figures refer to the leakages between the first pressurized chamber and the last chamber at low pressure.

Figure 14a,b, in particular, show the results in the gaps by increasing the pressure from 50 to 250 bar for oil temperatures of 50 and 80 °C, respectively, at the same pump speed. The velocity magnitude scale in all figures is always in the 0–13 m/s range.

In Figure 14a, by comparing graphs at the same oil temperature it is possible to distinguish the flow velocity magnitude profile through gaps. In fact, as consequence of the higher pressure drops between the two following chambers (from 50 to 250 bar), the flow velocity through the gap between the tooth and the pump housing increases and, as shown in pictures in Figure 13, flow becomes a jet.

Looking for example at Figure 14a, the flow velocity through the gap with a delta pressure of 50 bar is below 5 m/s. It increases in correspondence of the connection between the gap and the chamber at the lower pressure. This velocity profile is, of course, quite similar with a higher pressure drop, however the values increase. For a Δp of 150 bar the maximum velocity value registered in the analyzed fluid volume is of almost 9 m/s. It becomes 13 m/s for the delta pressure of 250 bar.

It is interesting to underline the effect of the T_{oil} on the flow velocity and as consequence, on the pump volumetric efficiency. For all the analyzed Δp, the velocity at 50 °C is lower than for an oil temperature of 80 °C. This depends by modification of the fluid properties with the temperature.

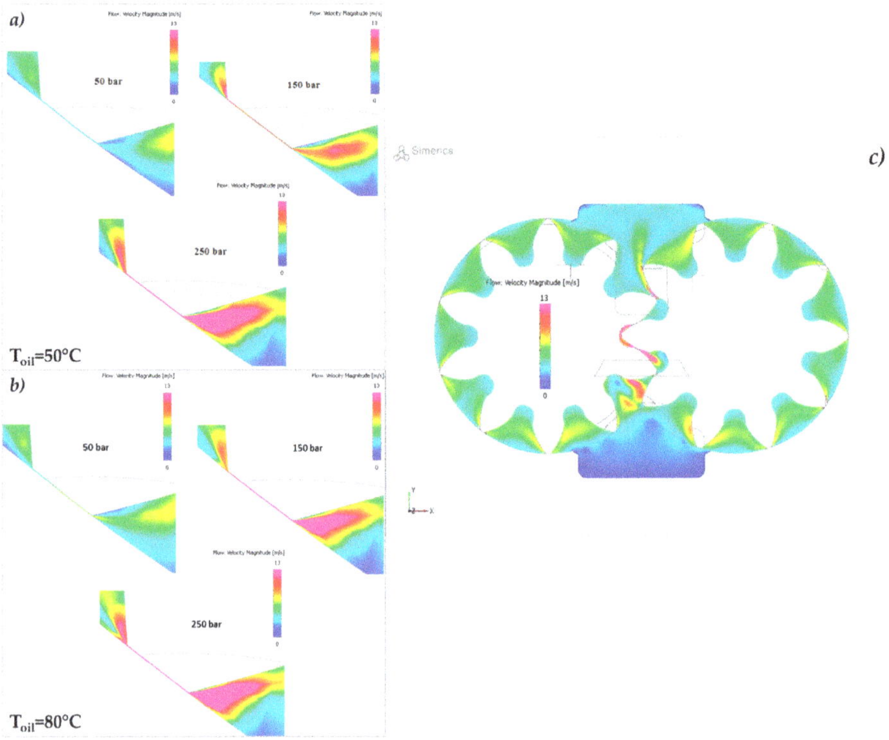

Figure 14. Velocity magnitude at (**a**) 50 °C; (**b**) 80 °C and (**c**) zoomed area.

In Figure 15, the pressure evolutions inside the chambers are diagrammed for a shaft revolution at the operating condition listed below:

- Oil temperature of 50 °C
- Pressure drop of 50 and 250 bar
- Pump speed of 1500 and 2500 rpm.

As is well known, it is also possible to obtain experimentally the results in Figure 15, however it is not easy. Therefore, the best solution is to obtain these important data with modeling techniques.

Data shown in graphs of Figure 15 allow evaluating the pressure evolution inside a chamber in rotation for each gear. In particular, the lines in red refer to the chambers of the gear 1, while the blue ones correspond to gear 2.

The graphs can be interpreted as follows. From the suction port the flow fills the generic chamber (obtained by coupling the two gears) and is compressed. Looking at Figure 15, it is clear that each chamber has a pressure evolution during the 360 degrees but the low-pressure zone has a lower angular extension. It is, in fact, more or less 90 degrees and depends by the pump geometry.

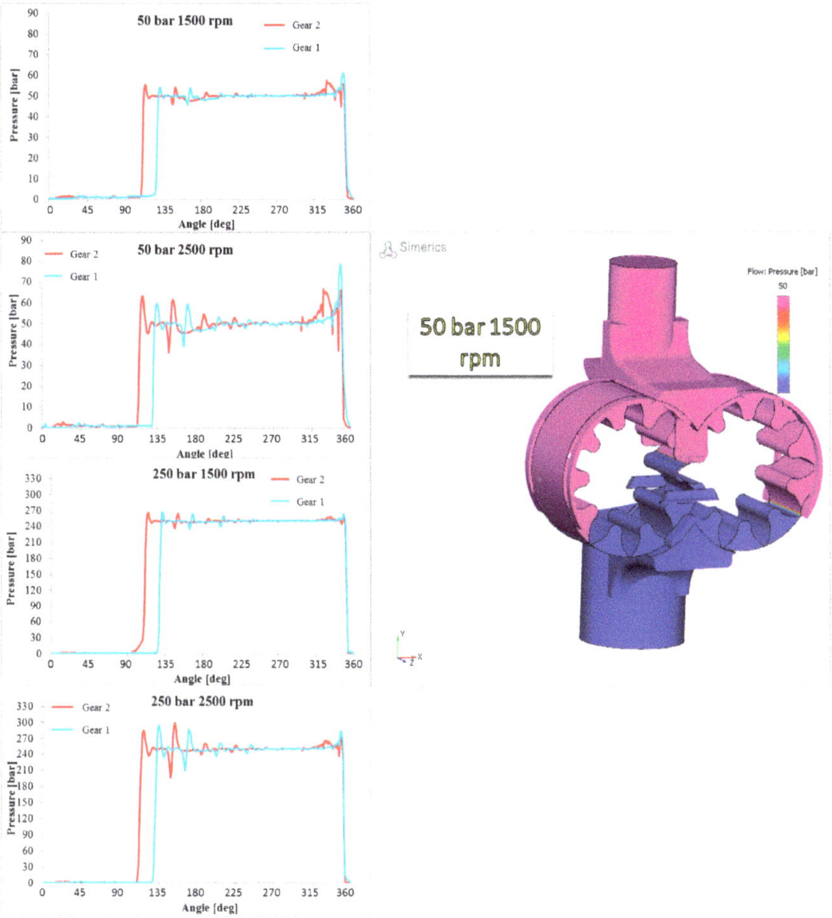

Figure 15. Pressure evolutions for two pressure levels at T_{oil} of 50 °C.

As expected, when the chamber is connected to the delivery port, it has an oscillating pressure around the delivery value (in this case 50 and 250 bar). The pressure is smoother around the mean value with a lower pump speed (1500 rpm). At 2500 rpm instead the oscillations around the mean value are high for both delivery pressures. The pressure spikes inside the chambers, at the higher delivery pressure level, are clear for both gears' fluid volumes. Those investigations, especially the evaluation of the spikes values, are really important for engineers to design or optimize a pump.

The results in Figure 15 are also of extreme interest by visualizing the pressure-axis in the range 0–3.5 bar and reducing the angle-axis to the only the degrees of connection with the pump suction side. In this way, it is possible to appreciate the behavior of the pressure when the chamber is connected to the suction port. These results are shown in Figure 16, providing a preliminary investigation of the cavitation phenomena. Cavitation, however, it analyzed in depth in Section 5.

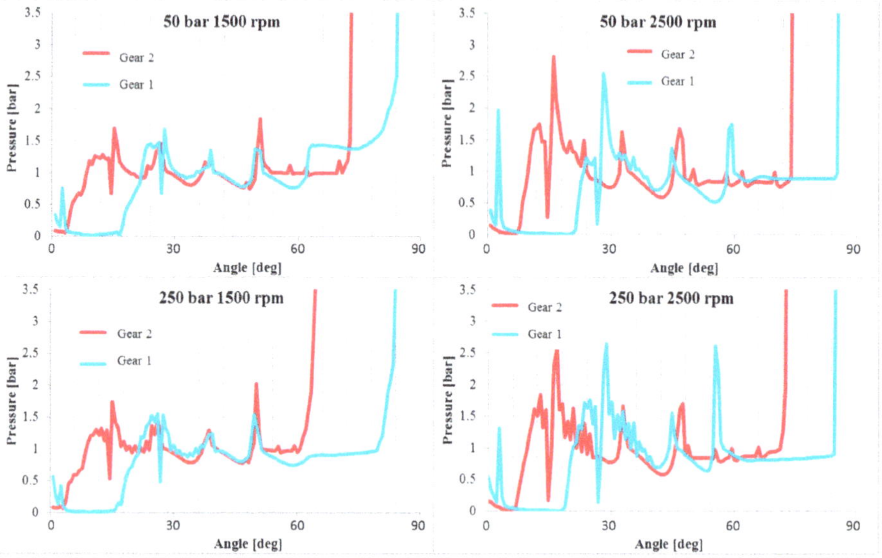

Figure 16. Pressure evolutions for two pressure levels in the range 0–3.5 bar, at T_{oil} of 50 °C.

However before studying the cavitation phenomena a study of the pressure ripples at the delivery side of the pump has been performed. For this reason the model has been implemented adding a supplementary volume in accordance with the experimental tests. This duct has the same geometry of the real one, installed on the test bench in order to correctly follow the experimental delivery pressure ripples. The geometry of the duct is shown in Figure 17a while in Figure 17b the final model is presented. In the model a monitoring point located at the same position of the pressure transducer has been added and a section set like the real lamination valve of the test bench has been added too. This section therefore simulates the lamination valve, and for this reason the entire model simulated with high accuracy the pressure waves inside the delivery duct.

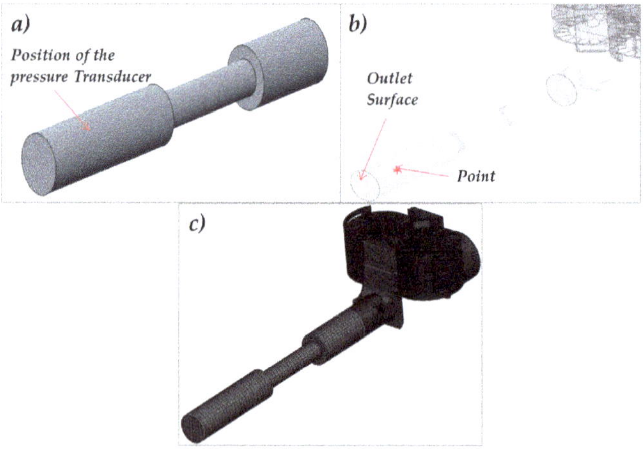

Figure 17. (**a**) Calibrated orifice geometry, (**b**) pressure ripples monitoring point; (**c**) mesh of the entire model.

During the tests a valve has been located close to the pump delivery in order to amplify the pressure ripples. As consequence, the model has been implemented by adding a calibrated orifice. The additional fluid volume has been inserted to replicate the real experimental setup. In Figure 17, the entire model of the pump is shown, including the fluid volume of the calibrated orifice. Simulations have been run with the model in Figure 17 and the model results have been compared with the experimental data.

The monitoring point for the simulation has been located in the same position of the pressure transducer installed on the test bench, as shown in Figure 17b; in that point the pressure ripple has been both measured and calculated. Comparisons have been done at two rotational speeds and three pressure levels at 50 °C (see Figure 17). The diagrams in Figure 18 have been normalized to a reference pressure.

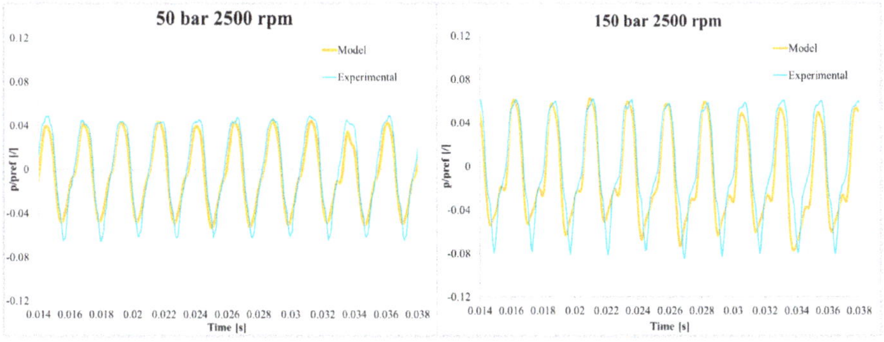

Figure 18. Pressure ripple comparison at 50 bar and 150 bar, 2500 rpm and 50 °C.

As said, the model results in Figure 18 have been obtained changing the delivery pressure at 2500 rpm. Comparison between the model and experimental data shows a good agreement in all the analyzed conditions, confirming the accuracy of the numerical model. The results, in fact, show that the timing and ripples have the same amplitude as the experimental data.

Pressure ripples show that the amplitude of the oscillation increases with the delivery pressure. In the graphs the y-axis has been fixed, therefore, looking at the ripples, it is clear that by increasing the delivery pressure the amplitude increases as well. This happens for both pump speeds.

5. Cavitation

In this section, the model results have been analyzed verifying if cavitation occurs under particular pump-operating conditions. The model, as already said in the corresponding section, includes the capability of prediction of the occurrence of cavitation. Thanks to the three-dimensional visualization of the total gas volume inside the pump, it is possible also to find when the cavitation occurs and where bubbles are located. The study of this phenomenon becomes important for engineers in the pump design and optimization phases.

In Figure 19, a front view of the pump is shown. The volume is colored as a function of the total gas fraction that varies in the range 0–1. The gas volume fraction is the volume fraction of the free non-condensable gas (NCG) in a liquid for a selected volume. In Figure 19, there are presented four sections relative to gears rotations of θ = 9°, 18°, 27° and 36°. The simulation has been done under fairly critical conditions for the cavitation: 2500 rpm and 250 bar.

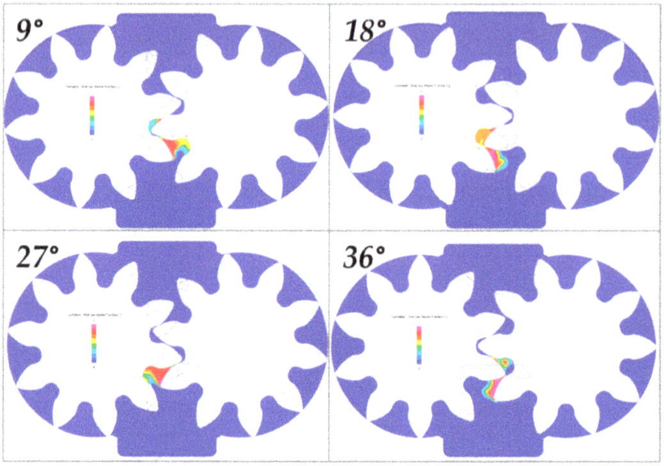

Figure 19. Distribution of the total gas fraction in a section view of the pump at 250 bar, 2500 rpm and 50 °C for a shaft rotation of θ = 9°, 18°, 27° and 36°.

Areas in blue correspond to a concentration of free gas equal to zero while areas colored in magenta are at a percentage of gas of 100%. As shown, the areas in magenta are located in areas where the rotating chamber, from the condition of minimum, starts to increase its volume. Therefore, in correspondence with the meshing area, an intense and sudden cavitation occur. As well known, this phenomenon causes noise and damage to parts and therefore must be avoided. Figure 20 shows two diagrams of the total gas volume fraction inside the chambers of gears. The total gas volume fraction, as said, varies in the range 0–100%.

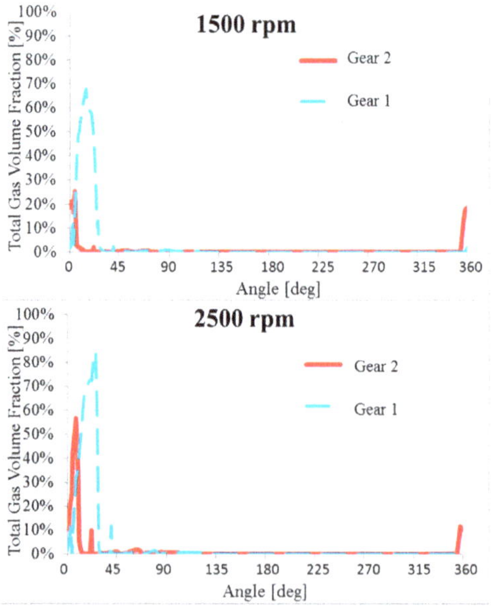

Figure 20. Total gas fraction behaviors at 1500 and 2500 rpm.

Signs of cavitation could be appreciable for values higher than 80%. Even if the total gas fraction in the chamber of the driving gear at 2500 rpm has a spike between 0° and 45° this pump demonstrates to not be subject to cavitation, even under critical operating conditions. However, the methodology described in this paper is demonstrated to be suitable because it offers the opportunity to fully study the complex fluid dynamics of external gear pumps. Results showing cavitation are really important for the optimization of a pump. Therefore, a CFD modeling analysis becomes fundamental in the design phase of components to achieve the best performance for all working conditions.

6. Conclusions

A three-dimensional CFD study a high-pressure external gear pump has been descried in this paper. The model has been realized based on a geometry of a pump manufactured by Casappa S.p.A. The model has been built up using a commercial code including accurate tools to predict the flow turbulence and cavitation. Leakages have been taken into account in order to correctly estimate the volumetric efficiency of the pump. The final model has demonstrated to achieve an accuracy close to 1%.

Model results have been compared with experimental data obtained on a dedicated test bench by the pump manufacturer. Thanks to the good agreement of the data the model has been run under some conditions (such as at high pressure value) which are very challenging from a modeling and numerical point of view. Simulations have found that at the higher pressure and speeds the pump has a tendency to cavitate. For this reason, monitoring points have been located inside the chamber volumes of both gears.

The pressure behavior underlined that the pressure, especially inside the driving gear chamber, goes below the oil saturation pressure. This has been confirmed by visualizing the total gas fraction distribution in sequential images in a pump cross-section. This study summarized a methodology able to completely describe the operation of pumps such as external gear pumps. All the working parameters have been analyzed and, where possible, compared with experimental data. Research has confirmed the importance of CFD, which can be a valuable instrument for engineers to study some working anomalies like cavitation.

Acknowledgments: This research is a result of a research collaboration between the Fluid Power Research Group (FPRG) of the Department of Industrial Engineering of the University of Naples "Federico II" and Casappa S.p.A. We appreciate the technical contribution from Federico Monterosso and Micaela Olivetti of OMIQ s.r.l.

Author Contributions: Emma Frosina and Manule Rigosi built up the model. Adolfo Senatore supervised the study. Emma Frosina wrote the paper.

Conflicts of Interest: The authors declare no conflict of interest.

Nomenclature

$c_{eq,k}$	Coefficient
CFD	Computational fluid dynamic
C_c	Cavitation model constant
C_e	Cavitation model constant
D_f	Diffusivity of vapor mass fraction
dS	Infinitesimal surface, m^2
f_v	Vapor mass fraction
f_g	Non-condensable gas mass fraction
G_t	Turbulent generation term
MGI	Mismatched grid interface
N_{driving}	Number of the teeth of the Gear 1
n	Pump velocity rotation, rpm
$p_{u,k}$	k-th inlet pressure, bar
$p_{d,k}$	k-th outlet pressure, bar
p_s	Oil supply pressure, Pa

Q_{teo}	Theoretical flow-rate, l/min
Q_{real}	Real flow-rate, l/min
R_i	Residual drop
R_e	Vapor generation rate
R_c	Vapor condensation rate
rpm	Pump speed
S'_{ij}	Strain tensor
T_{oil}	Oil temperature, °C
V	Pump displacement, cm^3/rev
Vi	i-th volume, m^3
Greek Letters	
β	Bulk modulus, bar
γ	Angle step, deg
ε	Turbulence dissipation
η_{vol}	Volumetric efficiency
Θ	Angle of rotation driving gear, deg
μ	Fluid viscosity, Pa·s
μ_t	Turbulent viscosity, Pa·s
v	Velocity, m/s
$\rho(p_k)$	Fluid density at pressure k, kg/m^3
ρ_g	Gas density, kg/m^3
ρ_l	Liquid density, kg/m^3
ρ_v	Vapor density, kg/m^3
τ	Stress tensor
θ	Angle of rotation driven gear, deg
ϕ	Shift angle, deg
Ω	Volume, m^3

References

1. Thiagarajan, D.; Dhar, S.; Vacca, A. Improvement of lubrication performance in external gear machines through micro-surface wedged gears. *Tribol. Trans.* **2017**, *60*, 337–348. [CrossRef]
2. Vacca, A.; Guidetti, M. Modelling and experimental validation of external spur gear machines for fluid power applications. *Simul. Model. Pract. Theory* **2011**, *19*, 2007–2031. [CrossRef]
3. Pellegri, M.; Vacca, M. A CFD-radial motion coupled model for the evaluation of the features of journal bearing in external gear machines. In Proceedings of the Symposium on Fluid Power and Motion Control FPMC2015, Chicago, IL, USA, 12–14 October 2015.
4. Borghi, M.; Milani, M.; Paltrinieri, F.; Zardin, B. Studying the axial balance of external gear pumps. In Proceedings of the SAE Commercial Vehicle Engineering Congress, Chicago, IL, USA, 1–3 November 2005.
5. Borghi, M.; Zardin, B.; Specchia, E. External gear pump volumetric efficiency: Numerical and experimental analysis. In Proceedings of the SAE 2014 World Congress and Exhibition, Detroit, MI, USA, 8–10 April 2009.
6. Mancò, S.; Nervegna, N. Simulation of an external gear pump and experimental verification. In Proceedings of the JHPS International Symposium on Fluid Power, Tokyo, Japan, 13–16 March 1989.
7. Koc, E.; Ng, K.; Hooke, C.J. An analysis of the lubrication mechanisms of the bush-type bearings in high pressure pumps. *Tribol. Int.* **1997**, *30*, 553–560. [CrossRef]
8. Borghi, M.; Milani, M.; Paltrinieri, F.; Zardin, B. Pressure transients in external gear pumps and motors meshing volumes. In Proceedings of the SAE Commercial Vehicle Engineering Congress, Chicago, IL, USA, 1–3 November 2005.
9. Strasser, W. CFD investigation of gear pump mixing using deforming/agglomerating mesh. *J. Fluids Eng.* **2007**, *129*, 476–484. [CrossRef]
10. Castilla, R.; Gamez-Montero, P.J.; Ertuürk, N.; Vernet, A.; Coussirat, M.; Codina, E. Numerical simulation of the turbulent flow in the suction chamber of a gear pump using deforming mesh and mesh replacement. *Int. J. Mech. Sci.* **2010**, *52*, 1334–1342. [CrossRef]

11. Yoon, Y.; Park, B.H.; Shim, J.; Han, Y.O.; Hong, B.J.; Yun, S.H. Numerical simulation of three-dimensional external gear pump using immersed solid method. *Appl. Therm. Eng.* **2017**, *118*, 539–550. [CrossRef]
12. Castilla, R.; Gamez-Montero, P.J.; del Campo, D.; Raush, G.; Garcia-Vilchez, M.; Codina, E. Three-dimensional numerical simulation of an external gear pump with decompression slot and meshing contact point. *ASME J. Fluids Eng.* **2015**, *137*, 041105. [CrossRef]
13. Senatore, A.; Buono, D.; Frosina, E.; Santato, L. Analysis and simulation of an oil lubrication pump for the internal combustion engine. In Proceedings of the ASME International Mechanical Engineering Congress and Exposition, San Diego, CA, USA, 13–21 November 2013.
14. Frosina, E.; Senatore, A.; Buono, D.; Pavanetto, M.; Olivetti, M.; Costin, I. Improving the performance of a two way flow control valve, using a 3D CFD modeling. In Proceedings of the ASME International Mechanical Engineering Congress and Exposition, Montreal, QC, Canada, 14–20 November 2014.
15. Frosina, E.; Senatore, A.; Buono, D.; Stelson, K.A.; Wang, F.; Mohanty, B.; Gust, M.J. Vane pump power split transmission: Three dimensional computational fluid dynamic modeling. In Proceedings of the ASME/BATH 2015 Symposium on Fluid Power and Motion Control, Chicago, IL, USA, 12–14 October 2015.
16. Frosina, E.; Buono, D.; Senatore, A.; Stelson, K.A. A modeling approach to study the fluid dynamic forces acting on the spool of a flow control valve. *J. Fluids Eng.* **2016**, *139*, 011103. [CrossRef]
17. Pellegri, M.; Vacca, A.; Frosina, E.; Senatore, A.; Buono, D. Numerical analysis and experimental validation of gerotor pumps: A comparison between a lumped parameter and a computational fluid dynamics-based approach. *Proc. Inst. Mech. Eng. Part C J. Mech. Eng. Sci.* **2016**, *1989–1996*, 203–210. [CrossRef]
18. Schleihs, C.; Viennet, E.; Deeken, M.; Ding, H.; Xia, X.; Lowry, S.; Murrenhoff, H. 3D-CFD simulation of an axial piston displacement unit. In Proceedings of the 9th International Fluid Power Conference, Aachen, Germany, 24–26 March 2014; pp. 332–343.
19. Vacca, A.; Franzoni, G.; Casoli, P. On the analysis of experimental data for external gear machines and their comparison with simulation results. In Proceedings of the IMECE2007 ASME International Mechanical Engineering Congress and Exposition, Design, Analysis, Control and Diagnosis of Fluid Power Systems, Seattle, WA, USA, 11–15 November 2007.
20. Borghi, M.; Zardin, B. Axial balance of external gear pumps and motors: Modelling and discussing the influence of elastohydrodynamic lubrication in the axial gap. In Proceedings of the IMECE2015 ASME International Mechanical Engineering Congress and Exposition, Advances in Multidisciplinary Engineering, Houston, TX, USA, 13–19 November 2015.
21. Zhou, J.; Vacca, A.; Casoli, P. A novel approach for predicting the operation of external gear pumps under cavitating conditions. *Simul. Model. Pract. Theory* **2014**, *45*, 35–49. [CrossRef]
22. Singhal, A.K.; Athavale, M.M.; Li, H.Y.; Jiang, Y. Mathematical basis and validation of the full cavitation model. *J. Fluids Eng.* **2002**, *124*, 617–624. [CrossRef]

energies

MDPI

Article

Numerical and Experimental Investigation of Equivalence Ratio (ER) and Feedstock Particle Size on Birchwood Gasification

Rukshan Jayathilake and Souman Rudra *

Department of Engineering and Sciences, University of Agder, Grimstad 4879, Norway; rukshanma@gmail.com
* Correspondence: souman.rudra@uia.no; Tel.: +47-3723-3036

Received: 13 July 2017; Accepted: 15 August 2017; Published: 19 August 2017

Abstract: This paper discusses the characteristics of Birchwood gasification using the simulated results of a Computational Fluid Dynamics (CFD) model. The CFD model is developed and validated with the experimental results obtained with the fixed bed downdraft gasifier available at the University of Agder (UIA), Norway. In this work, several parameters are examined and given importance, such as producer gas yield, syngas composition, lower heating value (LHV), and cold gas efficiency (CGE) of the syngas. The behavior of the parameters mentioned above is examined by varying the biomass particle size. The diameters of the two biomass particles are 11.5 mm and 9.18 mm. All the parameters investigate within the Equivalences Ratio (ER) range from 0.2 to 0.5. In the simulations, a variable air inflow rate is used to achieve different ER values. For the different biomass particle sizes, CO, CO_2, CH_4, and H_2 mass fractions of the syngas are analyzed along with syngas yield, LHV, and CGE. At an ER value of 0.35, 9.18 mm diameter particle shows average maximum values of 60% of CGE and 2.79 Nm^3/h of syngas yield, in turn showing 3.4% and 0.09 Nm^3/h improvement in the respective parameters over the 11.5 mm diameter biomass particle.

Keywords: Birchwood gasification; computational fluid dynamics; equivalence ratio; cold gas efficiency; syngas

---

1. Introduction

Among the available energy sources, biomass is envisaged to play a major role in the future energy supplement. It produces no net carbon emission, while being the fourth largest energy source available in the world [1–4]. Therefore, biomass has high potential in contributing to satisfy the future energy demands of the world. Further, it is seen that for countries where the economy is mostly based on agriculture, they can utilize the potential of biomass efficiently. To recover energy from biomass, either thermochemical or biochemical process can be used [5–7]. The gasification process is a thermochemical process which gives a set of gases as output, consisting of CO, CO_2, H_2, CH_4, and N_2 by converting organic or carbonaceous materials like coal or biomass [8–10]. In gasification, several types of reactors used, such as entrained flow, fixed bed, fluidized bed, and moving bed gasifiers [7,11–13].

The producer gas is a result of a set of endothermic and exothermic reactions. The required heat for the endothermic reactions is provided by the exothermic reactions [11,14]. Once the steady state is achieved, the gasifier could work in a certain temperature range. Then, the producer gas can be obtained, as long as the fuel is being fed [14,15].

It was proven that producer gas was more versatile and useful than the biomass [16]. The quality of the producer gas is established to be one of the most important aspects of the gasification which should be enhanced, as it is used to generate power [11,16]. Chemical reactions inside the gasifier consist of both homogeneous and heterogeneous reactions. There are some factors which are important for the improvement of the gasification process such as reactor temperature, reactor pressure, the flow rate of the gasifying agent, and inflow rate of the biomass to the gasifier [11,17]. To evaluate gasification characteristics of biomass, a thorough evaluation is required. With the recent development of the computational fluid dynamics (CFD) and numerical simulations, sophisticated and robust models can be developed to give more qualitative information on biomass gasification [17,18]. In addition, CFD can produce important data with a relatively low cost [19]. Thus, CFD has become popular and an often-used tool to examine gasification characteristics [18,20,21].

Previous researches on gasification process have been performed on a regular basis investigating a broad range of parameters and variables. Most of them focused on the thermodynamics of the process, syngas composition, energy and exergy output, effect of temperature, the efficiency of the process, etc. [22,23]. Moreover, numerous numerical simulations have been done on the flow patterns and turbulence, as well as the different mathematical approaches. Ali et al. [24] developed a simulation model to discuss the co-gasification of coal and rice straw blend to investigate the syngas and cold gas efficiency (CGE), noting an 87.5% CGE value during the work. Slezak et al. [25] discussed an entrained flow gasifier using a CFD model to study the effects of coal particle density and size variations, where two devolatilization models were used in their work. It concluded that a higher fixed carbon conversion and H_2 could be achieved with a mix of different partitioned coal. Rogel et al. [26] presented a detailed investigation on the use of Eulerian approaches on 1-D and 2-D CFD models to examine pine wood gasification in a downdraft gasifier. Syngas lower heating value, syngas gas composition, temperature profiles, and carbon conversion efficiency were investigated [26] for both models to conclude that these methods could be used effectively in determining the above-mentioned parameters. Sharma [17] developed a CFD model of a downdraft biomass gasifier to investigate the thermodynamic and kinetic modeling of char reduction reactions. Further, a conclusion has been made that char bed length is less sensitive to equilibrium predictions, while CO and H_2 component, the calorific value of product gas and the endothermic heat absorption rate in reduction zone are found to be sensitive to the reaction temperature. Wu et al. [27] developed a 2-D CFD model of a downdraft gasifier to study the high-temperature agent biomass gasification, and observed a syngas with a high concentration of H_2 with a limited need of combustion inside the gasifier. Janejreh et al. [28] discussed the evaluation of species and temperature distribution using a CFD model with k-ε turbulence model and Lagrangian particle models. The gasification process was assessed using the CGE. Ismail et al. [29] developed a 2-D CFD model by using the Eulerian-Eulerian approach. Coffee husks are used as the feedstock to model the gasification process of a fluidized bed reactor. The effect of Equivalence Ratio (ER) and moisture content on gasification temperature, syngas composition, CGE, and HHV were discussed. After the analysis, high moisture content of coffee husks was found as a negative impact on the CGE and Higher Heating Value (HHV) as the ER increased. Monteiro et al. [30] developed a comprehensive 2-D CFD model to examine the potential of syngas from gasification of Portuguese Miscanthus. The Eulerian-Eulerian approach was used to model the exchange of mass, energy, and momentum. For the work, the effect of ER, steam to biomass ratio and temperature on syngas quality assessed. ER was proven to be a positive impact on the Carbon Conversion Rate (CCE), where an adverse impact was shown on the syngas quality and LHV. Further, the syngas quality and gasification efficiency was found to be increased by the increasing temperature. In order to describe the gasification process of three Portuguese biomasses, Silva et al. [31] developed a 2-D CFD model. A k-ε model was used to model the gas phase, while an Eulerian-Eulerian approach was used to model the transport of mass, energy, and momentum. After the simulations, the highest CGE value was shown by vine-pruning residues, while a higher H_2/CO ratio was shown by both coffee husks and vine pruning residues over the forest residues. Couto et al. [32] developed a 2-D CFD model using the Eulerian-Eulerian approach to evaluate the gasification of municipal solid waste. The gasification temperature was defined to be vastly important in syngas heating value in conclusion. In comparison to the above research works, the novelty of this study lies in the field of the Birchwood gasification. In addition, the literature lacks the use of 3-D CFD models and experiments on the effect of wood particle size on CGE. In this article, a 3-D CFD model is developed to investigate the Birchwood gasification in a fixed bed downdraft gasifier. Both heterogeneous and homogeneous reactions are considered and variation of syngas composition, syngas yield, LHV of the syngas, and CGE are measured for two sizes of feedstock, as well as for ER. Fixed bed downdraft gasifiers have demonstrated to perform well in the ER range from 0.25 to 0.43. Therefore, in the present work, the variations of the gasification parameters are considered in the ER range from 0.2 to 0.5 [14,15]. For the experimental work, the downdraft gasifier is used, which is available at the University of Agder (UIA), Norway. In addition, the CFD model is developed as the same geometry of the gasifier used for the experiments. For the validation purposes, experiments are done with the same gasifier with one of the wood particle sizes (11.5 mm woodchip) as it is the only biomass particle size available for the experiments. Then, with the developed simulation model, both biomass particle sizes are simulated.

2. Feedstock Characteristics

For the study, Birchwood is used as the feedstock. The ultimate and proximate analysis results for the Birchwood are listed down in Table 1. The data is extracted from a previous study [33], which is done on Birchwood in the same lab and under the same conditions.

Table 1. Proximate and ultimate analysis results of Birchwood [33].

Type of Analysis	Physical or Chemical Property	Value
Proximate Analysis (Dry Basis)	Moisture %	7
	Volatiles %	82.2
	Fixed Carbon %	10.45
	Ash %	0.35
	LHV (MJ/kg)	17.9
Ultimate Analysis (Dry Basis)	Carbon %	50.4
	Hydrogen %	5.6
	Oxygen %	43.4
	Nitrogen %	0.12
	Sulphur %	0.017
	Chlorine %	0.019

According to the analysis above, Birchwood has a lower amount of moisture in comparison to some of the other wood types [33–35], which is an important aspect for gasification. Apart from the moisture content, Birchwood has shown a lower amount of ash, Cl, and S, while higher values of calorific value and volatile matter [34] has been indicated. It is obvious that Birchwood has a higher calorific value because of the higher amount of carbon content [33]. In addition, when a biomass consists of less ash, it can lead to a higher conversion process which has lesser slag [14]. Hence, Birchwood has some positive characteristics in the gasification perspective. According to some of the literature, K, Si, and Ca can have a small effect in determining gasification characteristics. However, in this work, those are not considered as they are not measured in the ultimate analysis [36].

3. Methods

In this section, first, the experimental setup and the experimental process is described. Then, a brief description of the equations which use for the calculations is presented. It is then followed by the numerical study and simulation process.

3.1. Experimental Study

For the experiment, initially, about 6 kg of biomass is fed into the gasifier along with 1 kg of charcoal. Charcoal is added to the system to help the ignition process. The simulations are also carried out following the same procedure. The average size of the Birchwood chip which was used for the experiments is approximately 11.5 mm × 11.5 mm. A schematic diagram of the system is shown in Figure 1.

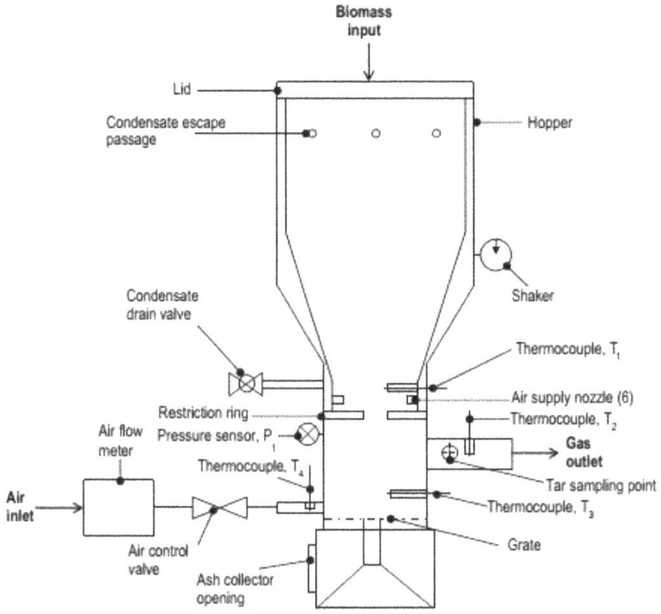

Figure 1. Schematic diagram of the gasifier system [34].

The hopper is a conically shaped stainless-steel structure with a volume of 0.13 m^3 and double-walled thick walls. It has a radius of 0.5 m. Biomass is fed through this hopper, and it is connected to an electrical shaker which can provide 15 shakes per minute (shakes/min). This shaking process helps to reduce or avoid some common problems, such as channeling and bridging. To drain out the condensed moisture, a drain valve is also attached to the bottom of the hopper.

The primary gasification reactor has a diameter of 0.26 m. It is 0.6 m in height. The air inlet system consists of six same size nozzles with equal spacing among each of a 5 mm diameter, and they are located from about 0.4 m from the bottom of the reactor. At the bottom of the reactor part, a reciprocally oscillating grate is connected and takes out the ash to the lower part of the ash collector. The grate is oscillating 30 s/min. To measure the weight loss during the gasification process, a mass scale locates at the bottom of the gasifier system. In addition, two pressure sensors and five K-type thermocouples are connected to the gasifier assembly to monitor and measure the parameters.

All the necessary experimental data is measured via a Lab view program which is specially designed to obtain data from the gasifier system. Further, to measure the composition of the syngas, a gas sample is sent to a gas analyzer, and throughout the experiment, changes in the composition of the syngas output is produced by the gas analyzer.

3.2. Theoretical and Numerical Study

In this study, the effect of biomass particle size and ER on the behavior of gasification parameters is examined using Birchwood as the feedstock. Thus, the variance of producer gas composition, producer gas yield, LHV of the syngas, and CGE is mainly assessed on two different values of biomass particle sizes.

Although downdraft gasifier demonstrates better performances at the ER ratio of 0.25, acceptable performances showed even close to the ER value of 0.43 [33,37]. Therefore, ER value range from 0.2 to 0.5 is selected in this study to investigate the best performance. ER is calculated using the following expression Equation (1) [8,15].

$$ER = \frac{\text{Actual air fuel ratio}}{\text{Stoichiometric air fuel ratio}} \tag{1}$$

here, the actual air-fuel ratio is calculated from the actual values measured from the experiment, while the stoichiometric air-fuel ratio is determined using Equation (2).

$$CH_aO_b + x(O_2 + 3.76N_2) \rightarrow yH_2O + zCO_2 + 3.76xN_2 \tag{2}$$

where a and b can be determined using the results of the ultimate analysis of the biomass (Birchwood). Here, y and z represent the stoichiometric coefficients of H_2O and CO_2, respectively, while x represent the stoichiometric coefficient of O_2 and 3.25 times of the N_2 amount. Further, the other parameters are calculated using the following Equation (3) to Equation (5) [33,38–40].

$$LHV_{syngas} = ((25.7 \times H_2) + 30 \times CO + (85.4 \times CH_4)) \times \frac{4.2}{1000} \tag{3}$$

Since the Hydro-Carbons (HC) higher than CH_4 are not considered for this study, they have not been used to calculate LHV_{syngas}.

$$\text{Producer gas yield } (V_{syngas}) = \frac{(Q_{air} \times 79)}{N_2 \times m_b} \tag{4}$$

$$\text{Cold gas efficiency (CGE)} = \frac{LHV_{syngas} \times V_{syngas}}{LHV_{biomass}} \times 100 \tag{5}$$

where LHV_{syngas} represents the dry base lower heating value of the syngas in MJ/Nm^3, V_{syngas} represents the syngas yield Nm^3/kg feed stock, Q_{air} represents the air input in Nm^3/h, m_b represents the biomass input in kg/h, $LHV_{biomass}$ represents the lower heating value of the feedstock in MJ/kg, and N_2 represents nitrogen mass fraction of the output gas.

In the simulation model, a spherical particle is used for modeling. Hence, spherical particles with a diameter of 11.5 mm (available wood chip size in the lab) and 9.18 mm are utilized for the simulations (a spherical particle which has the same surface area as an actual wood chip, with a diameter of 9.18 mm consider here).

In the presented work, the STAR CCM+ software package is used for the modeling and simulation. A 3D geometric model of the actual gasifier is designed to create the computational model. For the gasifier model, 0.01m mesh is used, while in the air inlets and air-fuel mixing region, a 0.001 m mesh has been used. Figure 2 shows the customized meshfor the gasifier (0.01 m) and for air inlets (0.001 m).

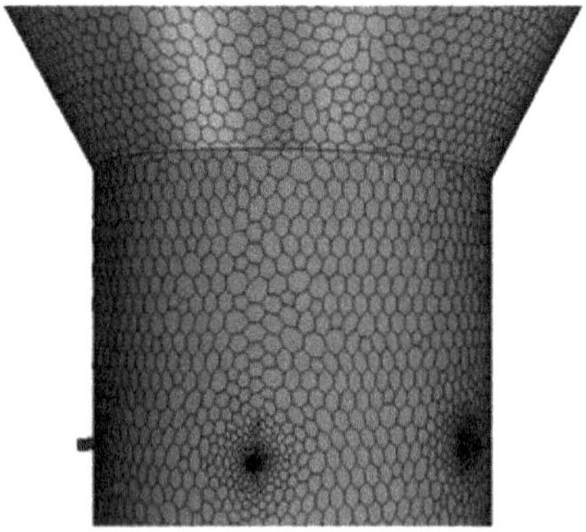

Figure 2. Meshed gasifier geometry and customized mesh at air inlets.

A segregated flow model is implemented to solve the transport equations. In addition, the Lagrangian particle modeling method is used to model the particle behavior (solid phase).

3.2.1. Assumptions

The underlying assumptions of the modeling process are summarized as follows:

- Steady flow is considered inside the gasifier.
- The flow inside the domain is considered as incompressible and turbulent.
- Spherical particles are used.
- Evenly distributed particle regime is utilized.
- The No-slip condition is imposed on inside wall surfaces.

3.2.2. Eulerian-Lagrangian Method

Reynolds averaged Navier-Stokes (RANS) equations are solved using a Eulerian-Lagrangian reference frame to solve the numerical scheme. The gas phase is considered as a reacting gas phase, and Lagrangian particle models are used to address the solid (particle) phase. Standard k-ε model is used for turbulence modeling. In addition, both finite rates of chemical reactions and eddy dissipation rates are used in the computational model. Moreover, for the modeling work, the conservation equations for transport of energy, momentum, and mass are used (Appendix A). Also, the default sub models available for moisture evaporation, devolatilization and char oxidation are employed (Appendix A). In the Eulerian-Eulerian approach, gas phase, as well as the solid phase, is considered as continuum [41]. In the Lagrangian approach, usually a large number of particles are tracked transiently. The method is started by solving the transient momentum equation for each particle. In order to calculate the trajectory of a particle, Newton's second law is used and written in a Lagrangian reference frame [41–44]. Momentum balance for particles is described in Equation (6),

$$m_p \frac{dV_p}{dt} = F_s + F_b \tag{6}$$

where V_p represents the velocity of the particle, m_p is the mass of the particle, and t is time. F_s and F_b represent the surface force and body force, respectively. When these forces are observed in depth, each of them is a composition of few forces available in the system. The following Equations (7) and (8) describe the two forces.

$$F_s = F_d + F_p + F_{vm} \tag{7}$$

$$F_b = F_g + F_u \tag{8}$$

where F_d, F_p, and F_{vm} represent the drag force on the particle, pressure force as a gradient in the static pressure, and the extra virtual mass added while the acceleration in the phase, respectively. F_g and F_u represent the gravitational force and the user defined force added to the system, respectively. In addition, the drag force can be defined as follows in Equation (9).

$$F_d = \frac{1}{2} C_D \rho A_p |V_s| V_s \tag{9}$$

where C_D is the drag coefficient, ρ is the density, A_p represents the surface area of the particle, and V_s is the slip velocity of the particle.

Using the Nusselt number (Nu_p), the heat transfer coefficient can be found for the Lagrangian particle, which is the energy model for the Lagrangian particle. This is shown in Equation (10),

$$h_p = \frac{Nu_p}{D_p} \cdot k \tag{10}$$

where k has represented the thermal conductivity of the phase and D_p is the particle diameter. Nusselt number (Nu_p) is expressed in the following Equation (11), where it is calculated by the Ranz-Marshall correlation and presented below in Equation (12).

$$Nu_p = 2\left(1 + 0.3Re_p^{\frac{1}{2}}Pr^{\frac{1}{3}}\right) \tag{11}$$

where Re represents the Reynolds number, while the expression Pr expresses the Prandtl number [45,46].

$$N_u = h\frac{d_p}{k} = a + cRe^m Pr^n \tag{12}$$

where h is the heat coefficient, d_p is the particle diameter, k represents the thermal conductivity of the gas, Re is the Reynolds number, and Pr represents the prandtl number. The letters a, c, m, and n are numerical constants which are determined by the flow field. According to the Ranz-Marshall correlation, at a fluid velocity (V) equal to zero, the heat transfer can only be affected by conduction. therefore, the values of a, c, m, and n become 2, 0.6, 0.5, and 0.33, respectively [47].

3.2.3. Reactions Modeling

In this work, the following approach is used to simulate the gasification process. This method has been used by [8,48]. In this study, it is assumed that the moisture is evaporated completely in the drying phase [49].

$$\text{Biomass (wet)} \rightarrow \text{Biomass (dry)} + H_2O \text{ (steam)} \tag{R1}$$

In the pyrolysis phase, the Biomass (dry) is converted into char, tar, ash, and volatile matter [49,50].

$$\text{Biomass (dry)} \rightarrow \text{volatiles} + \text{char} + \alpha\text{tar} + \text{ash} \tag{R2}$$

where, α is the stoichiometric coefficient of tar.

Here, the volatile matter is described as follows. It is assumed that the volatile matter consists of only CO, CO_2, H_2, and CH_4.

$$\text{Volatile} \rightarrow \alpha_1 CO + \alpha_2 CO_2 + \alpha_3 H_2 + \alpha_4 CH_4 \tag{R3}$$

where, $\alpha_1, \alpha_{2,}\alpha_{3,}$, and α_4 are the stoichiometric coefficients of CO, CO_2, H_2, and CH_4 respectively.

The following method is used to calculate the mass distribution in volatile matter and mass fractions in dry biomass. For the calculations, the data available in Table 1 is used. These values can be used to combine with the ultimate analysis. In this method, a set of linear equations is used to solve the C, O, and H components balance in the system [8,48].

$$\frac{12}{28}x + \frac{12}{44}y + \frac{12}{16}z = Y_{C,vol} = 0.449 \tag{13}$$

$$\frac{16}{28}x + \frac{32}{44}y = Y_{O,vol} = 0.488 \tag{14}$$

$$\frac{4}{16}z + w = Y_{H,vol} = 0.063 \tag{15}$$

here, x, y, z, and w are unknown mass fractions of CO, CO_2, CH_4, and H_2, respectively in volatile species after devolatilization. $Y_{C,vol}$, $Y_{O,vol}$, and $Y_{H,vol}$ are, respectively, the mass fractions of C, O, and H in biomass volatile.

In the above equations, the coefficients are the ratios of molar mass of each atom to the molar mass of each species. (As an example, in Equation (15), $\frac{4}{16} = \frac{M_H}{M_{CH_4}}$). Here, with the molar masses, the numerical values are rounded up for the simplification of the calculations. For finding the mass fractions $Y_{C,vol}$, $Y_{O,vol}$, and $Y_{H,vol}$, the following method is used along with the data from the ultimate analysis of the biomass.

$$\text{Volume fraction} = Y_{C\ total} - Y_{C\ fixed} + Y_O + Y_H = 0.89 \tag{16}$$

$$Y_{C,vol} = \frac{Y_{C\ total} - Y_{C\ fixed}}{\text{Volume fraction}} = 0.449 \tag{17}$$

$$Y_{O,vol} = \frac{Y_O}{\text{Volume fraction}} = 0.488 \tag{18}$$

$$Y_{H,vol} = \frac{Y_H}{\text{Volume fraction}} = 0.063 \tag{19}$$

According to the mass conservation,

total mass before devolatilization − mass of fixed carbon = total mass after devolatilization

Thus, according to the above criteria,

$$x + y + z + w = Y_{C,vol} + Y_{O,vol} + Y_{H,vol} = 1 \tag{20}$$

The ratio between CO and CO_2 is temperature dependent. In this study, the ratio has been determined as 2.43 according to [8]. For the calculations, the bed temperature is used as 907 K, which was obtained during the experiments. Below Equation (21) is used to determine the connection between CO and CO_2 at 907 K.

$$\frac{x}{y} = 2400e^{-\left(\frac{6234}{T}\right)} \tag{21}$$

$$\frac{x}{y} = 2.43 \tag{22}$$

The linear equations from Equation (9) to Equation (15) are solved using the MATLAB program, and the following values are obtained from the program.

$$x = 0.5605$$

$$y = 0.2306$$

$$z = 0.1945$$

$$w = 0.0144$$

According to the above results, 1 kg of volatile matter produce devolatilized products according to the following expression.

$$1\text{kg}_{vol} \rightarrow 0.5605\text{kg}_{CO} + 0.2306\text{kg}_{CO_2} + 0.1945\text{kg}_{CH_4} + 0.0144\text{kg}_{H_2} \tag{23}$$

Using the following molar mass values, the above Equation (23) can be used as a molar based equation.

$$M_{CO_2} = 44.008\ \text{kg/mol},\ M_{CO} = 28.009\ \text{kg/mol},\ M_{CH_4} = 16.042\ \text{kg/mol},\ M_{H_2} = 2.016\ \text{kg/mol}$$

$$1\text{kg}_{vol} \rightarrow 0.02CO + 0.00524CO_2 + 0.0072H_2 + 0.0122CH_4 \tag{24}$$

Therefore, according to the following (R4), the devolatilization of the volatiles is modeled.

$$0.0375CH_{1.69}O_{0.81} \rightarrow 0.02CO + 0.00524CO_2 + 0.0072H_2 + 0.0122CH_4 \tag{R4}$$

Based on the above devolatilization reaction (R4), the gasification process is modeled. In addition, for the modeling work, the following homogeneous and heterogeneous reactions are considered [8,28,51].

Water-gas shift reaction:

$$CO_{(g)} + H_2O_{(g)} \leftrightarrow CO_{2(g)} + H_{2(g)};\ \Delta H = +41.98\ \mathrm{kJ/mol} \tag{R5}$$

Methanation reaction:

$$CO_{(g)} + 3H_{2(g)} \leftrightarrow CH_{4(g)} + H_2O_{(g)};\ \Delta H = -227\ \mathrm{kJ/mol} \tag{R6}$$

Dry reforming reaction:

$$CH_{4(g)} + CO_{2(g)} \leftrightarrow 2CO_{(g)} + 2H_{2(g)};\ \Delta H = -793\ \mathrm{kJ/mol} \tag{R7}$$

Steam reforming reaction:

$$CH_{4(g)} + H_2O_{(g)} \leftrightarrow CO_{(g)} + 3H_{2(g)};\ \Delta H = +206\ \mathrm{kJ/mol} \tag{R8}$$

mbustion reaction:

$$C_{(s)} + O_{2(g)} \rightarrow CO_{2(g)};\Delta H = -408.8\ \mathrm{kJ/mol} \tag{R9}$$

Boudouard reaction:

$$C_{(s)} + CO_{2(g)} \rightarrow 2CO_{(g)};\ \Delta H = +172\ \mathrm{kJ/mol} \tag{R10}$$

Water-gas reaction:

$$C_{(s)} + H_2O_{(g)} \rightarrow CO_{(g)} + H_{2(g)};\ \Delta H = +131\ \mathrm{kJ/mol} \tag{R11}$$

Hydrogasification:

$$C_{(s)} + 2H_{2(g)} \rightarrow CH_{4(g)};\ \Delta H = +75\ \mathrm{kJ/mol} \tag{R12}$$

Carbon partial reaction:

$$2C_{(g)} + O_{2(g)} \leftrightarrow 2CO_{(g)};\ \Delta H = -246.4\ \mathrm{kJ/mol} \tag{R13}$$

4. Results and Discussions

In this section, the computational model validation is conducted by comparing the available experimental data and numerical simulation results. Moreover, the comparison of the two sizes of biomass particle sizes is carried out using the gasification characteristics.

4.1. Development of Mass Fractions of Considered Gases in the Syngas

A specially designed lab view program was used to measure all the gas component mass fractions throughout the experiment. From the obtained data, the variation of the mass fractions of the gas components during the experiment is presented below in Figure 3. For the tests, a biomass woodchip with an average size of 11.5 mm × 11.5 mm is used, which is available in the laboratory.

In the simulation model, for two iterations, one data point has been recorded. 600 iterations are used in the simulations, and 300 data points are recorded throughout the simulation. Using the obtained data points gas component mass fractions are graphed in Figure 4. For the simulations, two types of wood particle sizes are used. One of them is 11.5 mm in diameter, while the other is 9.18 mm in diameter. Below, Figure 4 was obtained using 11.5 mm diameter wood particle.

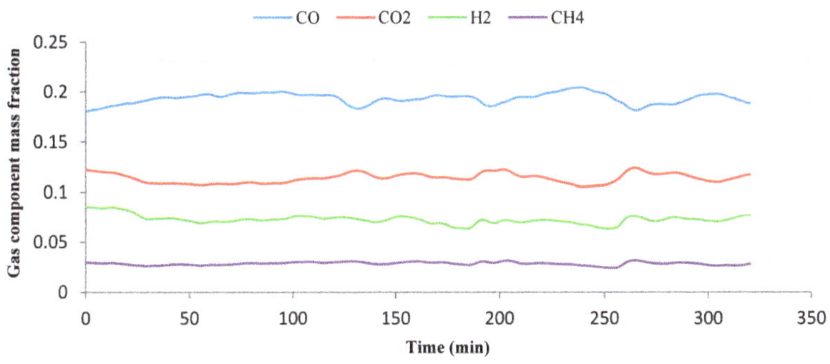

Figure 3. Mass fraction variation with time in the experimental study.

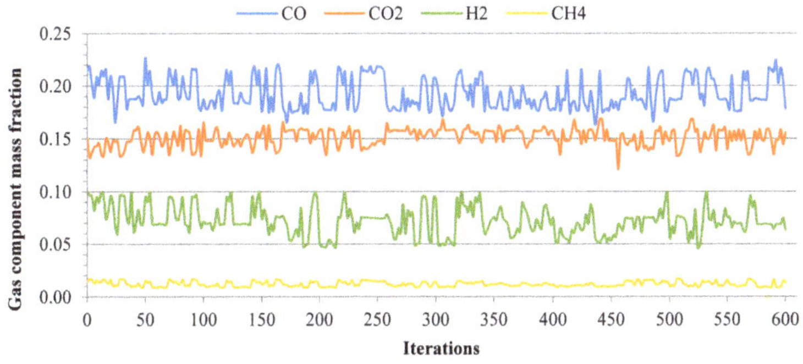

Figure 4. Mass fraction variation with iteration in the simulation model.

By comparing Figures 3 and 4, it shows the fluctuation in gas components mass friction in the simulation values, whereas in the experiment, it shows smooth variation. It can be due to the fact that the interval between two data points are bigger. However, an average value is used to plot graphs.

The data obtained from the model is also compared with some of the data collected from the literature (Study 1 [52], and Study 2 [53]), as well as with the data achieved during the experiment of the present work. Figure 5 shows the comparison of the predicted data along with the experimental data. According to the graph, predicted data shows good agreement with the literature and experimental data.

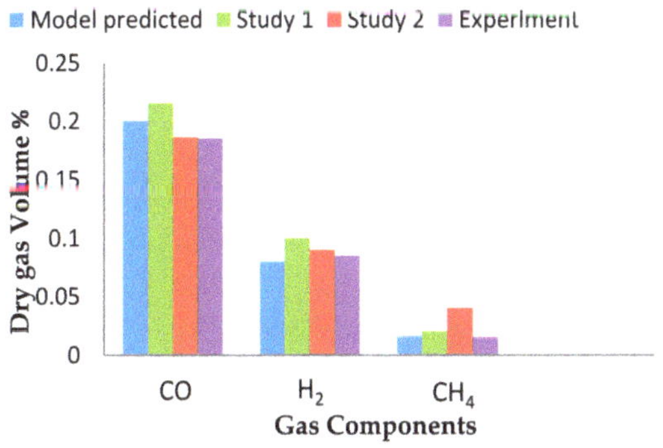

Figure 5. Composition of the syngas predicted by model compared with experimental results.

4.2. Effect of ER and Biomass Particle Size on Quality of Producer Gas

The variation of gas components in producer gas in both experiments and simulation models is illustrated in the following Figure 6. For the simulations, a biomass particle with a diameter of 11.5 mm and a biomass particle with a diameter of 9.18 mm were used. (The change of gas component contours in various ER values are illustrated in Appendix A (Figures A1–A3)).

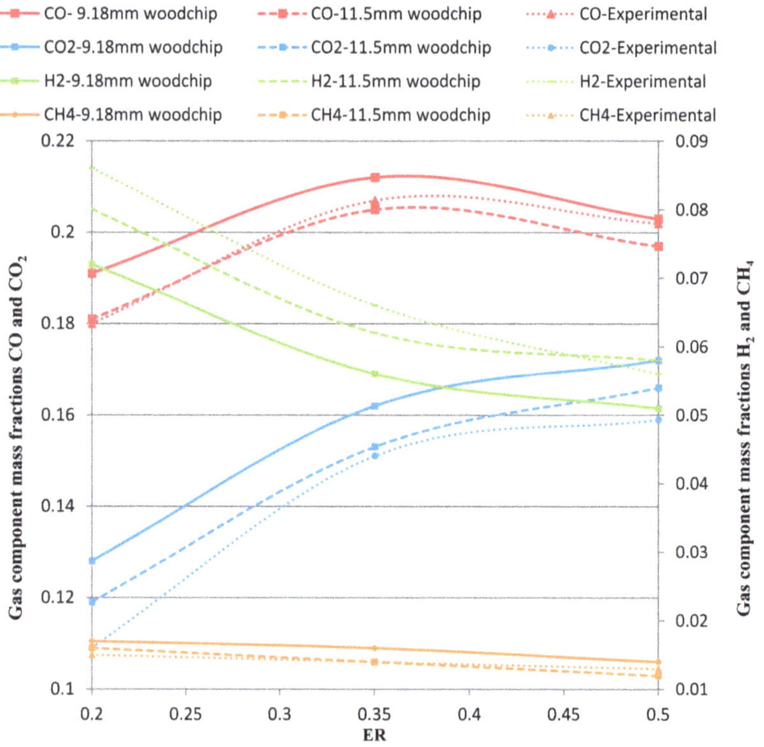

Figure 6. Variation of gas component mass fractions as a function of ER.

According to Figure 6, mass fractions of CO and CO_2 are increased when the ER is increased up to a certain level (ER = 0.2 to 0.35). Then, ER contributes to a reduction of the mass fraction of CO, while further increasing CO_2. This can be a result of the combined effect of Combustion reaction (R9), Boudouard reaction (R10), and Carbon partial reaction (R13). Therefore, in lower ER values, (R10) tends to produce more CO to the system at the expense of CO_2, while in high ER values, (R9) and (R13) convert more CO into CO_2 [2].

Rgearding CH_4 and H_2, increasing ER is effected on a decrease in both gas species. Nevertheless, H_2 is varied in a higher mass fraction range than of CH_4. This is mainly due to the effects of dry reforming (R7) and steam reforming (R8) reactions, which produce more H_2 to the system at the expense of CH_4.

Since a variable air inflow rate is used in the experiments, an average value of the air inflow rate is used for the calculation of ER values. However, a fixed air inflow rate is used in the simulations. Therefore, the impact of variable air inflow may have caused the deviation in the graphs of the experimental and simulation data. In the simulations, spherical shape wood particles are used, while wood chips with different shapes are used for the experiments. The difference of behavior of spherical wood particle and divergement shaped wood chips can also create the inequalities in the graphs. In addition, during the experiments, channeling and bridging can also have taken place [7]. The effect of that phenomenon may have also affected the experimental results.

According to Figure 6, all the considered gas component mass fractions, except H_2, are increased for the 9.18 mm wood particle over 11.5 mm wood particle. This pattern can be found in some of the literature [54,55]. When the particle size is getting smaller, the area to volume ratio of the wood particle is increased. This can result in an increased contact with the reactants [54–56]. Hence, the reaction rates can be increased and gasification reactions can also be improved. Further, the enhancement of hydrogasification (R12) is crucial for the behavior of the gas component mass fractions. When the biomass particle is getting smaller, it can enhance the reactivity and produce more CH_4 with R12.

4.3. Effect of Equivalences Ratio (ER) and Biomass Particle Size on Syngas Yield, Lower Heating Value (LHV), and Cold Gas Efficiency (CGE)

Figure 7 shows the variation of syngas yield, LHV and CGE of the syngas as a function of ER. According to Figure 7, the behavior of syngas yield could be an effect of higher volatilization [28].

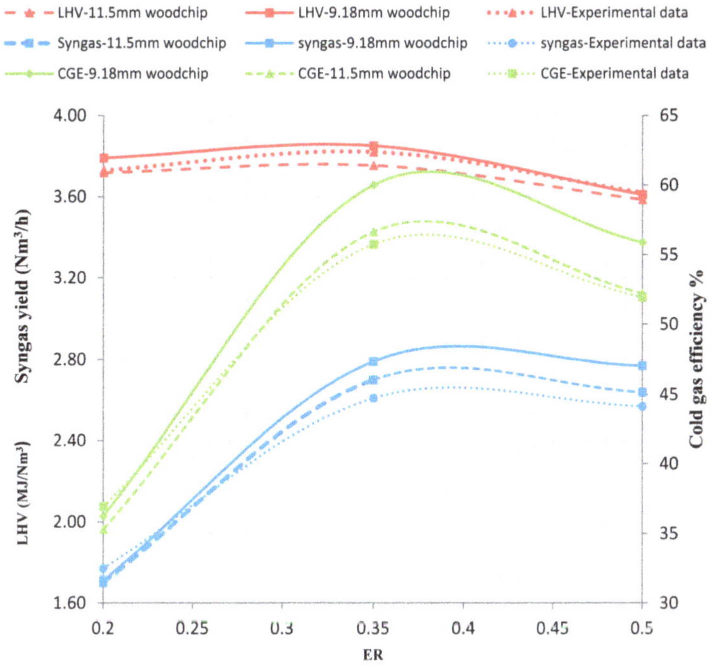

Figure 7. Variation of Syngas yield lower heating value (LHV) and cold gas efficiency (CGE) of syngas as a function of Equivalences Ratio (ER).

During the simulations, both biomass particles are evaluated in the same ER range. However, the gas yields with two separate sizes of biomass particles have shown a difference. This difference of syngas yield should be due to a different conversion. For the context of LHV of syngas, increasing ER has shown a positive effect within the range from 0.2 to 0.35. It is evident that, as the gas species mass fractions increase, LHV of syngas increases as well.

According to Figure 7, the smaller particle size has resulted in a significant increment of the LHV of the gas, although it has followed the same variation along the ER range as the bigger biomass particle. The increment of LHV can be mainly due to the increment of CH_4 with the smaller biomass particle. As mentioned before, LHV is a direct function of combustible gas species, and CH_4 has the highest contribution for the LHV among the gas species.

For CGE, a positive effect of increasing ER is observed from 0.2 to 0.35 of ER. Within this ER range, CGE could have been improved by increasing airflow. Increasing airflow (increasing ER) can contribute to creating a high composition of combustible gas species in syngas by enhancing the reaction chemistry of homogeneous and heterogeneous reactions. Therefore, as the gas species are increased up to 0.35 of ER, so does the CGE. As the gas species have shown a decrement in the mass fractions in the ER range from 0.35 to 0.5, CGE has also shown a decrement in its value. With more O_2 input to the system, the system can move towards combustion rather than gasification. This can be a result of the reduction of CGE after the ER value of 0.35. According to the graph in Figure 7, with the 9.18 mm biomass particle, CGE has shown an average maximum value of 60% at ER value of 0.35, compared to an average maximum value of 56.63 % with 11.5 mm diameter biomass particle.

For both particle sizes, same ER range is considered. Hence, an increase of LHV, syngas yield, and combustible gas spices in syngas with smaller biomass particle could be the reason for the increase of the CGE.

4.4. Temperature Profile

Temperature profile inside the gasifier in the simulation model is illustrated in Figure 8. In Figure 9, the temperature values recorded in the experiments and the simulation model are demonstrated in a graph. Figure 8 is mainly used to help understand the actual placement of the thermocouples. In the simulation model, seven temperature measurements are taken along the mid-axis, which is shown in Figure 9. However, in the experimental setup, only four thermocouples are available. Hence, to develop the curve, those four temperature measurement values are used. According to Figure 9, the maximum temperature is shown close to the air inlets in both experimental and simulation setups. In the experimental setup, the thermo couples are placed on the gasifier wall. Hence, the actual temperature can be slightly higher than what it is shown here.

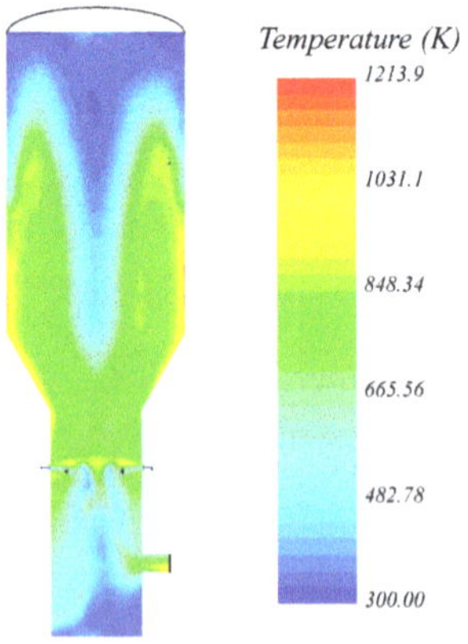

Figure 8. Temperature profile inside the gasifier.

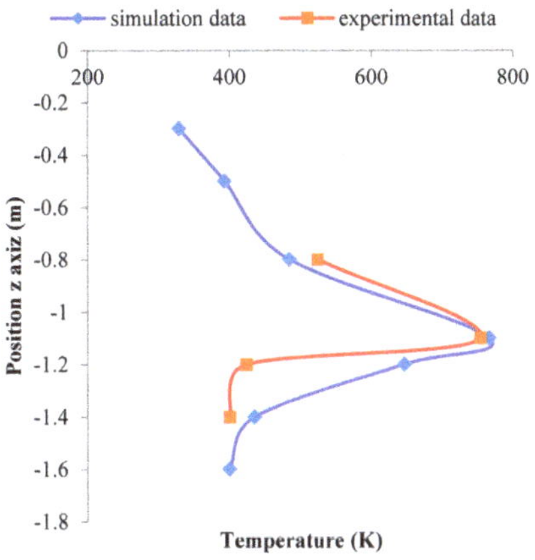

Figure 9. Temperature variation along the mid-axis of the gasifier simulation values vs. experimental values.

5. Conclusions

Based on the comparative study, it is concluded that most of the measured and calculated parameters (including syngas yield, LHV of syngas, and CGE) have shown an increasing trend as ER is increased up to 0.35, and when biomass particle size is reduced. Furthermore, the average value of CGE, for instance, reached a maximum value of 60% with the 9.18 mm diameter wood particle, and with a corresponding ER about 0.35. Moreover, 11.5 mm diameter wood particle has shown an average maximum value of 56.6% at the same ER value, which has been both an improvement and a significant output of this work. Reaching an average value of 2.79 Nm^3/h with a corresponding ER of 0.35, the average value of Syngas yield has shown an improvement of 0.09 Nm^3/h with the 9.18 mm diameter wood particle over 11.5 mm wood particle, which has been an important aspect in energy harvesting from the syngas. Therefore, as the result shows, an ER value close to 0.35 shows the best results in CGE, syngas yield, and LHV of the syngas. Furthermore, smaller biomass particles show improvements in the results. Therefore, according to our study, in order to obtain better outputs, ER should be kept close to 0.35. In the future, more simulations and experiments could be done with finer biomass particles. Further, more user defined functions could be used to improve the gasification process and to represent more realistic biomass particle shapes.

Acknowledgments: Authors would like to thank Department of Engineering and Sciences, UiA, for supporting to do the experiment at Bio-energy lab. In addition, we would appreciate the contribution and guidance given by Hans Jorgen Morch, CEO of CFD Marine AS, especially in obtaining the license for the STAR CCM+ software.

Author Contributions: Rukshan Jayathilake and Souman Rudra conceived and designed the experiments; Rukshan Jayathilake performed the experiments; Rukshan Jayathilake and Souman Rudra analyzed the data; Rukshan Jayathilake and Souman Rudra wrote the paper.

Conflicts of Interest: The authors declare no conflict of interest.

Appendix A

Appendix A.1. Conservation Equations

In order to develop the numerical model, conservation equations have been used, along with some other important equations. The equation for mass conservation is presented below in Equation (A1) [39],

$$\frac{\partial \rho}{\partial t} + \frac{\partial \rho v_i}{\partial x_i} = 0 \tag{A1}$$

where ρ is the density of the fluid, is the velocity tensor in the compact form in Einstein notation, t represents the time, and x_i represents the special first order tensor.

The momentum conservation equation is:

$$\rho\frac{\partial u_i}{\partial t}+\rho u_j\frac{\partial u_i}{\partial x_j}=\rho g_i-\nabla P+\frac{\partial}{\partial x}\left[\mu\left(\frac{\partial v_i}{\partial x}+\frac{\partial v_j}{\partial x}\right)\right] \tag{A2}$$

where ρ is the density of the fluid and g_i is the gravity force. P represents the pressure, while the dynamic viscosity is represented by μ.

The energy equation is presented below in Equation A3 [57],

$$\rho\left(\frac{\partial h}{\partial t}+u_i\frac{\partial h}{\partial x_i}\right)=\frac{\partial P}{\partial t}+u_i\frac{\partial P}{\partial x_i}+\frac{\partial}{\partial x_j}(k\nabla T)+\tau'_{ij}\frac{\partial u_i}{\partial x_j} \tag{A3}$$

where ρ represents the density of the fluid P represents the pressure, k represents the thermal conductivity, and τ'_{ij} represents the viscous stress tensor.

Appendix A.2. Submodels

Appendix A.2.1. Two Ways Coupling

When the Lagrangian particles are relatively smaller, a one-way coupling model is used, since the heat transfer and drag force are the only things which affect through continues phase. However, once the particle size is much larger compared to the volume cells, then a more advance model has to be introduced. Once the two-way coupling is used, an active particle or the Lagrangian phase can exchange mass momentum and energy with fluid phase [58–60].

Appendix A.2.2. Coal Combustion Model

Coal combustion model is a set of sub models consist of moisture evaporation, two step devolatilization, and char oxidation models. Thus, in the following part, each of these models will be discussed.

Appendix A.2.3. Moisture Evaporation Model

The evaporation model has been built on a small assumption. It assumes the moisture has covered the whole particle with a thin film. According to that, the moisture has to be evaporated before the volatiles are released because moisture is at the outermost layer. In this model, there are no properties associated. Here, the Ranz Marshall correlation has been used to calculate the relevant Nusselt and Sherwood numbers. In order to formulate this model, the Quasi-Steady single component droplet evaporation model has been applied to a water droplet [61].

Appendix A.2.4. Devolatilization Model

In order to simulate the volatile release from the coal particles, a devolatilization model has to be used. In STAR CCM+, there are two devolatilization models that can be found: the single step devolatilization model and two step devolatilization models [46]. In Equation (A4), the devolatilization model for n steps has been shown.

$$\text{rawcoal}\overset{k_{pn}}{\rightarrow}VM_{pa(\text{g})}+(1-VM_{pa})(Char)_p \tag{A4}$$

In the above equation, $(Char)_p$ represents the amount of char in the particle, while VM_{pa} represents the volatile matter in the proximate analysis. The kinetic rate for the above equation is shown in the below Equation (A5).

$$r_{vpn} = k_{pn} V M_{pa} \gamma_{cp} \tag{A5}$$

here, k_{pn} represents the reaction rate constant of the devolatilization reaction, and γ_{cp} is the coal mass fraction from the proximate analysis. Thus, the reaction rate constant k_{pn} is expressed in the following Equation (A6).

$$k_{pn} = A_{pn} \exp\left(-\frac{E_{pn}}{RT_p}\right) \tag{A6}$$

E_{pn} represents the activation energy of the nth reaction for the particle and A_{pn} is the pre-exponential factor for the reaction. T_p is the particle temperature, while R is the universal gas constant. For the modeling work a single step devolatilization model is used. Hence, the pre-exponent factor $A = 6 \times 10^{13}\ \mathrm{s}^{-1}$ and activation energy of $E = 2.5 \times 10^{8}\mathrm{J/kg\ mol}$ is used for the study, which are generally used as the pre-exponent and activation energy values for wood [62].

Appendix A.3. Figures

In the following Figure A1, change in CO contours against various ER values are shown.

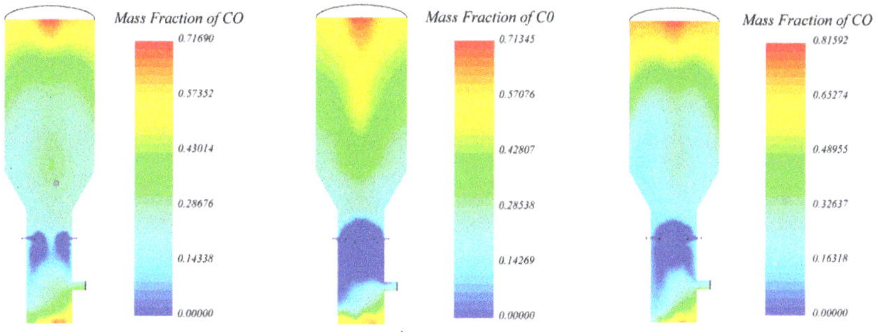

(**a**) (**b**) (**c**)

Figure A1. CO contours with ER values of (**a**) 0.2; (**b**) 0.35; (**c**) 0.5 from left to right.

In the following Figure A2, change in H_2 contours against various ER values are shown.

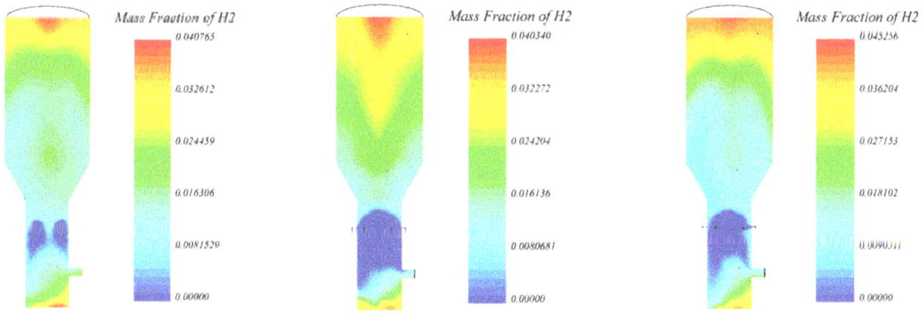

(**a**) (**b**) (**c**)

Figure A2. H_2 contours with ER values of (**a**) 0.2; (**b**) 0.35; (**c**) 0.5 from left to right.

In the following Figure A3, change in CO_2 contours against various ER values are shown.

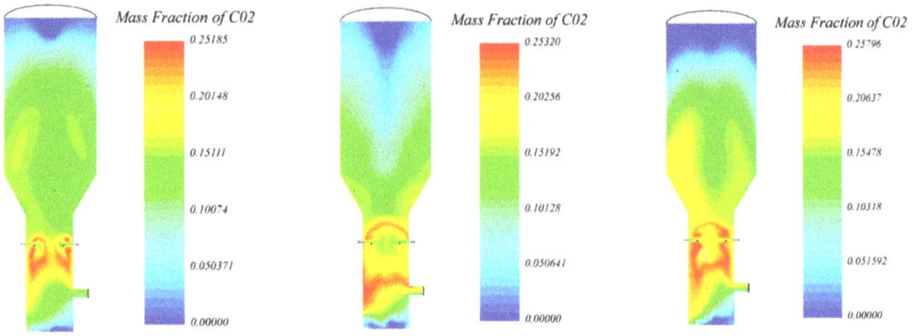

(**a**) (**b**) (**c**)

Figure A3. CO_2 contours with ER values of (**a**) 0.2; (**b**) 0.35; (**c**) 0.5 from left to right.

References

1. Caliandro, P.; Tock, L.; Ensinas, A.V.; Marechal, F. Thermo-economic optimization of a solid oxide fuel cell-gas turbine system fuelled with gasified lignocellulosic biomass. *Energy Convers. Manag.* **2014**, *85*, 764–773. [CrossRef]
2. Sarker, S.; Bimbela, F.; Sánchez, J.L.; Nielsen, H.K. Characterization and pilot scale fluidized bed gasification of herbaceous biomass: a case study on alfalfa pellets. *Energy Convers. Manag.* **2015**, *91*, 451–458. [CrossRef]
3. Saxena, R.C.; Adhikari, D.K.; Goyal, H.B. Biomass-based energy fuel through biochemical routes: A review. *Renew. Sustain. Energy Rev.* **2009**, *13*, 167–178. [CrossRef]
4. Neri, E.; Cespi, D.; Setti, L.; Gombi, E.; Bernardi, E.; Vassura, I.; Passarini, F. Biomass Residues to Renewable Energy: A Life Cycle Perspective Applied at a Local Scale. *Energies* **2016**, *9*, 922. [CrossRef]
5. Cotana, F.; Cavalaglio, G.; Coccia, V.; Petrozzi, A. Energy opportunities from lignocellulosic biomass for a biorefinery case study. *Energies* **2016**, *9*, 748. [CrossRef]
6. Cavalaglio, G.; Gelosia, M.; D'Antonio, S.; Nicolini, A.; Pisello, A.L.; Barbanera, M.; Cotana, F. Lignocellulosic ethanol production from the recovery of stranded driftwood residues. *Energies* **2016**, *9*, 634. [CrossRef]
7. Alauddin, Z.A.B.Z.; Lahijani, P.; Mohammadi, M.; Mohamed, A.R. Gasification of lignocellulosic biomass in fluidized beds for renewable energy development: A review. *Renew. Sustain. Energy Rev.* **2010**, *14*, 2852–2862. [CrossRef]
8. Basu, P. *Biomass Gasification and Pyrolysis: Practical Design and Theory*; Academic Press: Burlington, MA, USA, 2010.
9. Rudra, S.; Rosendahl, L.; Blarke, M.B. Process analysis of a biomass-based quad-generation plant for combined power, heat, cooling, and synthetic natural gas production. *Energy Convers. Manag.* **2015**, *106*, 1276–1285. [CrossRef]
10. Zaccariello, L.; Mastellone, M.L. Fluidized-bed gasification of plastic waste, wood, and their blends with coal. *Energies* **2015**, *8*, 8052–8068. [CrossRef]
11. Baruah, D.; Baruah, D.C. Modeling of biomass gasification: A review. *Renew. Sustain. Energy Rev.* **2014**, *39*, 806–815. [CrossRef]
12. James, R.A.M.; Yuan, W.; Boyette, M.D. The effect of biomass physical properties on top-lit updraft gasification of woodchips. *Energies* **2016**, *9*, 283. [CrossRef]
13. Shi, H.; Si, W.; Li, X. The concept, design and performance of a novel rotary kiln type air-staged biomass gasifier. *Energies* **2016**, *9*, 67. [CrossRef]
14. Rajvanshi, A.K.; Goswami, D.Y. *Biomass Gasification in Alternative Energy in Agriculture Vol. II*; CRC Press: Boca Raton, FL, USA, 1986.
15. Brown, R.C.; Stevens, C. *Thermochemical Processing of Biomass: Conversion into Fuels, Chemicals and Power*; Wiley series in Renewable Resources: Chichester, UK, 2011.

16. Sheth, P.N.; Babu, B.V. Experimental studies on producer gas generation from wood waste in a downdraft biomass gasifier. *Bioresour. Technol.* **2009**, *100*, 3127–3133. [CrossRef] [PubMed]
17. Sharma, A.K. Equilibrium and kinetic modeling of char reduction reactions in a downdraft biomass gasifier: A comparison. *Sol. Energy* **2008**, *82*, 918–928. [CrossRef]
18. Wang, Y.; Yan, L. CFD studies on biomass thermochemical conversion. *Int. J Mol. Sci.* **2008**, *9*, 1108–1130. [CrossRef] [PubMed]
19. Gerber, S.; Behrendt, F.; Oevermann, M. An Eulerian modeling approach of wood gasification in a bubbling fluidized bed reactor using char as bed material. *Fuel* **2010**, *89*, 2903–2917. [CrossRef]
20. Ku, X.; Li, T.; Løvås, T. CFD-DEM simulation of biomass gasification with steam in a fluidized bed reactor. *Chem. Eng. Sci.* **2015**, *122*, 270–283. [CrossRef]
21. Biglari, M.; Liu, H.; Elkamel, A.; Lohi, A. Application of scaling-law and CFD modeling to hydrodynamics of circulating biomass fluidized bed gasifier. *Energies* **2016**, *9*, 504. [CrossRef]
22. Patel, K.D.; Shah, N.K.; Patel, R.N. CFD Analysis of spatial distribution of various parameters in downdraft gasifier. *Procedia Eng.* **2013**, *51*, 764–769. [CrossRef]
23. Gerun, L.; Paraschiv, M.; Vijeu, R.; Bellettre, J.; Tazerout, M.; Gøbel, B.; Henriksen, U. Numerical investigation of the partial oxidation in a two-stage downdraft gasifier. *Fuel* **2008**, *87*, 1383. [CrossRef]
24. Ali, D.A.; Gadalla, M.A.; Abdelaziz, O.Y.; Hulteberg, C.P.; Ashour, F.H. Co-gasification of coal and biomass wastes in an entrained flow gasifier: Modelling, simulation and integration opportunities. *J. Nat. Gas Sci. Eng.* **2017**, *37*, 126–137. [CrossRef]
25. Slezak, A.; Kuhlman, J.M.; Shadle, L.J.; Spenik, J.; Shi, S. CFD simulation of entrained-flow coal gasification: Coal particle density/sizefraction effects. *Powder Technol.* **2010**, *203*, 98–108. [CrossRef]
26. Rogel, A.; Aguillon, J. The 2D Eulerian approach of entrained flow and temperature in a biomass stratified downdraft gasifier. *Am. J. Appl. Sci.* **2006**, *3*, 2068–2075. [CrossRef]
27. Wu, Y.; Zhang, Q.; Yang, W.; Blasiak, W. Two-dimensional computational fluid dynamics simulation of biomass gasification in a downdraft fixed-bed gasifier with highly preheated air and steam. *Energy Fuels* **2013**, *27*, 3274–3282. [CrossRef]
28. Janajreh, I.; Al Shrah, M. Numerical and experimental investigation of downdraft gasification of wood chips. *Energy Convers. Manag.* **2013**, *65*, 783–792. [CrossRef]
29. Ismail, T.M.; El-Salam, M.A.; Monteiro, E.; Rouboa, A. Eulerian-Eulerian CFD model on fluidized bed gasifier using coffee husks as fuel. *Appl. Therm. Eng.* **2016**, *106*, 1391–1402. [CrossRef]
30. Monteiro, E.; Ismail, T.M.; Ramos, A.; El-Salam, M.A.; Brito, P.S.D.; Rouboa, A. Assessment of the miscanthus gasification in a semi-industrial gasifier using a CFD model. *Appl. Therm. Eng.* **2017**, *123*, 448–457. [CrossRef]
31. Silva, V.; Monteiro, E.; Couto, N.; Brito, P.; Rouboa, A. Analysis of syngas quality from portuguese biomasses: An experimental and numerical study. *Energy Fuels* **2014**, *28*, 5766–5777. [CrossRef]
32. Couto, N.D.; Silva, V.B.; Monteiro, E.; Rouboa, A. Assessment of municipal solid wastes gasification in a semi-industrial gasifier using syngas quality indices. *Energy* **2015**, *93*, 864–873. [CrossRef]
33. Sarker, S.; Nielsen, H.K. Preliminary fixed-bed downdraft gasification of birch woodchips. *Int. J. Environ. Sci.* **2015**, *12*, 2119. [CrossRef]
34. Sarker, S.; Nielsen, H.K. Assessing the gasification potential of five woodchips species by employing a lab-scale fixed-bed downdraft reactor. *Energy Convers. Manag.* **2015**, *103*, 801–813. [CrossRef]
35. Atnaw, S.M.; Sulaiman, S.A.; Yusup, S. Influence of fuel moisture content and reactor temperature on the calorific value of syngas resulted from gasification of oil palm fronds. *Sci. World J.* **2014**, *2014*, 121908. [CrossRef] [PubMed]
36. Bouraoui, Z.; Jeguirim, M.; Guizani, C.; Limousy, L.; Dupont, C.; Gadiou, R. Thermogravimetric study on the influence of structural, textural and chemical properties of biomass chars on CO_2 gasification reactivity. *Energy* **2015**, *88*, 703–710. [CrossRef]
37. Zainal, Z.A.; Ali, R.; Lean, C.H.; Seetharamu, K.N. Prediction of performance of a downdraft gasifier using equilibrium modeling for different biomass materials. *Energy Convers. Manag.* **2001**, *42*, 1499–1515. [CrossRef]
38. Sarkar, M.; Kumar, A.; Tumuluru, J.S.; Patil, K.N.; Bellmer, D.D. Gasification performance of switchgrass pretreated with torrefaction and densification. *Appl. Energy* **2014**, *127*, 194–201. [CrossRef]
39. Syed, S.; Janajreh, I.; Ghenai, C. Thermodynamics equilibrium analysis within the entrained flow gasifier environment. *Int. J. Therm. Environ. Eng.* **2012**, *4*, 47–54. [CrossRef]

40. Renganathan, T.; Yadav, M.V.; Pushpavanam, S.; Voolapalli, R.K.; Cho, Y.S. CO_2 utilization for gasification of carbonaceous feedstocks: A thermodynamic analysis. *Chem. Eng. Sci.* **2012**, *83*, 159. [CrossRef]
41. Chiesa, M.; Mathiesen, V.; Melheim, J.A.; Halvorsen, B. Numerical simulation of particulate flow by the Eulerian-Lagrangian and the Eulerian-Eulerian approach with application to a fluidized bed. *Comput. Chem. Eng.* **2005**, *29*, 291–304. [CrossRef]
42. Zhang, Z.; Chen, Q. Comparison of the Eulerian and Lagrangian methods for predicting particle transport in enclosed spaces. *Atmos. Environ.* **2007**, *41*, 5236–5248. [CrossRef]
43. White, F.M. *Viscous Fluid Flow*; McGraw-Hill Higher Education: Boston, MA, USA, 2006.
44. Yu, X.; Hassan, M.; Ocone, R.; Makkawi, Y. A CFD study of biomass pyrolysis in a downer reactor equipped with a novel gas-solid separator-II thermochemical performance and products. *Fuel Process. Technol.* **2015**, *133*, 51–63. [CrossRef]
45. Aissa, A.; Abdelouahab, M.; Noureddine, A.; Elganaoui, M.; Pateyron, B. Ranz and Marshall correlations limits on heat flow between a sphere and its surrounding gas at high temperature. *Therm. Sci.* **2015**, *19*, 1521–1528. [CrossRef]
46. Luan, Y.T.; Chyou, Y.P.; Wang, T. Numerical analysis of gasification performance via finite-rate model in a cross-type two-stage gasifier. *Int. J. Heat Mass Transf.* **2013**, *57*, 558–566. [CrossRef]
47. Sharma, A.; Pareek, V.; Wang, S.; Zhang, Z.; Yang, H.; Zhang, D. A phenomenological model of the mechanisms of lignocellulosic biomass pyrolysis processes. *Comput. Chem. Eng.* **2014**, *60*, 231. [CrossRef]
48. Molcan, P.; Caillat, S. Modelling approach to woodchips combustion in spreader stoker boilers. In Proceedings of the 9th European Conference on Industrial Furnaces and Boilers, Estoril, Portugal, 26–29 April 2011.
49. Wu, Y.; Yang, W.; Blasiak, W. Energy and exergy analysis of high temperature agent gasification of biomass. *Energies* **2014**, *7*, 2107–2122. [CrossRef]
50. Allesina, G. Modeling of coupling gasification and anaerobic digestion processes for maize bioenergy conversion. *Biomass Bioenergy* **2015**, *81*, 444–451. [CrossRef]
51. Allesina, G.; Pedrazzi, S.; Tartarini, P. Modeling and investigation of the channeling phenomenon in downdraft stratified gasifers. *Bioresour. Technol.* **2013**, *146*, 704–712. [CrossRef] [PubMed]
52. Giltrap, D.L.; McKibbin, R.; Barnes, G.R.G. A steady state model of gas-char reactions in a downdraft biomass gasifier. *Sol. Energy* **2003**, *74*, 85–91. [CrossRef]
53. González-Vázquez, M.P.; García, R.; Pevida, C.; Rubiera, F. Optimization of a bubbling fluidized bed plant for low-temperature gasification of biomass. *Energies* **2017**, *10*, 306. [CrossRef]
54. Zhang, C. Numerical modeling of coal gasification in an entrained-flow gasifier. *ASME Int. Mech. Eng. Congr. Expo.* **2012**, *6*, 1193–1203. [CrossRef]
55. Lv, P.; Chang, J.; Wang, T.; Fu, Y.; Chen, Y.; Zhu, J. Hydrogen-rich gas production from biomass catalytic gasification. *Int. J. Hydrogen Energy* **2009**, *34*, 1260–1264. [CrossRef]
56. Hernández, J.J.; Aranda-Almansa, G.; Bula, A. Gasification of biomass wastes in an entrained flow gasifier: Effect of the particle size and the residence time. *Fuel Process. Technol.* **2010**, *91*, 681–692. [CrossRef]
57. Cornejo, P.; Farias, O. Mathematical modeling of coal gasification in a fluidized bed reactor using an Eulerian Granular description. *Int. J. Chem. React. Eng.* **2011**, *9*. [CrossRef]
58. Discrete Element method (DEM) in STAR CCM+. Available online: https://mdx2.plm.automation.siemens.com/presentation/discrete-element-method-dem-star-ccm (accessed on 10 August 2017).
59. Alletto, M.; Breuer, M. One-way, two-way and four-way coupled LES predictions of a particle-laden turbulent flow at high mass loading downstream of a confined bluff body. *Int. J. Multiph. Flow* **2012**, *45*, 70–90. [CrossRef]
60. Benra, F.K.; Dohmen, H.J.; Pei, J.; Schuster, S.; Wan, B. A comparison of one-way and two-way coupling methods for numerical analysis of fluid-structure interactions. *J. Appl. Math.* **2011**, *2011*, 853560. [CrossRef]

61. Kent, J.C. Quasi-steady diffusion-controlled droplet evaporation and condensation. *Appl. Sci. Res.* **1973**, *28*, 315–360. [CrossRef]
62. Belosevic, S. Modeling approaches to predict biomass co-firing with pulverized coal. *Open Thermodyn. J.* **2010**, *4*, 50–70. [CrossRef]

MDPI AG
St. Alban-Anlage 66
4052 Basel, Switzerland
Tel. +41 61 683 77 34
Fax +41 61 302 89 18
http://www.mdpi.com

Energies Editorial Office
E-mail: energies@mdpi.com
http://www.mdpi.com/journal/energies

www.ingramcontent.com/pod-product-compliance
Lightning Source LLC
Chambersburg PA
CBHW051847210326
41597CB00033B/5812
9783038426684